Cheeney
1983 July

THE
OXFORD ENGINEERING SCIENCE SERIES

General Editors

A. L. CULLEN

F. W. CRAWFORD   J. W. HUTCHINSON

W. H. WITTRICK   L. C. WOODS

THE MATHEMATICAL THEORY OF PLASTICITY

BY

R. HILL

OXFORD
AT THE CLARENDON PRESS

Oxford University Press, Walton Street, Oxford OX2 6DP
Oxford London Glasgow
New York Toronto Melbourne Auckland
Kuala Lumpur Singapore Hong Kong Tokyo
Delhi Bombay Calcutta Madras Karachi
Nairobi Dar es Salaam Cape Town
and associated companies in
Beirut Berlin Ibadan Mexico City Nicosia

OXFORD is a trade mark of Oxford University Press

Published in the United States by
Oxford University Press, New York

ISBN 0 19 856162 8

First published 1950
Reprinted 1956, 1960, 1964, 1967, 1971
First issued in paperback 1983

All rights reserved. No part of this publication may be reproduced, stored in a retrieval system, or transmitted, in any form or by any means, electronic, mechanical, photocopying, recording, or otherwise, without the prior permission of Oxford University Press

This book is sold subject to the condition that it shall not, by way of trade or otherwise, be lent, re-sold, hired out or otherwise circulated without the publisher's prior consent in any form of binding or cover other than that in which it is published and without a similar condition including this condition being imposed on the subsequent purchaser

Printed in Hong Kong

PREFACE

ALTHOUGH the first mathematical account of the plastic straining of metals was outlined about eighty years ago by Lévy and Saint-Venant, significant advances did not follow until after 1920. The theory of plasticity still remains a young science, with few serious students. I have written the book in the hope that it will attract engineers and applied mathematicians to a field which well rewards study and research, and which is finding an increasing application in metal technology.

To give an added unity to the book, the examples chosen to illustrate the applications of the theory are arranged as a progression from simple to complex boundary-value problems. Many of these relate to processes for the shaping of metals, such as rolling and drawing, not only on account of my own interests but also because it is here that the plasticity of a metal is demonstrated most strikingly. I have been especially concerned to provide the data needed by research workers in industry, and wherever it seems justifiable I have introduced approximations to simplify and condense the results of the analysis. Rather than cover the whole field cursorily, I have preferred to take selected topics and make the treatment of these as complete and factual as present knowledge appears to warrant.

The reader is assumed to have an acquaintance with the elementary theory of elasticity. Two useful mathematical techniques which are probably unfamiliar to many engineers are explained in appendixes: these are suffix notation, together with the summation convention, and the solution of hyperbolic differential equations. The first can be mastered at once and is an indispensable shorthand for a thorough discussion of stress-strain relations. The second appears in many branches of mechanics, and its methods usually appeal for their simplicity and directness.

In an account of such a rapidly expanding subject, with a literature that is often vague and contradictory, it is inevitable that there are many stages in the argument where I have relied on my own judgement. Moreover, while the first draft was being prepared, several serious gaps in the theory became apparent and these I have done my best to fill. There should be no difficulty in recognizing statements for which I alone am responsible: any quoted result from the work of others is accompanied by a reference to the original source, to the best of my knowledge of papers appearing in scientific journals before the autumn of 1949.

It is a pleasure to record my appreciation of the ready co-operation given me by the officers of the Clarendon Press: their contribution speaks for itself. My task has been made immeasurably easier by my wife, without whose help and encouragement I would have come much less near the book I set out to write.

May 1950

R. H.

CONTENTS

I. INTRODUCTION 1
   1. Definition of the subject 1
   2. Historical outline 2
   3. Physical background 4
   4. The stress-strain curve 8

II. FOUNDATIONS OF THE THEORY 14
   1. The ideal plastic body 14
   2. The criterion of yielding 15
   3. Strain-hardening 23
   4. The complete stress-strain relations 33
   5. The Lévy–Mises and Reuss equations 38
   6. The Hencky stress-strain equations 45
   7. Other theories 48

III. GENERAL THEOREMS 50
   1. The plastic potential 50
   2. Uniqueness of a stress distribution under given boundary conditions . 53
   3. Extremum and variational principles 60

IV. THE SOLUTION OF PLASTIC-ELASTIC PROBLEMS. I . 70
   1. Introduction 70
   2. Theory of Hohenemser's experiment 71
   3. Combined torsion and tension of a thin-walled tube . 74
   4. Combined torsion and tension of a cylindrical bar . 75
   5. Compression under conditions of plane strain . . 77
   6. Bending under conditions of plane strain . . . 79
   7. Bending of a prismatic beam 81
   8. Torsion of a prismatic bar 84
   9. Torsion of a bar of non-uniform section . . . 94

V. THE SOLUTION OF PLASTIC-ELASTIC PROBLEMS. II . 97
   1. The expansion of a spherical shell 97
   2. The expansion of a cylindrical tube 106
   3. Theory of the autofrettage process 114
   4. Expansion of a cylindrical cavity in an infinite medium . 125

VI. PLANE PLASTIC STRAIN AND THE THEORY OF THE SLIP-LINE FIELD 128
   1. Assumption of a plastic-rigid material 128
   2. The plane strain equations referred to Cartesian coordinates . 129

3. The plane strain equations referred to the slip-lines . . 132
4. Geometry of the slip-line field 136
5. The numerical calculation of slip-line fields. . . . 140
6. The numerical calculation of the velocity distribution . . 149
7. Analytic integration of the plane strain equations . . . 151
8. Discontinuities in the stress 157

VII. TWO-DIMENSIONAL PROBLEMS OF STEADY MOTION . 161
1. Formulation of the problem 161
2. Sheet-drawing 163
3. Ironing of a thin-walled cup 178
4. Sheet-extrusion 181
5. Piercing 186
6. Strip-rolling 188
7. Machining 206
8. Flow through a converging channel 209

VIII. NON-STEADY MOTION PROBLEMS IN TWO DIMENSIONS. I 213
1. Geometric similarity and the unit diagram 213
2. Wedge-indentation 215
3. Compression of a wedge by a flat die 221
4. Expansion of a semi-cylindrical cavity in a surface . . . 223
5. Compression of a block between rough plates . . . 226

IX. NON-STEADY MOTION PROBLEMS IN TWO DIMENSIONS. II 237
1. Introduction 237
2. Formulation of the problem 238
3. Yielding of notched bars under tension 245
4. Plastic yielding round a cavity 252
5. Indentation and the theory of hardness tests . . . 254

X. AXIAL SYMMETRY 262
1. Fundamental equations 262
2. Extrusion from a contracting cylindrical container . . . 263
3. Compression of a cylinder under certain distributed loads . . 265
4. Cylindrical tube under axial tension and internal pressure . . 267
5. Tube-sinking 269
6. Stress distribution in the neck of a tension specimen . . 272
7. Compression of a cylinder between rough plates . . . 277
8. Relations along slip-lines and flow-lines 278

XI. MISCELLANEOUS TOPICS 282
1. Deep-drawing 282
2. General theory of sheet-bending 287

3. Plane strain of a general plastic material 294
   4. The theory of plane plastic stress, with applications . . 300
   5. Completely plastic states of stress in a prismatic bar . . 313

XII. PLASTIC ANISOTROPY 317
   1. The yield criterion 318
   2. Relations between stress and strain-increment . . . 320
   3. Plastic anisotropy of rolled sheet 321
   4. Length changes in a twisted tube 325
   5. The earing of deep-drawn cups 328
   6. Variation of the anisotropic parameters during cold-work . . 332
   7. Theory of plane strain for anisotropic metals . . . 334

APPENDIXES 341

AUTHOR INDEX 351

SUBJECT INDEX 354

I
INTRODUCTION

1. Definition of the subject

'THEORY of plasticity' is the name given to the mathematical study of stress and strain in plastically deformed solids, especially metals. This follows the well-established precedent set by the 'theory of elasticity', which deals with methods of calculating stress and strain in elastically deformed solids, and not, as a literal interpretation suggests, with the physical explanation of elasticity. The relation of plastic and elastic properties of metals to crystal structure and cohesive forces belongs to the subject now known as 'metal physics'.

The theory of plasticity takes as its starting-point certain experimental observations of the macroscopic behaviour of a plastic solid in uniform states of combined stress. The task of the theory is twofold: first, to construct explicit relations between stress and strain agreeing with the observations as closely and as universally as need be; and second, to develop mathematical techniques for calculating non-uniform distributions of stress and strain in bodies permanently distorted in any way. At the present time metals are the only plastic solids for which there is enough data to warrant the construction of a *general* theory. For this reason the theory is related specifically to the properties of metals, though it *may* apply to other potentially plastic materials (e.g. ice, clay, or rock).

By contrast with many other plastic solids the most striking attribute of a metal is its capacity for cold-work. At ordinary temperatures, and under favourable applied stresses, a dimensional change of twentyfold can easily be obtained with a ductile metal, for example by compressing or shearing a cylinder of copper. More severe strains are enforced locally when a metal billet is extruded or pierced. The theory of plasticity is, therefore, especially concerned with technological forming processes such as the rolling of strip, extrusion of rods and tubes, drawing of wire, and deep-drawing of sheet. The purpose of the analysis is to determine the external loads, the power consumption, and the non-uniform strain and hardening due to the cold-working. Intense plastic strains are also produced locally in many standard mechanical tests of metals, for instance indentation by a conical die, the bending of a notched bar, or the extension of a tensile specimen past the necking point. A rational account of the physical significance of these tests requires a knowledge of the state of stress and the extent of the plastic zone.

At the other extreme, where the subjects of elasticity and plasticity meet, a typical application is to predict the critical loading which just causes a structural member to yield plastically at its weakest point. Between these extremes come problems where the plastic and elastic strains are of a similar order of magnitude, as in a beam partly overstrained by bending or twisting, or in a pressure vessel strengthened by an initial permanent expansion.

2. Historical outline

The scientific study of the plasticity of metals may justly be regarded as beginning in 1864. In that year Tresca published a preliminary account of experiments on punching and extrusion, which led him to state that a metal yielded plastically when the maximum shear stress attained a critical value. Criteria for the yielding of plastic solids, mainly soils, had been proposed before, for example by Coulomb (1773), and had been applied by Poncelet (1840) and Rankine (1853) to problems such as the calculation of earth-pressure on retaining walls; there appears, however, to have been no earlier important investigation for metals. Tresca's yield criterion was applied by Saint-Venant to determine the stresses in a partly plastic cylinder subjected to torsion or bending (1870) and in a completely plastic tube expanded by internal pressure (1872) (the first step towards the solution for a partly plastic tube was taken by Turner in 1909). Saint-Venant also set up a system of five equations governing the stresses and strains in two-dimensional flow, and, recognizing that there is no one-one relation between stress and total plastic strain, postulated that the directions of the maximum shear strain-*rate* coincided at each moment with the directions of the maximum shear stress. In 1871 Lévy, adopting Saint-Venant's conception of an ideal plastic material, proposed three-dimensional relations between stress and rate of plastic strain.

There seems to have been no further significant advance until the close of the century when Guest investigated the yielding of hollow tubes under combined axial tension and internal pressure, and obtained results broadly in agreement with the maximum shear-stress criterion. During the next decade many similar experiments were performed, mainly in England, with slightly differing conclusions. Various yield criteria were suggested, but for many metals, as later and more accurate work was to show, the most satisfactory was that advanced by von Mises

(1913) on the basis of purely mathematical considerations; it was interpreted by Hencky some years afterwards as implying that yielding occurred when the elastic shear-strain energy reached a critical value. Von Mises also independently proposed equations similar to Lévy's.

Between the two wars the subject was actively developed by German writers. In 1920 and 1921 Prandtl showed that the two-dimensional plastic problem is hyperbolic, and calculated the loads needed to indent a plane surface and a truncated wedge by a flat die. Parallel experiments by Nadai were in accord with these calculations, but it has been shown recently that Prandtl's work is defective in certain respects. The general theory underlying Prandtl's special solutions was supplied in 1923 by Hencky, who also discovered simple geometrical properties of the field of slip-lines in a state of plane plastic strain. It was some time, however, before the equations governing the variation of the velocity of flow along slip-lines were obtained (Geiringer, 1930) and even longer before the correct approach to the solution of plane problems was clarified (1945–9). In 1923 Nadai investigated, both theoretically and experimentally, the plastic zones in a twisted prismatic bar of arbitrary contour. The effective application of plastic theory to technological processes began in 1925 when von Karman analysed, by an elementary method, the state of stress in rolling. In the following year Siebel, and soon afterwards Sachs, put forward similar theories for wire-drawing.

It was not until 1926, when Lode measured the deformation of tubes of various metals under combined tension and internal pressure, that the Lévy–Mises stress-strain relations were shown to be valid to a first approximation. However, Lode's results indicated certain divergences, and these were afterwards confirmed by the more controlled experiments of Taylor and Quinney (1931). The theory was now generalized in two important directions: first by Reuss (1930) who made allowance for the elastic component of strain, following an earlier suggestion by Prandtl; second by Schmidt (1932) and Odquist (1933) who showed, in slightly different ways, how work-hardening could be brought within the framework of the Lévy–Mises equations. The first generalization was broadly confirmed by experiments of Hohenemser (1931–2), and the second by investigations of Schmidt. Thus, by 1932, a theory had been constructed, reproducing the main plastic and elastic properties of an isotropic metal at ordinary temperatures, and substantially in accord with observation. However, from then until the early 1940's little progress was made in the solution of special problems. Further generalizations were formulated (for example, by von Mises in 1928 and by Melan in 1938), but mathematical expediency and lack of accurate data combined to render them, for the time being, academic.

Meanwhile a rival theory proposed by Hencky in 1924 was favoured for its analytic convenience in problems where the plastic strain was small, despite its conflict with experience in establishing a one-one relation between stress and strain. This theory was given prominence by Nadai in his book on plasticity (1931), and was afterwards extensively employed by the Russian school (1935 onwards). Hencky's equations lead to approximately correct results only for certain loading-paths, but many writers have applied them without discrimination.

The war stimulated research in England and America, through problems such as the calculation of the stresses in autofrettaged gun-barrels and of the forces resisting a shot penetrating armour plate. Since then the subject has been intensively studied in many countries, and the advances made are such that the present book is largely an account of the work of the five years 1945–9.

3. Physical background

During the construction of the theory frequent reference is made to the plastic properties of metal single crystals and polycrystalline aggregates. It is assumed that the reader has a general knowledge of these, and only a brief résumé of the relevant properties is given here. For broader and more detailed accounts, presented from other standpoints, specialist works on metal physics should be consulted.†

(i) *Single crystals*. In a freshly grown metal crystal, isolated from external disturbances and of the highest purity, the atoms are disposed in equilibrium under their mutual forces in a regular three-dimensional array, with a periodic structure characteristic of the metal. Most of the well-known metals have a lattice structure which is either face-centred cubic (copper, aluminium, lead, silver, gold), body-centred cubic (alpha iron, vanadium, tungsten, and the alkali metals), or hexagonal close-packed (zinc, magnesium, cadmium). According to current theory, the cohesive forces binding the atoms together are such that a perfect metal crystal could sustain, with only slight displacements of the atoms from their normal positions, very much greater applied stresses than an actual

† C. S. Barrett, *Structure of Metals* (McGraw-Hill Book Co., 1943); W. Boas, *An Introduction to the Physics of Metals and Alloys* (John Wiley & Sons, Inc., New York, 1947); A. H. Cottrell, *Theoretical Structural Metallurgy* (Edward Arnold & Co., London, 1948); C. F. Elam, *Distortion of Metal Crystals* (Clarendon Press, Oxford, 1935); W. Hume-Rothery, *Atomic Theory for Students of Metallurgy* (Institute of Metals, London, 1946); E. Schmid and W. Boas, *Kristallplastizität* (Julius Springer, Berlin, 1935); F. Seitz, *The Physics of Metals* (McGraw-Hill Book Co., 1943).

crystal is observed to do. Thus, in a freshly prepared crystal, sensitive measurements show that elastic (or reversible) deformation has ceased after a *macroscopic* shear strain of order 10^{-4}, whereas a perfect crystal should be capable of an elastic shear strain of order 10^{-1}. The discrepancy is attributed to faults or disturbances in the lattice structure, formed during growth or introduced by subsequent handling. It is thought that the faults are separated by distances of the order of 1,000 atomic spacings, and that each extends over a volume containing perhaps 100–1,000 atoms. Certain of the faults are considered to be of a kind that 'weaken' the crystal, and are known as 'dislocations'. Mathematical studies of conjectural atomic arrangements in a dislocation indicate that a very small applied stress would cause it to move (as a geometrical entity) through an otherwise perfect crystal. The resultant effect of the passage of a dislocation is a relative displacement of the parts of the crystal bordering its path by an amount equal to one or two atomic spacings. In this way the movement through the crystal of many dislocations produces an overall strain without affecting the main lattice structure; the substantial preservation of the structure during plastic deformation is confirmed by X-ray examination. That the strain is plastic and irreversible is attributed to the 'trapping' of the dislocations at other faults which are not mobile under the external stresses. Additional dislocations are thought to be created or liberated during the deformation, but despite this a continually increasing stress is usually needed to enforce plastic strain (strain-hardening); the movement of free dislocations is progressively impeded by the local disordering of the lattice at points where trapped dislocations accumulate. The increase in potential energy of the deformed crystal is only a small fraction (of order one-tenth) of the work done by the applied stress; the remainder appears as vibrational energy of the atoms in or near moving dislocations (whose speed must be close to that of sound) and is ultimately dissipated as heat throughout the crystal.

If the external temperature is sufficiently great, the activation energy needed to move existing dislocations may also be provided by thermal agitation; the effect of the applied stress is mainly to give direction to the resultant flow (transient creep). If the stress is removed, and the crystal is held at a sufficiently high temperature for a certain period, thermal fluctuations assist the atoms over their potential barriers towards the original regular array, which is the configuration of greatest stability. Ultimately the imperfections created during the previous deformation are removed, and the crystal is said to have been 'annealed'.

In the theory of plasticity the strain is regarded as macroscopically uniform, but on a microscopic scale it is known that the plastic distortion is largely confined to narrow bands (slip-lines) which extend through the crystal and are presumably created by the passage, along closely grouped planes, of large numbers of dislocations, many of which become trapped or mutually locked. These bands are perhaps some 100 atoms thick, and the planes to which they are parallel are known as slip-, or glide-, planes; they are often the crystallographic planes most densely packed with atoms. The average spacing of the slip-bands depends on the amount and rate of strain, and on the metal, but is normally of the order of 10,000 atomic distances. The lattice between the slip-bands is still virtually perfect and only distorted elastically; the strain there is greater than the overall strain at the elastic limit since the applied stress has increased because of the hardening; moreover, since the slip-bands extend over only a small proportion of the total volume, even after quite large strains, the elastic moduli of the crystal as a whole are little affected by the plastic deformation.

When a crystal in the form of a wire is stretched under tension the cross-section becomes elliptical. Macroscopically, the deformation may be described by saying that the crystal has undergone a shear in a certain direction over a certain set of parallel slip-planes, together with a rotation bringing these planes more nearly parallel to the axis of the wire. Only a limited number of active slip-directions have ever been observed. In face-centred cubic metals at ordinary temperatures there are apparently only four possible slip-planes (the octahedral planes), in each of which there are three possible slip-directions (lying in the cubic planes); slip occurs in a body-centred cubic metal in many more ways, but in a hexagonal metal only over the basal planes and along the digonal axes. The tension needed to deform the wire plastically varies greatly with the orientation of the crystal to the axis, but it is found that a slip-direction is activated only when a certain critical value (the yield stress) is attained by the component of shear stress acting over the slip-plane and in the slip-direction; this is of order 100 gm./mm.^{2} in an annealed crystal at ordinary temperatures. The yield stress is approximately the same for all the different slip-directions in a crystal in a given state, and is independent of the type of test (for example, whether tension or compression). For ordinary strain-rates and for temperatures where creep is negligible, the yield stress is a function mainly of the amount of previous plastic distortion. In particular, the same relationship is obtained between shear stress and shear strain irrespective of which set of slip-planes is operating, and, if the applied stress is subsequently changed so that another set of planes is operated, the new shear-hardening curve is a continuation of the previous one. It appears, therefore, that the disordering of the lattice affects all planes equally whether or not they are active; this phenomenon is known as latent hardening. When a gradually increasing stress is applied to a crystal (either in its original state or plastically deformed), it is observed that the active slip-planes are those on which the critical value of the shear stress is first attained. Double slip begins when the rotation of the crystal brings another set of planes into the position where the corresponding shear stress is equal to the current yield value; several sets can be operated simultaneously under combined stresses.

(ii) *Polycrystalline aggregate.* A metal, in its generally used form, is a compact aggregate of crystal grains with varying shapes and orientations, each grain having grown from a separate nucleus in the melt. The metal may be considered macroscopically isotropic when the orientations are randomly distributed and when the average dimensions of the individual crystals are small compared with the dimensions of the whole specimen (for example, 10^{-3}–10^{-2} cm. compared with 1–100 cm.). Nevertheless, the properties of an aggregate are not always simply statistical averages of the properties of a single crystal, taken over all orientations. While this is approximately true of properties which depend mainly on the bulk structure, such as the coefficient of thermal expansion or the elastic moduli, it is not necessarily true of plastic phenomena.

Theory and experiment suggest that the transition from one orientation to another in neighbouring grains takes place through a layer only a few atoms thick. In this transition zone, or grain boundary, the atoms take up equilibrium positions which are a compromise between the normal positions in each of the two lattices. These atoms have a higher free energy than atoms within the grains, and are consequently main agents of viscous flow and intercrystalline fracture at high temperatures. They are also thought to be potential sources of dislocations, but at the same time hindrances to the passage of others, in greater or lesser degree. Being the centre of a stress concentration extending over many atoms, a dislocation arriving at a boundary can activate slip on a skew plane in a neighbouring crystal, and this should be easier when many slip-planes of near orientation are available, as in face-centred or body-centred cubic metals. Indeed, the shear-hardening curve of a polycrystalline cubic metal does not exceed the mean curve for single crystals of arbitrary orientation by more than a factor of two or three. Where, as in a hexagonal metal, only one set of slip-planes can be activated, the polycrystal is as much as one hundred times as hard as a single crystal, despite the intervention of other and more favourable mechanisms of plastic deformation (such as twinning). The existence of such mechanisms in hexagonal metals permits the grains to maintain contact (though not their absolute identities, since grain boundaries are unlikely to comprise the same atoms throughout a process of plastic distortion). Contact could not be preserved only by slipping over the basal planes in each grain, since a (uniform) plastic strain generally requires the simultaneous operation of five independent sets of planes if it is to be produced by slip alone. Sufficient sets are, on the other hand, available in cubic metals, and in these it is probable that the macroscopic strain in each crystal resembles that of the aggregate as a whole.

If, then, the specimen is subjected to a *large* distortion of one kind (for instance, monotonic compression) the same crystallographic directions in each grain are gradually rotated towards a common axis. A preferred orientation is thereby created, and the specimen becomes increasingly anisotropic. When a plastically deformed specimen is unloaded, residual stresses on a microscopic scale remain, due mainly to the different states of stress existing in the variously oriented crystals before unloading. If a different loading is now applied such residual stresses must influence the plastic yielding. For example, if the previous strain was a uniform extension and the specimen is then reloaded in compression in the opposite direction, it is observed that yielding (of the specimen as a whole) occurs at a much reduced stress. This is known as the Bauschinger effect, and in so far as it is absent from single crystals of pure metals it is attributable to a particular kind of residual stress due to the grain boundaries.

4. The stress-strain curve

The stress-strain curve of a polycrystalline metal under any simple loading has the shape shown diagrammatically in the figure. The three most usual types of test are the tension of a rod, the compression of a short cylindrical block, and the twisting of a thin-walled tube. If special precautions are observed (for example, with regard to axiality of loading, efficiency of lubrication, and isotropy of material), the macroscopic strain can be made effectively uniform except where an instability occurs. The result of such a test is represented by plotting the mean stress σ (tensile, compressive or shear), acting over the current cross-sectional area, against some measure of the total strain. If the length of a tensile specimen is increased from l_0 to l the amount of deformation is customarily measured either as $\epsilon = \ln(l/l_0)$ (called logarithmic or natural strain) or as $e = (l-l_0)/l_0$ (engineering or conventional strain); it is evident that $\epsilon = \ln(1+e)$. ϵ and e are approximately identical when the change in length is small, as in the elastic range, but ϵ becomes progressively less than e as the strain increases. If the height of a compression specimen is reduced from h_0 to h, the two analogous measures of the deformation are defined to be $\epsilon = \ln(h_0/h)$ and $e = (h_0-h)/h_0$; here, however, $\epsilon = \ln\{1/(1-e)\}$. As the height of the specimen approaches

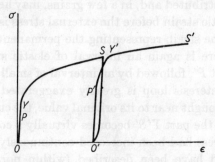

zero, $\epsilon \to \infty$ but $e \to 1$. In torsion the only commonly used measure is the engineering shear strain $\gamma = r\theta$, where r is the tube radius and θ is the twist in radians per unit length. It is shown later how the stress-strain curves from different tests of the same material can be correlated on the basis of a *universal* definition of strain.

The main characteristics of a stress-strain curve will now be recapitulated, logarithmic strain being meant in the case of tension or compression. When a gradually increasing load is applied, the specimen first deforms elastically and regains its original dimensions if the load is removed. The stress at which the strain ceases to be proportional to the applied stress is known as the proportional limit (P in the figure); its value depends somewhat on the sensitivity of the measuring apparatus. The elastic range generally extends beyond the proportional limit, and the stress at which an appreciable permanent deformation is observed is known as the yield stress (Y in the figure); the location of this point, too, depends partly on the apparatus, since the transition from elastic to fully plastic behaviour is not immediate owing to the successive yielding of the crystal grains. In most metals the stress must be continually raised to enforce further plastic strain, but a progressively smaller increment of stress is normally needed to produce a given increment of strain.

In other words, the rate of strain-hardening falls steadily and the curve bends over more and more. An actual stress-strain curve for annealed copper is shown in Fig. 5 on p. 28; curves for any one metal prepared in different ways may vary by some 10 per cent. There is evidence to show that at very large strains (up to the point of fracture) the rate of hardening approaches a small constant value (perhaps zero), due possibly to spontaneous reordering of the heavily distorted lattice.

If, following a certain plastic deformation, the stress is reduced from its current value at S the change in length is at first elastic. However, owing to the different orientations of the grains the residual stresses are not uniformly distributed and, in a few grains, may act so as to produce a very small plastic strain before the external stress is entirely removed (at O'). OO' is the strain representing the permanent change of shape. On reloading there is again an interval of elastic strain, with a new proportional limit P', followed by an interval of small plastic strain (the width of the hysteresis loop is greatly exaggerated in the diagram). As the stress is brought near to its original value, the curve bends sharply over near Y' and the part $Y'S'$ becomes virtually a continuation of YS. Indeed, if the stress had been increased continuously from S the same curve SS' would have been described (within normal experimental error); the state of the specimen at Y' after unloading and reloading differs from what it would have been (had the stress increased continuously) only by the additional disordering of the lattice during the hysteresis loop $SO'Y'$, and this is quickly overshadowed by further deformation. S, and similarly any point on the curve, can therefore be regarded, for practical purposes, as the yield stress of the specimen strained by the amount OO'. The stress-strain curve $O'Y'S'$ for such a pre-strained metal is distinguished from the curve for the annealed metal by a higher yield stress and a more rapid subsequent bend.

In a few metals, such as annealed mild steel and certain alloys of aluminium, the stress falls abruptly after the yield stress Y. This is then known as the upper yield-point, and the stress which is needed to enforce further plastic deformation is known as the lower yield-point. In a tensile test on annealed mild steel a several per cent. change in length occurs without a significant increase in stress above the lower yield-point. The fall in stress is observed to coincide with the propagation of a Lüders' band across the specimen. This is a lamellar zone of plastic distortion, inclined at about 45° to the axis, and in which the macroscopic strain appears to be a simple shear; the remainder of the specimen is still only strained elastically. During the lower yield-point extension other bands usually appear. The deformation does not become uniform, nor does the stress-strain curve begin to rise, until the bands have spread through the whole specimen. Pre-strained mild steel does not show the yield-point drop unless it has been rested for a certain time (strain-ageing). It seems that the phenomenon is caused by solute atoms, of carbon or some impurity, which arrange themselves interstitially in the lattice under thermal activation in such a way that the normal yielding of the solvent metal is retarded.† The fall in stress at the yield-point is attributed to the freeing of dislocations, which are then able to travel through the lattice in the ordinary way under the lower yield-point stress. Since the first Lüders' band is initiated at a point where there is a local concentration of stress of unknown magnitude (due to a slight geometrical or structural non-uniformity), the upper yield-point varies greatly with the conditions of the test. The lower yield-point is, on the other hand, fairly reproducible and in annealed mild steel is about 10–12 tn./in.² (in tension), while the upper yield-point may exceed this by 20 or 30 per cent. in a suitably designed test at normal temperatures and rates of strain.

In a compression test the applied load steadily increases during plastic deformation since both the stress and the cross-sectional area increase. If A_0 is the initial area and A is the area when the height is h, then $Ah = A_0 h_0$ for constant volume (neglecting the small elastic change). The load is therefore

$$L = \sigma A = \sigma A_0 h_0/h = \sigma A_0/(1-e).$$

The graph of load against strain has an upward inflexion and rises without limit as the cylinder is reduced to a thin disk. In a tensile test the load is
$$L = \sigma A = \sigma A_0 l_0/l = \sigma A_0/(1+e),$$

which reaches a maximum and then decreases when the rate of diminution of area outweighs the rate of hardening. The maximum load is given by $dL = 0$, or
$$\sigma\, dA + A\, d\sigma = 0.$$

Combining this with $$l\, dA + A\, dl = 0,$$

we obtain
$$-\frac{dA}{A} = \frac{d\sigma}{\sigma} = \frac{dl}{l} = d\epsilon = \frac{de}{1+e}.$$

† A. H. Cottrell, 'Report of Bristol Conference', *Phys. Soc.* (1948), 30, has proposed a possible mechanism depending on the tendency of solute atoms to cluster round dislocations.

Thus, the load is a maximum when

$$\frac{d\sigma}{d\epsilon} = \sigma, \qquad (1)$$

or

$$\frac{d\sigma}{de} = \frac{\sigma}{1+e}. \qquad (2)$$

It follows from the last equation that the maximum load corresponds to the point of contact of the tangent to the (σ, e) curve from the point -1 on the e axis. A normal tensile test becomes unstable when the load reaches its maximum; the specimen 'necks' locally while the remainder of the specimen recovers elastically under the decreasing load. Since the deformation and distribution of stress is then non-uniform, the test does not provide a direct measure of the tensile stress-strain curve after necking. However, a correction can be made for this effect if the geometry of the neck is measured during the test (see Chapter X, Section 6). Alternatively, stress-strain behaviour at large strains can be found by a compression test provided that barrelling is avoided by adequate lubrication; even so, a small correction for the effect of friction at the ends may be necessary (Chapter X, Section 7).

Several empirical formulae have been proposed for fitting stress-strain curves. An early one, due to Ludwik,‡ is the power law

$$\sigma = a + b\epsilon^c, \qquad (3)$$

where a, b, c are arbitrary constants. When (3) is fitted to data for an annealed metal up to a strain of order 0·2, the tendency is to underestimate the actual stress where ϵ is small and to overestimate where ϵ is large, since a power law does not fall away sufficiently rapidly. In many metals (particularly high carbon and alloy steels) the (σ, ϵ) curve has an approximately constant slope at very large strains.§ The curve for such a metal, when heavily pre-strained, is closely represented by $\sigma = a + b\epsilon$ ($c = 1$). In some metals a more successful formula for moderate strains is

$$\sigma = a + (b-a)(1 - e^{-c\epsilon}), \qquad (4)$$

where e denotes the exponential constant. This was proposed independently by Voce and Palm.† Although $\sigma \to b$ as $\epsilon \to \infty$, b is unlikely

‡ P. Ludwik, *Elemente der technologischen Mechanik* (Berlin, 1909).

§ C. W. MacGregor, *Timoshenko Anniversary Volume* (New York, Macmillan, 1938); *Journ. App. Mech.* **6** (1939), A-156; *Proc. Am. Soc. Test. Mat.* **40** (1940), 508; *Journ. Franklin Inst.* **238** (1944), 111. See also P. W. Bridgman, loc. cit.

† E. Voce, *Journ. Inst. Metals*, **74** (1948), 537; see also discussion on p. 760. J. H. Palm, *Applied Scientific Research*, A-2 (1949), 198. A somewhat similar relation was proposed by M. Reiner, *Proc. 6th Int. Cong. App. Mech.* (Paris), 1946.

to be the saturation stress in a metal where the actual rate of hardening becomes vanishingly small. Similarly, $\sigma = a$ when $\epsilon = 0$, but the value of a giving best fit over a certain strain-range does not necessarily coincide with the initial yield stress; the formula fails, of course, for very small or purely elastic strains.

II
FOUNDATIONS OF THE THEORY

1. The ideal plastic body

WITH present knowledge the element of time cannot be adequately incorporated in a mathematical account of the plasticity of metals. The theory to be described is valid only at temperatures for which recovery, creep, and thermal phenomena generally, can be neglected. The absence of a high-temperature theory is due partly to insufficient data, but more to the circumstance that the stress needed to enforce plastic flow at a given temperature is not a single-valued function of the rate of strain and the total distortion (as measured conventionally).† Preliminary attempts to include thermal effects within the framework of the low-temperature theory have always assumed a fictitious equation of state.‡

Another class of phenomena which is secondary in many applications, and which will be disregarded here, arises from non-uniformity on a microscopic scale. The principal examples are the Bauschinger effect and the hysteresis loop in unloading and reloading; these are due, as we have seen, to differential hardening of the variously oriented crystals during straining. The neglect of these phenomena in the theory does not prevent the inclusion of anisotropy due to preferred orientation, in which local uniformity can, in principle, be preserved (plastic anisotropy is treated in Chap. XII). It is true that anisotropy and the Bauschinger effect are generally found together, but they arise from different causes and the Bauschinger effect can be removed by a mild annealing while the preferred orientation is retained.§

Size effects are also neglected; they may be due to grain size, inclusions, or the difficulty of nucleation of a slip process.‖ In the following sections other properties will be ascribed to the ideal plastic body.

---

† E. Orowan, *West of Scotland Iron and Steel Inst.* February 1947; J. E. Dorn, A. Goldberg, T. E. Tietz, *Metals Technology*, Tech. Pub. 2445, (1948).

‡ P. Ludwik, *Elemente der technologischen Mechanik*, (Julius Springer, Berlin, 1909). J. H. Hollomon, *Metals Technology*, Tech. Pub. 2304, September 1946; *Trans. Am. Inst. Min. Met. Eng.*, 171 (1947), 535. J. H. Hollomon and C. Zener, *Journ. App. Phys.* 17 (1946), 82. J. H. Hollomon and J. D. Lubahn, *Phys. Rev.* 70 (1946), 775. J. D. Lubahn, *Journ. App. Mech.* 14 (1948), A–299.

§ G. Sachs, *Zeits. Ver. deut. Ing.* 71 (1927), 1511; G. Sachs and H. Shoji, *Zeits. Phys.* 45 (1927), 776. See also C. F. Elam, *Distortion of Metal Crystals*, p. 90 (Oxford, Clarendon Press, 1934).

‖ W. L. Bragg, *Nature*, 149 (1942), 511; E. Orowan, J. F. Nye, and W. J. Cairns, Ministry of Supply, Armament Research Department, Theoretical Research Report 16/45.

2. The criterion of yielding

(i) *General considerations.* It is supposed that if an element of the ideal solid is plastically deformed, and then unloaded, it will recover elastically and in such a way that the change of strain depends linearly on the change of stress. Unless the previous working has been severe the macroscopic elastic behaviour of a metal can only be slightly affected by distortions of the lattice. During unloading, elastic recovery is limited by the plastic yielding of favourably oriented grains, but there is evidence to show that the elastic modulus, calculated from the initial slope of the unloading curve, remains invariant.† The elastic constants of the ideal solid are therefore assumed to retain the same values provided they are defined with respect to the current shape of the element. Moreover, it is supposed that an element recovers its original shape when reloaded along the same path to the initial state of stress, and that there is no hysteresis loop. The reloading may, however, be carried out with other combinations of stress, each of which will eventually produce a further plastic strain if all the components of stress are increased monotonically. For example, if a thin cylindrical tube has been uniformly strained in tension and then partly unloaded, we may inquire what torque must be added to cause a further permanent distortion. A law defining the limit of elasticity under any possible combination of stresses is known as a criterion of yielding. The concept of a yield criterion is not, of course, restricted merely to loading directly from the annealed state, as is sometimes thought.

It is supposed, for the present, that the material is isotropic. Since plastic yielding can then depend only on the magnitudes of the three principal applied stresses, and not on their directions, any yield criterion is expressible in the form

$$f(J_1, J_2, J_3) = 0, \qquad (1)$$

where J_1, J_2, and J_3 are the first three invariants of the stress tensor σ_{ij}. They are defined in terms of the principal components of stress σ_1, σ_2, σ_3, by the relations

$$J_1 = \sigma_1 + \sigma_2 + \sigma_3; \qquad J_2 = -(\sigma_1\sigma_2 + \sigma_2\sigma_3 + \sigma_3\sigma_1); \qquad J_3 = \sigma_1\sigma_2\sigma_3. \qquad (2)$$

The principal stresses are the roots of the cubic equation

$$\lambda^3 - J_1\lambda^2 - J_2\lambda - J_3 = 0.$$

The stress may be specified either by the three principal components or by the three tensor invariants. However, although any function of the

---

† J. V. Howard and S. L. Smith, *Proc. Roy. Soc.* A, **107** (1925), 113; R. W. Mebs and D. J. McAdam, Jr., *Nat. Adv. Comm. Aero.*, Tech. Note 1100, March 1947.

invariants can be expressed in terms of the principal stresses, it is not true that any function of the principal stresses is a possible yield criterion; only functions symmetrical in the three principal stresses are permissible. The function f is characteristic of the state of the element immediately before unloading, and hence depends on the whole mechanical and heat treatment of the metal since it was last in the fully annealed condition. It will later be necessary to settle the precise manner in which f varies with the amount of pre-strain, but for the moment this question is left open.

An immediate simplification of (1) can be achieved by using the experimental fact that the yielding of a metal is, to a first approximation, unaffected by a moderate hydrostatic pressure or tension, either applied alone or superposed on some state of combined stress.† Supposing this to be strictly true for the ideal plastic body, it follows that yielding depends only on the principal components $(\sigma'_1, \sigma'_2, \sigma'_3)$ of the deviatoric, or reduced, stress tensor

$$\sigma'_{ij} = \sigma_{ij} - \sigma\delta_{ij}, \qquad (3)‡$$

where $\sigma = \tfrac{1}{3}\sigma_{ii}$ is the hydrostatic component of the stress. The principal components are not independent, since $\sigma'_1 + \sigma'_2 + \sigma'_3$ is identically zero. The yield criterion now reduces to the form

$$f(J'_2, J'_3) = 0, \qquad (4)$$

where

$$\left. \begin{aligned} J'_2 &= -(\sigma'_1\sigma'_2 + \sigma'_2\sigma'_3 + \sigma'_3\sigma'_1) = \tfrac{1}{2}(\sigma'^2_1 + \sigma'^2_2 + \sigma'^2_3) = \tfrac{1}{2}\sigma'_{ij}\sigma'_{ij}, \\ J'_3 &= \sigma'_1\sigma'_2\sigma'_3 = \tfrac{1}{3}(\sigma'^3_1 + \sigma'^3_2 + \sigma'^3_3) = \tfrac{1}{3}\sigma'_{ij}\sigma'_{jk}\sigma'_{ki}. \end{aligned} \right\} \qquad (5)$$

A further restriction must be imposed in view of the supposition that the ideal plastic body does not show a Bauschinger effect, so that the magnitude of the yield stress is the same in tension and compression. This, as we have seen, is true of metal single crystals where reversing the applied stress only changes the sense of the constituent shears and not the operative glide-planes; the same is true of a polycrystal after internal stresses due to the differential crystal orientations have been removed by annealing. More generally, when an element is unloaded from a plastic stress state σ_{ij} and then reloaded to the state $-\sigma_{ij}$, keeping the ratios of the stress components constant throughout, it is assumed that the element is deformed only elastically and is finally again on the point

† M. Polanyi and E. Schmid, *Zeits. Phys.* **16** (1923), 336. P. W. Bridgman, *Metals Technology*, Tech. Pub. 1782 (1944); *Trans. Am. Inst. Min. Met. Eng.* **162** (1945), 569; *Rev. Mod. Phys.* **17** (1945), 3; *Journ. App. Phys.* **17** (1946), 201; *Trans. Am. Soc. Met.* **40** (1948), 246.

‡ See Appendix I for an explanation of tensor notation and the summation convention.

of yielding. Since J_3' changes sign when the stresses are reversed, it follows that f must be an even function of J_3'.

(ii) *A geometrical representation of stress.* It is helpful to introduce here a geometrical representation of stress. For present purposes a state of stress is completely specified by the values of the three principal components, so that any stress-state may be represented by a bound vector in a three-dimensional space where the principal stresses are taken as Cartesian coordinates.† In Fig. 1 **OS** is the vector $(\sigma_1, \sigma_2, \sigma_3)$, while

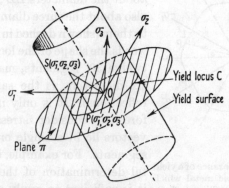

FIG. 1. Geometrical representation of a plastic state of stress in $(\sigma_1, \sigma_2, \sigma_3)$ space, where $\sigma_1, \sigma_2, \sigma_3$ are the principal components. A yield criterion, which is independent of the hydrostatic component of stress, is represented by a curve C in the plane Π whose equation is $\sigma_1 + \sigma_2 + \sigma_3 = 0$.

OP is the vector representing the deviatoric stress $(\sigma_1', \sigma_2', \sigma_3')$. **OP** always lies in the plane Π whose equation is $\sigma_1 + \sigma_2 + \sigma_3 = 0$, while **PS**, representing the hydrostatic component (σ, σ, σ) of the stress, has direction cosines $(1/\sqrt{3}, 1/\sqrt{3}, 1/\sqrt{3})$ and is perpendicular to Π. Now the yield criterion (4), for a particular state of the metal, can be regarded as a surface in this space. Since the yielding is independent of the hydrostatic component of stress, it is evident that this surface is a right cylinder with generators perpendicular to Π and cutting it in some curve C. It is sufficient to discuss possible forms of the curve C, and to consider only stress-states whose hydrostatic component is zero. In Fig. 2 Π is the plane of the paper; the yield locus C and the orthogonal projections of the axes of reference are shown. The locus may be convex or concave to the origin, but obviously not so that a radius cuts it twice.

† B. P. Haigh, *Engineering*, **109** (1920), 158; H. M. Westergaard, *Journ. Franklin Inst.* **189** (1920), 627. See also W. M. Baldwin, Jr., *Metals Technology*, Tech. Pub. 1980, (1946); W. W. Sokolovsky, *Doklady Akad. Nauk S.S.S.R.* **61** (1948), 223.

Now if $(\sigma_1, \sigma_2, \sigma_3)$ is a plastic state, so also is $(\sigma_1, \sigma_3, \sigma_2)$, since the element is isotropic. The locus is therefore symmetrical about LL', and similarly about MM' and NN'. This is equivalent to saying that the yield criterion is a function of the tensor invariants. If, for any point on the locus, a radius is drawn through the origin (representing unloading with constant stress ratios), it must meet the locus again at the same distance from the origin, since there is no Bauschinger effect. Hence, the locus is symmetrical, not only about the diameters LL', MM', NN', but also about the three diameters orthogonal to them (shown dashed in Fig. 2). In other words the shape of the locus in each of the twelve 30° segments, marked off by the six diameters, is the same (apart from reflections). It is only necessary, therefore, to consider stress states whose vectors lie in a single one of the twelve segments. For example, in an experimental determination of the yield criterion, it is sufficient to apply only those stress systems whose stress vectors lie in a selected segment. This may be otherwise expressed in terms of a parameter μ introduced by Lode:†

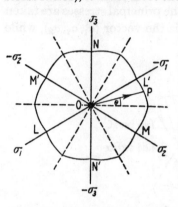

FIG. 2. General appearance of a yield locus for an isotropic metal which does not show a Bauschinger effect.

$$\mu = \frac{2\sigma_3 - \sigma_1 - \sigma_2}{\sigma_1 - \sigma_2}. \qquad (6)$$

It can be shown by simple geometry that

$$\mu = -\sqrt{3}\tan\theta, \qquad (7)$$

where θ defines the position of a stress vector **OP**, as in Fig. 2.‡ The yield locus can be completely determined by applying stress systems such that θ covers the range from 0° to 30°, μ varying from 0 to -1. When $\mu = 0$, $\sigma_3 = \frac{1}{2}(\sigma_1 + \sigma_2)$; the state of stress is then a pure shear $\frac{1}{2}(\sigma_1 - \sigma_2, \sigma_2 - \sigma_1, 0)$, together with a hydrostatic stress $\frac{1}{2}(\sigma_1 + \sigma_2)$. When $\mu = -1$, $\sigma_2 = \sigma_3$; the state of stress is a uniaxial stress $(\sigma_1 - \sigma_2, 0, 0)$ together with a hydrostatic stress σ_2. Thus the extreme values $\mu = 0$, -1, correspond respectively to pure shear and pure tension (or compres-

† W. Lode, *Zeits. Phys.* **36** (1926), 913; also *Zeits. ang. Math. Mech.* **5** (1925), 142.

‡ If (x, y) are the Cartesian coordinates of P with respect to the dashed line $\theta = 0$ and the line ON, then
$$x = (\sigma_2 - \sigma_1)/\sqrt{2}; \qquad y = (2\sigma_3 - \sigma_1 - \sigma_2)/\sqrt{6}.$$
Since $\mu = -3(\sigma_1' + \sigma_2')/(\sigma_1' - \sigma_2')$ we obtain from (7): $\sigma_1'/\sigma_2' = (\tan\theta + \sqrt{3})/(\tan\theta - \sqrt{3})$. From this equation the ratios $\sigma_1' : \sigma_2' : \sigma_3'$ can be found when the direction of OP is given.

sion), while the intermediate range may be covered by combined torsion and tension of a thin-walled tube. If σ is the tensile stress and τ the shear stress acting over a plane transverse to the tube axis, the principal stresses are

$$\sigma_1 = \tfrac{1}{2}\sigma + (\tfrac{1}{4}\sigma^2 + \tau^2)^{\frac{1}{2}}, \qquad \sigma_2 = \tfrac{1}{2}\sigma - (\tfrac{1}{4}\sigma^2 + \tau^2)^{\frac{1}{2}}, \qquad \sigma_3 = 0.$$

μ is therefore equal to $-\sigma/(\sigma^2 + 4\tau^2)^{\frac{1}{2}}$.

Another convenient method of covering the range is to stress a thin cylindrical tube under combined tensile load L and internal pressure P. Here the mean radial stress component σ_2 is negligible and the state of stress is effectively biaxial, namely a circumferential stress $\sigma_1 = Pa/t$ and an axial stress $\sigma_3 = L/2\pi at$, where t is the wall thickness and a is the tube radius. μ is then equal to $L/\pi a^2 P - 1$, and so $L/\pi a^2 P$ must be varied between $+1$ and 0. The state of stress corresponding to the first limit may be realized in a tube with closed ends under internal pressure, while the second limit corresponds to a tube stressed only by internal pressure with its ends free to contract.

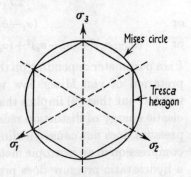

Fig. 3. Yield loci corresponding to the criteria of von Mises (circle) and Tresca (regular hexagon).

(iii) *The yield criteria of Tresca and von Mises. Experimental evidence.* The first investigation of a yield criterion appears to be the long series of experiments by Tresca (1864), in which he measured the loads required to extrude metals through dies of various shapes. The distribution of stress in the plastic region was of course far from uniform, and Tresca's attempts to analyse it were very crude. However, he concluded† that yielding occurred when the maximum shear stress reached a certain value. In this he was probably influenced by a slightly more general law for the failure of soils, suggested many years earlier by Coulomb.‡ Tresca's yield criterion is

$$\sigma_1 - \sigma_3 = \text{constant}, \tag{8}$$

where $\sigma_1 \geqslant \sigma_2 \geqslant \sigma_3$. It may alternatively be written in terms of the invariants J_2', J_3', but the result is complicated and not very useful. The yield locus in the plane diagram is a regular hexagon (Fig. 3).

† H. Tresca, *Comptes Rendus Acad. Sci. Paris*, **59** (1864), 754; and **64** (1867), 809; *Mém. Sav. Acad. Sci. Paris*, **18** (1868), 733; and **20** (1872), 75 and 281. See also J. Boussinesq, *Comptes Rendus Acad. Sci. Paris*, **166** and **167** (1918), for several articles dealing with the theory of Tresca's experiments.
‡ C. A. Coulomb, *Mém. Math. et Phys.* **7** (1773), 343.

Most of the various yield criteria that have been suggested for metals are now only of historic interest, since they conflict with later experiments in predicting that a hydrostatic stress *always* influences yielding. The two simplest which do not have this fault are the criterion of Tresca, just described, and the criterion due to von Mises† (1913). Von Mises suggested that yielding occurred when J_2' reached a critical value, or, in other words, that the function f in (4) did not involve J_3'. His criterion can be written in the alternative forms

$$\left. \begin{array}{l} 2J_2' = \sigma_{ij}'\sigma_{ij}' = \sigma_1'^2+\sigma_2'^2+\sigma_3'^2 = 2k^2, \\ \text{or} \quad (\sigma_1-\sigma_2)^2+(\sigma_2-\sigma_3)^2+(\sigma_3-\sigma_1)^2 = 6k^2, \\ \text{or} \quad (\sigma_x-\sigma_y)^2+(\sigma_y-\sigma_z)^2+(\sigma_z-\sigma_x)^2+6(\tau_{yz}^2+\tau_{zx}^2+\tau_{xy}^2) = 6k^2. \end{array} \right\} \quad (9)$$

k is a parameter depending on the amount of pre-strain. A physical interpretation of von Mises' law was suggested by Hencky‡ (1924), who pointed out that (9) implies that yielding begins when the (recoverable) elastic energy of distortion reaches a critical value. Thus a hydrostatic pressure does not cause yielding since it produces only elastic energy of compression in an isotropic metal. In some anisotropic metals, however, a hydrostatic pressure does produce elastic distortion, but presumably no yielding if it is of moderate amount. For this reason alone Hencky's interpretation does not seem to have a general physical significance. In fact, of course, several interpretations of (9) are possible. For example, as Nadai has proposed,§ (9) implies that yielding begins when the shear stress acting over the octahedral planes reaches a certain value. The octahedral planes are those having the same relation to the axes of principal stress as the faces of an octahedron have to its cubic axes. The octahedral shear stress has the value $\sqrt{\tfrac{2}{3}}k$ at the yield-point. Von Mises' criterion was anticipated, to some extent, by Huber‖ (1904) in a paper†† in Polish which did not attract general attention until nearly twenty years later. Huber distinguished two cases depending on whether the hydrostatic component of the stress was a tension or a compression. If the latter, he proposed that yielding‡‡ was determined by the elastic energy of distortion; if the former, by the total elastic energy (as suggested universally by Beltrami and Haigh§§). Von Mises and

† R. von Mises, *Göttinger Nachrichten, math.-phys. Klasse*, (1913), 582.
‡ H. Hencky, *Zeits. ang. Math. Mech.* 4 (1924), 323.
§ A. Nadai, *Journ. App. Phys.* 8 (1937), 205.
‖ Apparently also by Clerk Maxwell; letter to W. Thomson, 18 Dec. 1856.
†† M. T. Huber, *Czasopismo techniczne, Lemberg*, 22 (1904), 81.
‡‡ It is not clear whether Huber was thinking of plastic yielding or brittle rupture.
§§ E. Beltrami, *Rend. Ist. Lomb.* 18 (1885), 704; B. P. Haigh, *Brit. Ass. Reports*, Section G, 1919.

Schleicher† independently generalized (9) by replacing k by an arbitrary function of $(\sigma_1+\sigma_2+\sigma_3)$. This criterion could be used for materials whose yielding is influenced by the hydrostatic component of stress; it includes the laws of Huber and Beltrami as particular cases.

By setting $\sigma_1 = -\sigma_2$, $\sigma_3 = 0$, in (9), k may be identified with the maximum shear stress in yielding in a state of pure shear, which may, for example, be obtained by twisting a thin cylindrical tube. The yield stress Y in uniaxial tension is then $\sqrt{3}k$, as may be seen on substituting $\sigma_1 = Y$, $\sigma_2 = \sigma_3 = 0$. Thus the Mises criterion predicts that the maximum shear stress in pure torsion is greater by a factor $2/\sqrt{3}$, or about 1·155, than that in pure tension. Tresca's criterion, on the other hand, predicts that they are equal. This is the most significant difference between the two criteria, and the one most easily tested by experiment. A general comparison is best made in terms of the respective yield loci in the plane diagram. It has been mentioned that the Tresca locus is a hexagon, while it is obvious from the first relation in (9) that the Mises locus is a circle of radius $\sqrt{2}k$ or $\sqrt{\frac{2}{3}}Y$. By suitably choosing the values of k and the constant in (8), the two criteria can be made to agree with each other and with experiment for a single state of stress. This may be selected arbitrarily; it is conventional to make the circle pass through the corners of the hexagon by taking the constant in (8) to be Y, the yield stress in simple tension. The loci then differ most for a state of pure shear, where the Mises criterion gives a yield stress $2/\sqrt{3}$ times that given by the Tresca criterion. For most metals von Mises' law fits the data more closely than Tresca's, but it frequently happens that Tresca's is simpler to use in theoretical applications. If the latter is adopted, a greater overall accuracy can be obtained by taking the constant in (8) as mY, where m is an empirically assigned number lying between 1 and 1·155. The maximum error in the stress components for *a given stress state* (specified by a certain value of μ) is then never more than 8 per cent.‡ However, the comparison of the distributions of stress in a body, corresponding to the two yield criteria, is made at *a given point of the body*, and the discrepancy there can be very much greater since the stress states are generally not identical. On the other hand, when the stress states in the particular application cover only a limited arc of the circle, it may be possible to choose m so that the error is nowhere more than 1 or 2 per cent.

† F. Schleicher, *Zeits. ang. Math. Mech.* 5 (1925), 478; 6 (1926), 199; R. von Mises, ibid. 6 (1926), 199 (discussion of paper by Schleicher).

‡ It is frequently stated that the error is only less than 15 per cent., m automatically being taken as unity.

The yield criterion of von Mises has been shown to be in excellent agreement with experiment for many ductile metals, for example copper, nickel, aluminium, iron, cold-worked mild steel, medium carbon and alloy steels. The influence of the intermediate principal stress on yielding, and the corresponding failure of Tresca's criterion, was first clearly shown in the work of Lode† (1925), who stressed tubes of iron, copper,

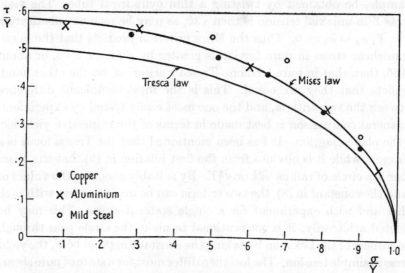

Fig. 4. Experimental results of Taylor and Quinney from combined torsion and tension tests, each metal being work-hardened to the same state for all tests. The Mises law is $\sigma^2 + 3\tau^2 = Y^2$, while the Tresca law is $\sigma^2 + 4\tau^2 = Y^2$, where σ = tensile stress, τ = shear stress, Y = tensile yield stress.

and nickel, under combined tension and internal pressure. The substantial accuracy of the Mises law was afterwards demonstrated by the work of Taylor and Quinney‡ (1931), Lessells and MacGregor§ (1940), and Davis‖ (1945). As an example, the results of Taylor and Quinney are given in Fig. 4. Better agreement could occasionally be obtained by adding a small correction term in J_3', but in view of other differences between the ideal plastic body and a real metal, it would hardly be worthwhile in practical applications.

For the upper yield-point of annealed mild steel Tresca's law appears

† W. Lode, *Zeits. ang. Math. Mech.* **5** (1925), 142; *Zeits. Phys.* **36** (1926), 913; *Forschungsarbeiten des Vereines deutscher Ingenieure*, **303** (1927).
‡ G. I. Taylor and H. Quinney, *Phil. Trans. Roy. Soc.* A, **230** (1931), 323.
§ J. M. Lessells and C. W. MacGregor, *Journ. Franklin Inst.* **230** (1940), 163.
‖ E. A. Davis, *Trans. Am. Soc. Mech. Eng.* **62** (1940), 577; **65** (1943), A–187. See also Miller and Edwards, *Journ. Am. Petr. Inst.* (1939), 483; Marin and Stanley, *Journ. Am. Weld. Soc., Weld. Res. Suppl.* **19** (1940), 748.

to fit the data better than Mises'. However, the situation is confused by the sensitivity of the upper yield-point to the conditions of testing, for example, eccentricity of loading, non-uniformity of the specimen, stress concentration in the fillets. Morrison,† in particular, has advocated caution in accepting many of the early observations on mild steel contained in the work of Guest‡ (1900), Scoble§ (1906), Seely and Putnam|| (1919), and Ros and Eichinger†† (1926). In very careful experiments with annealed mild steel Morrison has found that the criterion of yielding appears to vary with the absolute size of the specimen. It has been suggested by Cook,‡‡ and by Cook and Robertson,§§ that the yield stress in mild steel is somewhat higher when the stress distribution is not uniform (as in bending, or in a tube expanded by pressure). This may be true for metals in general, since there are theoretical reasons for supposing that a certain minimum volume of plastic material has to be produced before slip can be initiated.

Several attempts|||| have been made to derive the yield criterion of a polycrystal from the observed behaviour of a face-centred cubic single crystal. No really satisfactory theory, taking account of the mutual constraints between the grains, has yet been proposed.

3. Strain-hardening

(i) *Dependence of the yield locus on the strain-history.* The yield locus for a given state of the metal must depend in some complicated way on the whole of the previous process of plastic deformation since the last annealing. It will be supposed that the crystals remain randomly oriented, so that isotropy is preserved (the anisotropy developed during cold work is usually negligible so long as the total strain is not too large,

† J. L. M. Morrison, *Proc. Inst. Mech. Eng.* **142** (1940), 193; see also discussion in **144**, 33.
‡ J. J. Guest, *Phil. Mag.* **50** (1900), 69; see, however, Guest's most recent conclusions: *Proc. Inst. Auto. Eng.* **35** (1940), 33.
§ W. A. Scoble, *Phil. Mag.* **12** (1906), 533.
|| F. B. Seely and W. J. Putnam, *Univ. Illinois Eng. Exp. Bull.* Series 115, (1919).
†† M. Ros and A. Eichinger, *Proc. 2nd Int. Cong. App. Mech.* Zürich (1926), 315. See also W. Mason, *Proc. Inst. Mech. Eng.* **4** (1909), 1205; A. J. Becker, *Univ. Illinois Eng. Dept. Bull.* **85** (1916), 84.
‡‡ G. Cook, *Engineering*, **132** (1931), 343; *Phil. Trans. Roy. Soc.* A, **230** (1931), 103; *Proc. Roy. Soc.* A, **137** (1932), 559; *Trans. Engineers and Shipbuilders in Scotland*, **81** (1937), 371.
§§ G. Cook and A. Robertson, *Engineering*, **92** (1911), 783; A. Robertson and G. Cook, *Proc. Roy. Soc.* A, **88** (1913), 462. See also D. Morkovin and O. Sidebotham, *Univ. Illinois. Eng. Exp. Bull.*, Series 372, (1947); J. A. Pope, *Engineering*, **164** (1947), 284.
|||| G. Sachs, *Zeits. Ver. deut. Ing.*, **72** (1928), 734; H. L. Cox and D. G. Sopwith, *Proc. Phys. Soc.* **49** (1937), 134; U. Dehlinger, *Zeits. Metallkunde*, **35** (1943), 182; N. K. Snitko, *Journ. Tech. Phys. (Russian)*, **18** (1948), 857.

for example less than 30 per cent. reduction in rolling). In searching for a mathematical formulation of strain-hardening we shall introduce a broad simplification at the outset. It is assumed that no matter by what strain-path a given stress state is reached the final yield locus is the same. This cannot be true of a real metal since the yield locus depends on the distribution of microscopic internal stresses, and these in turn depend on the strain-history. The latent hardening of non-active glide-planes in a single crystal suggests that the yield locus would be a function only of the final applied stress were it not for *additional* internal stresses induced during the previous deformation by the non-uniform orientation of the crystal grains. It appears probable, however, that since a mild annealing is known to remove the additional internal stresses following a simple strain-path and responsible for the Bauschinger effect, it should also remove those remaining after a complex path. Since we are disregarding the Bauschinger effect we must, for consistency, postulate that the yield locus for the ideal body is uniquely determined by the final plastic state of stress. The yield loci for two states, reached by different strain-paths, then either do not have any point in common, or are entirely coincident so that the states are identical as regards plastic yielding. There are, of course, many strain-paths which produce a given final state of hardening, defined, say, by the value of the yield stress in tension. For all of these the yield loci are identical, the yield stresses being equal under any other stress system. Otherwise expressed, there are only a single infinity of distinct states, and all can be obtained by, for example, different amounts of pure tension.

During continued deformation the shape of the yield locus may change, and it should not be tacitly assumed that it merely increases in size. Thus, in an experimental investigation of the yield criterion, it would be wrong to assume the same relation to be defined by data from combined stressing after different amounts of pre-strain in tension, the data being represented non-dimensionally in terms of the tensile yield stress. Possible changes in shape of the yield locus have not been investigated in detail. The experimental data for mild steel, reviewed in the last section, suggests that the yield locus changes over from a hexagon to a circle with progressive cold-work.† However, for other steels, and for copper and aluminium, von Mises' criterion appears to fit the data equally well no matter what the degree of pre-strain. The loci are then always circles, expanding steadily during continued loading. This is

† A possible explanation has been suggested by G. I. Taylor, *Proc. Roy. Soc.* A, **145** (1934), 1.

the simplest and most attractive hypothesis conceivable for the ideal plastic body, and it will accordingly be used in all subsequent applications.

It now remains to relate the radius $\sqrt{2k}$ of the Mises circle to the plastic deformation since the last annealing. The obvious association, marked by the term 'work-hardening', between the work used to produce plastic glide, and the hardening created by the glide, suggests the hypothesis that the degree of hardening is a function *only* of the total plastic work, and is otherwise independent of the strain-path. This does not, of course, imply that the strain energy associated with the lattice distortions responsible for hardening is equal to the external plastic work; in fact, as we saw in Chapter I, only a small fraction of the external work appears to be used in permanently distorting the lattice. Whether the hypothesis is sound is for future theoretical investigation to show, probably by considerations of the production, arrangement, and movement of lattice imperfections.† At present the assumption is justified by its simplicity and approximate agreement with observation.

We now state the hypothesis mathematically. Let $d\epsilon_{ij}$ represent an infinitesimal increment of strain during the continued loading of an element of material. $d\epsilon_{ij}$ is measured with respect to the current configuration, and is defined by the equation

$$d\epsilon_{ij} = \frac{1}{2}\left[\frac{\partial}{\partial x_j}(du_i) + \frac{\partial}{\partial x_i}(du_j)\right], \qquad (10)$$

where du_i ($i = 1, 2, 3$) is the vector representing the incremental displacement of a point whose current position-vector is x_i referred to Cartesian coordinates fixed in the element. A part of this strain is recoverable on removing the added stress $d\sigma_{ij}$. This is the elastic component of the strain, defined by

$$d\epsilon_{ij}^e = \frac{d\sigma'_{ij}}{2G} + (1-2\nu)\delta_{ij}\frac{d\sigma}{E}, \qquad (11)$$

† G. I. Taylor, *Proc. Roy. Soc.* A, **145** (1934), 362 and 388, has shown that the parabolic type of stress-strain curve of a f.c.c. crystal in shear can be obtained from suitable assumptions about the distribution of dislocations. In later papers, *Journ. Inst. Metals*, **62** (1938), 307, and *Timoshenko Anniversary Volume*, p. 218 (Macmillan, New York, 1938), he has calculated the tensile stress-strain curve of polycrystalline aluminium in terms of the observed shear-hardening of a single crystal. While the agreement with the measured curve is good, there is little evidence as to the validity of individual assumptions, for example concerning operative glide-planes and the mutual constraints between grains. Most data concerning the macroscopic behaviour of grains in an aggregate relates to surface phenomena, to which Taylor's theory is not meant to apply. An extension of the theory to general states of stress has not yet been made. See also A. Köchendörfer, *Metallforschung*, **2** (1947), 173.

where $d\sigma'_{ij}$ and $d\sigma$ have the meanings assigned in (3), and E, G, and ν are Young's modulus, modulus of shear, and Poisson's ratio, respectively. The external work dW per unit volume done on the element during the strain $d\epsilon_{ij}$ is $\sigma_{ij}d\epsilon_{ij}$, of which a part $dW_e = \sigma_{ij}d\epsilon^e_{ij}$ is recoverable elastic energy. The remainder is called the plastic work per unit volume, and is, therefore, by definition

$$dW_p = dW - dW_e = \sigma_{ij}(d\epsilon_{ij}-d\epsilon^e_{ij}) = \sigma_{ij}d\epsilon^p_{ij}, \qquad (12)$$

where $d\epsilon^p_{ij} = d\epsilon_{ij} - d\epsilon^e_{ij}$ is called the plastic strain-increment. dW_p is essentially positive since plastic distortion is an irreversible process, in the thermodynamic sense. Since the compressibility under stresses exceeding the elastic limit is of the same order of magnitude as the elastic compressibility, the total plastic work per unit volume expended during a certain finite deformation is, with neglect of a term of order σ/E,†

$$W_p = \int \sigma_{ij}\,d\epsilon^p_{ij}, \qquad (13)$$

where the integral is taken over the actual strain-path from some initial state of the metal. The hypothesis that the radius of the Mises circle is a function only of W_p may then be written as

$$\bar{\sigma} = +\sqrt{\tfrac{1}{2}\{(\sigma_1-\sigma_2)^2+(\sigma_2-\sigma_3)^2+(\sigma_3-\sigma_1)^2\}^{\frac{1}{2}}} = \sqrt{\tfrac{3}{2}}\{\sigma'_{ij}\sigma'_{ij}\}^{\frac{1}{2}} = F(W_p), \qquad (14)$$

where, to follow the accepted convention, $\bar{\sigma}$ is written for Y or $\sqrt{3}k$.‡ $\bar{\sigma}$ is variously known as the generalized stress, effective stress, or equivalent stress. The assumption that hardening occurs if, and only if, plastic work is done is evidently consistent with the fact that purely elastic strains do not produce hardening. There is, however, one restriction which (14) appears to impose on possible plastic strains. We have previously seen that the yield criterion may normally be assumed independent of the hydrostatic component of stress. For a similar range of pressure (and tension) this appears to be true also of the plastic deformation and the resultant hardening. If we take this to be a further property of the ideal plastic body, as we shall, it must for consistency be supposed that no plastic work is done by the hydrostatic component of the applied stress. This implies that there is no plastic or irrecoverable change in volume. According to experiment this is true to a very close approxi-

† Here, and elsewhere in the theory of plastic deformation, σ/E is neglected in comparison with unity, just as in the theory of elasticity. Thus, for copper, E is about 12×10^{11} dyne/cm.², while σ will usually be less than 5×10^9 dyne/cm.².

‡ More generally one could assume $W_p = \phi(J'_2, J'_3)$ when the yield locus is not a circle. The function ϕ must be consistent with the requirement that the locus steadily expands with increasing W_p.

mation,† and so no practical limitation is placed on the applicability of (14). Stated otherwise, changes in volume during plastic deformation are elastic, and for our ideal plastic body $d\epsilon_{ii}^p = 0$. If the deviatoric strain-increment is defined as

$$d\epsilon'_{ij} = d\epsilon_{ij} - \delta_{ij} d\epsilon, \qquad d\epsilon = \tfrac{1}{3} d\epsilon_{ii},$$

we have from (11) and (12):

$$\left. \begin{aligned} d\epsilon &= \frac{(1-2\nu)}{E} d\sigma, \\ d\epsilon'_{ij} &= d\epsilon_{ij}^p + \frac{d\sigma'_{ij}}{2G}. \end{aligned} \right\} \qquad (15)$$

and

The increment of plastic work may then be written

$$dW_p = \sigma_{ij}\left(d\epsilon'_{ij} - \frac{d\sigma'_{ij}}{2G}\right) = \sigma'_{ij}\left(d\epsilon'_{ij} - \frac{d\sigma'_{ij}}{2G}\right).$$

(ii) *Comparison of stress-strain curves in tension, compression, and torsion.* It is not clear when the relation (14) was first introduced into the literature, but Taylor and Quinney‡ (1931), and Schmidt§ (1932), proposed a hypothesis to which (14) reduces when the elastic strain-increment can be neglected. This is permissible except where the stresses are changing very rapidly with increasing strain, for example where the rate of work-hardening is of order E. The function F can be determined from the stress-strain curve of a cylindrical rod in tension (until necking intervenes). $\bar{\sigma}$ is then just the applied tensile stress σ, while

$$W_p = \int \sigma\left(\frac{dl}{l} - \frac{d\sigma}{E}\right) = \int_{l_0}^{l} \frac{\sigma\, dl}{l} - \frac{\sigma^2}{2E},$$

where l is the current length of the rod, and l_0 its initial length. Thus

$$\sigma = F\left(\int_{l_0}^{l} \frac{\sigma\, dl}{l} - \frac{\sigma^2}{2E}\right).$$

---

† W. E. Alkins, *Journ. Inst. Metals*, **23** (1920), 381; D. Hanson and M. A. Wheeler, ibid. **45** (1931), 229. Usually the density first rises, as holes and cracks are closed, and afterwards falls. Decreases of less than 0·5 per cent. were observed even after 90 per cent. reduction by rolling. The absence of any positive change for single crystals suggests that the decrease is caused by the 'opening-up' of grain boundaries as dislocations accumulate there. The experimental difficulties are so great that probably only the direction of the change, and not its magnitude, is significant.
‡ G. I. Taylor and H. Quinney, *Phil. Trans. Roy. Soc.* A, **230** (1931), 323.
§ R. Schmidt, *Ingenieur-Archiv*, **3** (1932), 215.

If σ is plotted against $\ln(l/l_0) - \sigma/E$, the argument of F is simply the area under this curve up to the ordinate σ.

In uniform compression of a short cylindrical specimen, the applied pressure

$$p = F\left(\int_h^{h_0} \frac{p\,dh}{h} - \frac{p^2}{2E}\right),$$

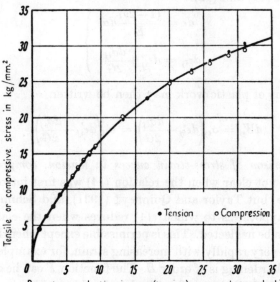

Fig. 5. Experimental data of Ludwik and Scheu, showing that the tension and compression stress-strain curves of annealed copper coincide when the stress is plotted against fractional reduction in area in tension and against fractional reduction in height in compression.

where h_0 and h are the initial and current heights. It is evident that the pressure p is the same function of $\ln(h_0/h)$ as σ is of $\ln(l/l_0)$. In other words, (14) implies that the tensile and compressive stress-strain curves coincide when true stress is plotted against the so-called logarithmic strain, but *not* when plotted against the conventional or engineering strains $(l-l_0)/l_0$, $(h_0-h)/h_0$. More generally, it is implied that the curves coincide when σ and p are plotted against the same functions of l/l_0 and h_0/h, respectively; for example, fractional reduction in area $(1-l_0/l)$ in tension and fractional reduction in height $(1-h/h_0)$ in compression. This has been found to be true in loading from the annealed state (thus avoiding any Bauschinger effect) provided the compression plates are efficiently lubricated.† An example is shown in Fig. 5 which reproduces Ludwik and Scheu's results for copper. The comparison of the tension and compression curves on the basis of logarithmic strain (sometimes called also natural strain) was first suggested by Ludwik.‡ It is apparent that a comparison could not be based on the engineering strain (except for small strains) since when this is unity the length of a tension specimen has only been doubled, while the compression specimen has been totally flattened.

Consider, as a further example, the twisting of a thin cylindrical tube. A line on the tube, originally parallel to the axis, is distorted into a helix. If ϕ is the angle made with the axial direction by the helix, the work per unit volume done by the shear stress τ in a further small twist $d\phi$ is $\tau\, d(\tan\phi)$, since $d(\tan\phi)$ is the relative displacement of two transverse sections at unit distance apart. Now $\bar{\sigma} = \sqrt{3}\tau$ and so

$$\sqrt{3}\tau = F\left[\int_0^\phi \tau\, d(\tan\phi) - \frac{\tau^2}{2G}\right].$$

Thus $\sqrt{3}\tau$ is the same function of $(\tan\phi - \tau/G)/\sqrt{3}$ as σ is of $\ln(l/l_0) - \sigma/E$. This is confirmed by experiment, both in the initial part of the stress-strain curve where the elastic component of strain is not negligible,§ and also for a further range of strains where the elastic component may be disregarded.‖ For logarithmic strains greater than about 0·2, however, the torsion curve falls increasingly below the tensile curve. This seems to be due to anisotropy which is possibly developed to a greater extent in torsion than in tension at corresponding strains, owing to transverse planes being directions of maximum shear-strain throughout the twisting. It is probable, too, that a part of the divergence may be attributable to non-uniformity introduced by necking in tension. For more complex strain-paths the state of stress must be measured experimentally in order to evaluate W_p and to test the truth of the hypothesis. Schmidt (loc. cit., p. 27) verified its approximate validity under combined torsion and tension. There seems to be no published data for other stress systems. In the next section we shall consider theoretical expressions relating the state of stress to the strain-path; these expressions can be used in conjunction with (14) to determine the work-hardening in applications.

† M. P. Malaval, *Revue de Métallurgie*, **20** (1923), 46; *Comptes Rendus Acad. Sci. Paris*, **176** (1923), 488; P. Ludwik and A. Scheu, *Stahl u. Eisen*, **45** (1925), 373; F. Körber and A. Eichinger, *Mitt. Kais. Wilh. Inst. Eisenf.* **26** (1943), 37.
‡ P. Ludwik, *Elemente der technologischen Mechanik* (Julius Springer, Berlin, 1909).
§ W. M. Shepherd, *Proc. Inst. Mech. Eng.* **159** (1948), 95.
‖ C. Zener and J. H. Hollomon, *Journ. App. Phys.* **17** (1946), 2.

(iii) *An alternative hypothesis for strain-hardening.* Another hypothesis, less natural than (14) but used more frequently, relates $\bar{\sigma}$ to a certain measure of the total plastic deformation. A quantity $\overline{d\epsilon^p}$, known as the generalized or equivalent plastic strain-increment, is defined by the equation

$$\overline{d\epsilon^p} = +\sqrt{\tfrac{2}{3}}\{d\epsilon_{ij}^p d\epsilon_{ij}^p\}^{\tfrac{1}{2}}. \tag{16}$$

By comparison with the definition of $\bar{\sigma}$ in (14) it will be seen that, apart from a numerical factor, $\overline{d\epsilon^p}$ is the same invariant function of the components of the plastic strain-increment tensor as $\bar{\sigma}$ is of the components of the deviatoric stress tensor (remembering that $d\epsilon_{ij}'^p = d\epsilon_{ij}^p$, since $d\epsilon_{ii}^p = 0$). The equivalent strain $\int \overline{d\epsilon^p}$, integrated over the strain-path, then provides a measure of the plastic distortion. It is assumed that

$$\bar{\sigma} = H\left(\int \overline{d\epsilon^p}\right), \tag{17}$$

where H is a certain function depending on the metal concerned.† The use of strain-increments is dictated by the consideration that the amount of hardening is obviously not determined only by the difference between the initial and final shapes of an element. If it were, an element would not be hardened by an extension followed by an equal compression, nor in fact by any strain-path restoring its original shape. This is disproved by experiment, which shows that a definite contribution to the hardening is made by every plastic distortion (apart from Bauschinger effects). To be satisfactory, then, the measure of total strain must involve the summation of some continually positive quantity over the whole strain-path. The measure adopted in (17) is by no means the only possible one, but it is perhaps the simplest and most natural, and satisfies all obvious requirements. A similar relation was proposed by Odquist‡ (1933), in terms of $\overline{d\epsilon}$, the analogous expression in the components of the increment of total strain. This is equivalent to (17) only when the elastic strains are negligible.

The numerical factor in the definition of $\overline{d\epsilon^p}$ has been chosen so that the function H is identical with the relation between true stress and logarithmic plastic strain in a tension or compression test. In torsion, $\int \overline{d\epsilon^p}$ is equal to $(\tan\phi - \tau/G)/\sqrt{3}$, in the previous notation, and the graph of $\sqrt{3}\tau$ plotted against $(\tan\phi - \tau/G)/\sqrt{3}$ coincides with the stress-strain curve in tension. In this case (17) leads to the same result as (14); this, of course, is not necessarily true in general. However, it will be shown in the

---

† A more general assumption would be $\int \overline{d\epsilon^p} = \psi(J_2', J_3')$, when the yield locus is not a circle.
‡ F. K. G. Odquist, *Zeits. ang. Math. Mech.* **13** (1933), 360. For other accounts see C. Zener and J. H. Hollomon, *Trans. Am. Soc. Met.* **33** (1944), 163; L. R. Jackson, *Metals Technology*, Tech. Pub. 2072 (1946).

next section that for many metals the dependence of the state of stress on the strain-path is such that the two hypotheses always lead to approximately the same results. In fact, Schmidt's data (mentioned previously) agrees equally well, within experimental error, with both (17) and (14).

It is worth noting that an explicit expression for the integral in (17) can also be obtained when the principal axes of successive strain-increments do not rotate relatively to the element, and, further, when the components of any strain-increment bear constant ratios to one another. It is supposed, too, that elastic strain-increments can be neglected, so that $\overline{d\epsilon}^p$ and $\overline{d\epsilon}$ are identical. Let $d\epsilon_1$, $d\epsilon_2$, and $d\epsilon_3$, be the principal components of an increment of strain. Since there is no change of volume, we may write
$$d\epsilon_1 : d\epsilon_2 : d\epsilon_3 = 1 : \alpha : -(1+\alpha),$$
where α is constant throughout the strain-path. Then
$$\overline{d\epsilon} = \sqrt{\tfrac{2}{3}}(d\epsilon_1^2 + d\epsilon_2^2 + d\epsilon_3^2)^{\frac{1}{2}} = \frac{2}{\sqrt{3}}(1+\alpha+\alpha^2)^{\frac{1}{2}} d\epsilon_1$$
if $d\epsilon_1$ is positive; if $d\epsilon_1$ is negative, the negative radical must be taken. Let axes of reference be chosen in each element of the material coinciding with the principal axes of strain-increment, whose directions in the element have been assumed constant. By the definition in (10), $d\epsilon_1$ is equal to $\partial(du_1)/\partial x_1$ where x_1 is measured along the first principal axis from the centre of the element. Similar expressions may be written for $d\epsilon_2$ and $d\epsilon_3$. If x_1^0 is the value of x_1 before the element is deformed, $u_1 = x_1 - x_1^0$, and so, regarding x_1^0 as an independent variable,
$$d\epsilon_1 = \frac{\dfrac{\partial}{\partial x_1^0}(du_1)}{\dfrac{\partial x_1}{\partial x_1^0}} = \frac{d\!\left(\dfrac{\partial u_1}{\partial x_1^0}\right)}{1 + \dfrac{\partial u_1}{\partial x_1^0}}.$$

Hence
$$\int d\epsilon_1 = \ln\!\left(1 + \frac{\partial u_1}{\partial x_1^0}\right) = \ln\!\left(\frac{\partial x_1}{\partial x_1^0}\right) = \epsilon_1, \text{ say.}$$

Therefore
$$\int \overline{d\epsilon} = \frac{2}{\sqrt{3}}(1+\alpha+\alpha^2)^{\frac{1}{2}}\epsilon_1.$$

ϵ_2 and ϵ_3 can also be defined by similar expressions, and we then have
$$\epsilon_1 : \epsilon_2 : \epsilon_3 = 1 : \alpha : -(1+\alpha); \qquad \epsilon_i = \ln\!\left(\frac{\partial x_i}{\partial x_i^0}\right); \tag{18}$$

and
$$\int \overline{d\epsilon} = \sqrt{\tfrac{2}{3}}(\epsilon_1^2 + \epsilon_2^2 + \epsilon_3^2)^{\frac{1}{2}} = \bar{\epsilon}, \text{ say.} \tag{19}$$

$\bar{\epsilon}$ may be thought of as an equivalent total strain, being the same function of the ϵ_i as $\overline{d\epsilon}$ is of the $d\epsilon_i$. The quantities defined in (18) are evidently analogous to the logarithmic strain $\ln(l/l_0)$ already defined for simple tension, but modified to take account of possible non-uniformity of the deformation. If there is a reversal of the strain-path (α retaining the same value) the integration must be restarted from the new origin, so that $\overline{d\epsilon}$ is always taken positive. It should also be carefully noted that, although quantities $\int d\epsilon_i$ can always be formed, even where the principal axes rotate relatively to the element, they cannot generally be evaluated explicitly, as in (18), nor do they possess any geometrical significance.

If, then, only strain-paths of this special type are considered, and if elastic strain-increments are neglected, (17) is equivalent to

$$\bar{\sigma} = H(\bar{\epsilon}), \tag{20}$$

where $\bar{\epsilon}$ is the quantity defined in terms of the logarithmic finite strains by (19). The function H is just the relation between true stress σ and logarithmic strain $\ln(l/l_0)$ in uniaxial tension (for which $\alpha = -\tfrac{1}{2}$). The work-hardening hypothesis in the restricted form (20) was first stated by Ros and Eichinger[†] (1929), though they did not limit its use to strain-paths where the strain ratios are constant.[‡] Even so, their experimental results on the deformation of mild steel tubes by internal pressure and axial load were consistent with (20) to within about 5 per cent.[§] Later tests by Davis[‖] on annealed copper tubes under internal pressure and axial load confirmed the hypothesis to a similar accuracy; the strain ratios were again not held strictly constant. Many more experiments are desirable to examine further the precise ranges of validity of (14) or (17). However, it seems safe to use either hypothesis in practical

† M. Ros and A. Eichinger, *Metalle Diskussionsbericht No. 34 der Eidgenossen Materialprüfungsanstalt*, Zürich, 1929; *Proc. 3rd Int. Cong. App. Mech.*, Stockholm, **2** (1930), 254.

‡ The reader is warned that some writers have applied (20) in cases where the principal axes rotate relatively to the element (e.g. torsion), in conjunction with definitions of the ϵ_i which lead to results at variance with (17); see, for example, A. Nadai, *Journ. App. Phys.* **8** (1937), 205; E. A. Davis, ibid. 213. Such a procedure is legitimate mathematically, but its relevance for the deformation of metals depends on its physical appropriateness and agreement with experiment. At present there seems to be no general theory worth considering as an alternative to (14) or (17).

§ The mild steel was annealed, and so better agreement could not be expected if the yield locus varies from a hexagon to a circle.

‖ E. A. Davis, *Trans. Am. Soc. Mech. Eng.* **65** (1943), A-187; *Journ. App. Mech.* **12** (1945), A-13. See also W. T. Lankford, J. R. Low, and M. Gensamer, *Trans. Am. Inst. Min. Met. Eng.* **171** (1947), 574; D. M. Cunningham, E. G. Thomsen, and J. E. Dorn, *Proc. Am. Soc. Test. Mat.* **47** (1947), 546. This paper also includes an account of an experiment in which the stress-ratios were varied, namely by expanding a tube under internal pressure while applying a constant axial tension. In this connexion see also H. E. Davis and E. R. Parker, *Journ. App. Mech.* **15** (1948), 201; W. R. Osgood, ibid. **69** (1947), A-147; S. J. Fraenkel, ibid. **15** (1948), 193.

applications, where random variations in the properties of the material may well exceed small inaccuracies of theory.

4. The complete stress-strain relations

(i) *General relations between stress and strain.* In the last section a theory was formulated to describe the hardening of a metal during plastic distortion. Suppose, now, that an element of metal has been cold-worked by taking it along a certain strain-path, and that its state of hardening is represented by some yield locus. If the stress is increased to a point just outside the yield locus, a further increment of plastic strain must be enforced. The associated increment of elastic strain is known from (11) directly in terms of the stress increment, but we have not yet considered how to calculate the increment of plastic strain. It is true that two conditions restricting possible plastic strains are already known: the plastic volume change must be zero, and the magnitude of the strain is governed, through (14) or (17), by the position of the new yield locus. Four more relations must be found, however, since there are six components of the strain-increment tensor to be determined.

The choice of possible relations is severely restricted by the consideration that no plastic strain can occur during any increment of stress for which the stress-point remains on the same yield locus. This follows from the definition of a yield locus, and in fact is implied by equations (14) and (17) when the locus is a circle. Such stress changes have been well called 'neutral', since they constitute neither loading nor unloading.† In the special, and rather trivial, case when the principal axes rotate, but the reduced principal stresses retain the same values, the stress-point stays where it is. Now according to the postulate in Section 3 (i) the yield loci depend on the strain-history through a single parameter only. We suppose, therefore, that the yield criterion (4) can be put into the form
$$f(J'_2, J'_3) = c, \qquad (21)$$
where f does not depend on the strain-history, which enters only through the parameter c. A general neutral change $d\sigma_{ij}$ is then such that
$$df = \frac{\partial f}{\partial \sigma_{ij}} d\sigma_{ij} = \frac{\partial f}{\partial J'_2} dJ'_2 + \frac{\partial f}{\partial J'_3} dJ'_3 = 0.$$
The condition that $d\epsilon_{ij}^p$ is zero for a neutral change of stress is satisfied by assuming that
$$d\epsilon_{ij}^p = G_{ij} df,$$
where G_{ij} is a symmetric tensor. The G_{ij} are supposed to be functions of the stress components and possibly of the previous strain-history, but not

† G. H. Handelman, C. C. Lin, and W. Prager, *Quart. App. Math.* 4 (1947), 397.

of the stress-increment. The significance of this last assumption should be carefully studied: it means that the ratios of the components of the plastic strain-increment are functions of the current stress but not of the stress-increment. The assumption is suggested by the following considerations. In a crystal grain a plastic strain-increment is produced by a combination of shears along certain slip directions, depending on the orientation of the grain and its external constraint; for the operation of such a glide-system a certain state of stress is needed, and hence, as a statistical average over all grains, a definite macroscopic stress. The stress-increment enters only in determining the magnitude of the strain-increment, as measured by the invariant $\overline{d\epsilon^p}$. As we shall see later, all experimental determinations of the stress-strain relations have been carried out under conditions where any effect of the stress-increment would be negligible. Consequently, there is at present no evidence that the stress-increment influences the ratios of the components of the plastic strain-increment.

The functions G_{ij} must satisfy two necessary conditions: (i) the restriction $G_{ii} = 0$ must be imposed to ensure zero plastic volume change; (ii) the principal axes of the plastic strain-increment tensor, and so of G_{ij}, must coincide with the principal stress axes, since the element is isotropic. These conditions can be satisfied with sufficient generality by choosing

$$G_{ij} = h\frac{\partial g}{\partial \sigma_{ij}},$$

where g and h are scalar functions of the invariants J'_2 and J'_3, and possibly also of the strain-history. We now have the equations

$$d\epsilon^p_{ij} = h\frac{\partial g}{\partial \sigma_{ij}}df, \qquad (22)$$

which were apparently first used by Melan† in 1938. With the definitions of J'_2 and J'_3 in (5), equation (22) may be written

$$d\epsilon^p_{ij} = h\left(\frac{\partial g}{\partial J'_2}\sigma'_{ij} + \frac{\partial g}{\partial J'_3}t'_{ij}\right)df, \qquad (22')$$

where $t'_{ij} = \sigma'_{ik}\sigma'_{kj} - \tfrac{2}{3}J'_2\delta_{ij}$ is the deviator of the square of the reduced stress. If g and h are chosen, this set of equations determines the plastic strain-increment corresponding to a given stress-increment. A new state of hardening is thereby produced, represented by a slightly different yield locus. The new locus is calculable from the change in the parameter c, which we may regard as varying in a known way along the

† E. Melan, *Ingenieur-Archiv*, 9 (1938), 116.

strain-path. Since this locus must naturally pass through the new stress-point, it follows that g and h cannot both be chosen arbitrarily but must satisfy a certain necessary condition. This we now proceed to find. The opposite standpoint may, of course, equally well be taken: the relations between strain-increment, stress, and stress-increment, may be laid down arbitrarily, and the way in which the parameter c varies with the stress- and strain-history will then be implied.

(ii) *Geometrical representation of the plastic strain-increment.* Sufficient generality is retained if (22) is specialized further by assuming that the ratios of the plastic strain components depend only on the ratios of the reduced stress components and not on their absolute magnitudes. In other words, states of stress specified by the same value of μ, applied to the element after varying amounts of pre-strain, always produce the same ratios of the incremental plastic strain components. This has usually been tacitly assumed in experimental work;† it is probable only so long as the element remains isotropic and deforms by the same basic atomic processes. g can then be taken to be a homogeneous function of the stress components, and independent of the strain-history. The surface $g = $ constant in $(\sigma_1, \sigma_2, \sigma_3)$ space is a cylinder of uniform section, cutting the plane Π orthogonally in some curve Γ. Now the plastic strain-increment can also be represented in the same space by a free vector $2G(d\epsilon_1^p, d\epsilon_2^p, d\epsilon_3^p)$, where the factor $2G$ is introduced to obtain the dimensions of stress. This vector lies in Π since $d\epsilon_1^p + d\epsilon_2^p + d\epsilon_3^p = 0$. The equations (22) may then be interpreted as stating that the vector representing the plastic strain-increment is parallel to the normal to Γ at the point of intersection with the stress vector (we must obviously suppose that Γ is met only once by any radius from the origin). Now, with a single crystal, the reversal of the sign of the applied stress does not change the operative glide-planes but only the sense of the respective shears; hence the strain-increment is changed only in sign. This would also be true of a polycrystal were it not for those internal stresses produced by the previous straining as a result of the differential crystal orientations. Since we are neglecting the effects of these, the slope of the curve Γ must be the same at opposite ends of a diameter. Hence, the function g must be an even function of the stresses, and therefore of J_3'. Furthermore, if the material is isotropic, the effect of interchanging σ_1 and σ_2 is merely to interchange $d\epsilon_1$ and $d\epsilon_2$; thus, Γ is symmetrical with respect to the three axes. It follows that Γ must,

---

† See, however, S. J. Fraenkel, *Journ. App. Mech.* 15 (1948), 193. Lode, also, measured ν in various states of hardening for a given μ.

like the yield locus C, be identical in each of the 30° segments of Fig. 2.

Fig. 6 shows a typical 30° segment of Π, cut off by the stress vectors $\mu = 0$ and $\mu = -1$. The absolute size of Γ is, of course, immaterial. The plastic strain-increment vector **RQ'** is parallel to the normal at the point R where the stress vector **OP** meets Γ (**RQ'** is a free vector and can be placed anywhere in the plane). If ψ denotes the angle between

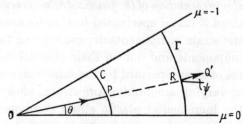

Fig. 6. Geometrical representation of the stress and the plastic strain-increment by vectors in the plane Π.

RQ' and the radius $\mu = 0$, a parameter ν (not to be confused with Poisson's ratio) is defined by the relation

$$\nu = -\sqrt{3}\tan\psi = \frac{2d\epsilon_3^p - d\epsilon_1^p - d\epsilon_2^p}{d\epsilon_1^p - d\epsilon_2^p}. \tag{23}$$

The definition of ν is analogous to (6) for μ, and was also introduced by Lode. A plastic strain-increment is completely specified when ν is given (for the present discussion we do not need to know the directions of the common principal axes of the stress and plastic strain-increment tensors). In pure tension ($\mu = -1$) it follows from symmetry that $\nu = -1$ for an isotropic element. For the ideal body the strain corresponding to a pure shear stress ($\mu = 0$) must be a pure shear strain ($\nu = 0$). This may be shown by the following argument. Suppose that the strain $(d\epsilon_1, d\epsilon_2, d\epsilon_3)$ corresponds to the shear stress $(\sigma, -\sigma, 0)$. If, as we assume, the directional effects of internal stresses are negligible, the reverse of this stress, namely $(-\sigma, \sigma, 0)$, produces the strain $(-d\epsilon_1, -d\epsilon_2, -d\epsilon_3)$. But the stress $(-\sigma, \sigma, 0)$ is also obtainable from $(\sigma, -\sigma, 0)$ by interchanging the axes, and hence, in an isotropic material, produces the strain

$$(d\epsilon_2, d\epsilon_1, d\epsilon_3).$$

Thus $(-d\epsilon_1, -d\epsilon_2, -d\epsilon_3)$ is equivalent to $(d\epsilon_2, d\epsilon_1, d\epsilon_3)$; this is only possible if $d\epsilon_1 = -d\epsilon_2$ and $d\epsilon_3 = 0$, that is, if the strain is a pure shear.†

† It should be noted that it is only necessary to assume that the reversal of the stress reverses the strain when the stress is a pure shear. Thus g need not be an even function; it is sufficient if $\partial g/\partial J_3'$ is zero when $J_3' = 0$.

Ideally, a thin tube twisted by couples should therefore not change its length; if a change is observed it is due either to anisotropy resulting from a preferred orientation, or to internal stresses, or to a combination of both. If the tube is isotropic the sense of the length change is the same if the torque is reversed; if the sense or relative magnitude is altered, the tube must be anisotropic. Returning to the curve Γ, it follows, for reasons of continuity, that it must cut orthogonally the radii bounding the 30° segments. Apart from this, Γ may conceivably be of any shape provided that the normals at two points in the same segment are not parallel, for reasons of uniqueness. The increment of plastic work per unit volume may also be found in terms of geometrical quantities:

$$dW_p = \sigma_1' d\epsilon_1^p + \sigma_2' d\epsilon_2^p + \sigma_3' d\epsilon_3^p = \frac{\mathbf{OP.RQ'}}{2G}.$$

But
$$|\mathbf{OP}| = \sqrt{(\sigma_1'^2 + \sigma_2'^2 + \sigma_3'^2)} = \sqrt{\tfrac{2}{3}}\,\bar\sigma,$$
and
$$|\mathbf{RQ'}| = 2G\sqrt{(d\epsilon_1^{p^2} + d\epsilon_2^{p^2} + d\epsilon_3^{p^2})} = 2G\sqrt{\tfrac{3}{2}}\,\overline{d\epsilon}^p. \quad (24)$$

Hence
$$dW_p = \bar\sigma\, \overline{d\epsilon}^p \cos(\psi - \theta). \qquad (25)$$

It will be supposed that the shape of the yield locus is also preserved during progressive deformation, implying that f is a homogeneous function of the stresses which is independent of strain-history. By convention we choose the sign of f so that the condition for continued loading of a plastic element is $df > 0$. The function h is determined from the way in which the element work-hardens. For our ideal plastic body we assume that the state of hardening depends only on the total plastic work so that the yield criterion (21) is

$$f(J_2', J_3') = F\!\left(\int \sigma_{ij}\, d\epsilon_{ij}^p\right). \qquad (26)$$

This is a generalization of (14). F is a monotonically-increasing positive function and can be determined from the stress-strain curve in simple tension, once f has been assigned. Equation (22) can now be written as

$$d\epsilon_{ij}^p = hF'\,\frac{\partial g}{\partial \sigma_{ij}}(\sigma_{kl}\,d\epsilon_{kl}^p),$$

where F' is the derivative of F with respect to its argument. Multiplying through by σ_{ij} and summing:

$$\sigma_{ij}\,d\epsilon_{ij}^p = hF'\sigma_{ij}\frac{\partial g}{\partial \sigma_{ij}}(\sigma_{kl}\,d\epsilon_{kl}^p) = nF'gh\sigma_{kl}\,d\epsilon_{kl}^p,$$

by Euler's theorem for homogeneous functions, n being the degree of g. Thus
$$nF'gh = 1.$$

This is the condition on g and h that we were seeking. On eliminating h the relations for the plastic strain-increment become

$$d\epsilon_{ij}^p = \frac{1}{ng}\frac{\partial g}{\partial \sigma_{ij}}\frac{df}{F'} \quad (df \geqslant 0). \tag{27}$$

The functions f, g, and F are to be regarded as known, either calculated from experimental data or arbitrarily prescribed. The equations (27) are a complete statement of plastic behaviour during continued loading. If the stress-increment is such that $df < 0$ the element unloads, and the change of strain is no longer governed by (27) but by the elastic equations (11). Further plastic deformation does not occur until the stress-point again lies on the yield locus defined by the stress from which unloading began.

5. The Lévy–Mises and Reuss equations

(i) Theoretical speculation about the relation between stress and strain originated in 1870 with Saint-Venant's treatment of plane plastic strain.[†] With great physical insight Saint-Venant proposed that the principal axes of the strain-increment (and not the total strain) coincided with the axes of principal stress. Saint-Venant did not discuss the dependence of μ on ν, either for plane strain or more generally.[‡] A general relationship between the ratios of the components of the strain-increment and the stress ratios was first suggested by Lévy (1871).[§] Lévy's work remained largely unknown outside his own country, and it was not until the same equations were suggested independently by von Mises[||] in 1913 that they became widely used as the basis of plasticity theory. The Lévy–Mises equations, as they are known, may be expressed in the form

$$\frac{d\epsilon_x}{\sigma_x'} = \frac{d\epsilon_y}{\sigma_y'} = \frac{d\epsilon_z}{\sigma_z'} = \frac{d\gamma_{yz}}{\tau_{yz}} = \frac{d\gamma_{zx}}{\tau_{zx}} = \frac{d\gamma_{xy}}{\tau_{xy}},$$

or, more compactly, as

$$d\epsilon_{ij} = \sigma_{ij}'\, d\lambda, \tag{28}$$

where $d\lambda$ is a scalar factor of proportionality.[††] Since Lévy and von

[†] B. de Saint-Venant, *Comptes Rendus Acad. Sci. Paris*, **70** (1870), 473; *Journ. Math. pures et app.* **16** (1871), 308; *Comptes Rendus Acad. Sci. Paris*, **74** (1872), 1009 and 1083.

[‡] In applications to particular problems Saint-Venant used Tresca's yield criterion and supposed the work-hardening to be zero.

[§] M. Lévy, *Comptes Rendus Acad. Sci. Paris*, **70** (1870), 1323; *Journ. Math. pures et app.* **16** (1871), 369. See I. Todhunter and K. Pearson, *A History of the Elasticity and Strength of Materials* (Cambridge, 1893), vol. ii, part i, 165 et seq. for a critical review of the work of Saint-Venant and Lévy.

[||] R. von Mises, *Göttinger Nachrichten, math.-phys. Klasse* (1913), 582.

[††] The reader should guard against a facile analogy with the equations for a Newtonian viscous fluid: $\dot{\epsilon}_{ij} = \sigma_{ij}'/2\mu$ where the viscosity μ is a material constant, and the rate of strain is directly linked with the applied stress. Although (28) can be written as $\dot{\epsilon}_{ij} = \dot{\lambda}\sigma_{ij}'$ the relations between stress and strain are still independent of time since they are

Mises used the total strain-increment, and not the plastic strain-increment, the equations are strictly applicable only to a fictitious material in which the elastic strains are zero. Accordingly Young's modulus must be regarded as infinitely large, the material remaining rigid when unloaded. The extension of the Lévy–Mises equations to allow for the elastic component of the strain was carried out by Prandtl† (1924) for the plane problem, and in complete generality by Reuss‡ (1930). Reuss assumed that

$$d\epsilon_{ij}^p = \sigma_{ij}' \, d\lambda. \tag{29}$$

It is evident that these equations are equivalent to the combined statements that the principal axes of stress and plastic strain-increment are coincident, and that $\mu = \nu$. Regarded from the geometrical representation of Fig. 6, this means that **RQ'** is parallel to **OP**. The curve Γ is then a circle, and $g = J_2'$, $\partial g/\partial \sigma_{ij} = \sigma_{ij}'$. Equations (27) become

$$d\epsilon_{ij}^p = \frac{\sigma_{ij}' \, df}{2J_2' F'}, \quad \text{with } f = F\left(\int \bar{\sigma} \, \overline{d\epsilon^p}\right) \text{ from (25)}.$$

If this is combined with von Mises' yield criterion $f = \bar{\sigma} = \sqrt{(3J_2')}$, so that (26) reduces to (14), we obtain

$$d\epsilon_{ij}^p = \frac{3\sigma_{ij}' \, d\bar{\sigma}}{2\bar{\sigma}^2 F'}, \qquad F' = \frac{1}{\bar{\sigma}} \frac{d\bar{\sigma}}{\overline{d\epsilon^p}}.$$

But since $\bar{\sigma} = F\left(\int \bar{\sigma} \, \overline{d\epsilon^p}\right)$ it follows that $\bar{\sigma}$ is a function only of $\int \overline{d\epsilon^p}$. In other words, when $\mu = \nu$, the work-hardening hypotheses (14) and (17) are equivalent. From the definition of the function H in (17), F' is equal to $H'/\bar{\sigma}$ where H' is the slope of the equivalent stress/plastic strain curve. For simplicity it is better to write the last equation in terms of H rather than F:

$$d\epsilon_{ij}^p = \frac{3\sigma_{ij}' \, d\bar{\sigma}}{2\bar{\sigma} H'}. \tag{30}$$

Alternatively, this may be split into the two statements

$$\frac{d\epsilon_{ij}^p}{\overline{d\epsilon^p}} = \frac{3\sigma_{ij}'}{2\bar{\sigma}}, \qquad \bar{\sigma} = H\left(\int \overline{d\epsilon^p}\right). \tag{31}$$

From (15) the complete stress-strain relations are

$$\left.\begin{array}{l} d\epsilon_{ij}' = \dfrac{3\sigma_{ij}' \, d\bar{\sigma}}{2\bar{\sigma} H'} + \dfrac{d\sigma_{ij}'}{2G} \quad (d\bar{\sigma} \geqslant 0), \\[2mm] d\epsilon_{ii} = \dfrac{(1-2\nu)}{E} d\sigma_{ii}. \end{array}\right\} \tag{32}$$

dimensionally homogeneous. Also λ is certainly not a material constant, but varies during the deformation.

† L. Prandtl, *Proc. 1st Int. Cong. App. Mech.*, Delft, (1924), 43.
‡ A. Reuss, *Zeits. ang. Math. Mech.* **10** (1930), 266.

All the applications of plasticity theory to be described in this book are based on these equations.

In many problems it is necessary to introduce a further simplification in order to avoid mathematical difficulties. Work-hardening is neglected altogether, so that the yield locus remains unchanged during the range of strain considered. $\bar{\sigma}$ is then a constant (usually replaced by Y, the tensile yield stress) and $H' = 0$. In tension the material deforms elastically up to a yield-point Y and then extends under constant stress. This is the ideal material of classical theories. From (31) the equations for the plastic strain-increment are

$$d\epsilon_{ij}^p = \frac{3\sigma'_{ij}}{2Y}\overline{d\epsilon^p}.$$

These equations are a complete statement since they carry the implication that $\bar{\sigma} = Y$ (as may be verified by squaring each side and summing). Many writers introduce the proportionality factor $d\lambda$, in which case two statements are needed:

$$d\epsilon_{ij}^p = \sigma'_{ij}\, d\lambda; \qquad \sigma'_{ij}\sigma'_{ij} = 2Y^2/3 = 2k^2.$$

On combining this with the elastic component of the strain, we obtain equations originally due to Reuss:

$$\left.\begin{aligned} d\epsilon'_{ij} &= \sigma'_{ij}\, d\lambda + \frac{d\sigma'_{ij}}{2G}, \\ d\epsilon_{ii} &= \frac{(1-2\nu)}{E}\, d\sigma_{ii}, \\ \sigma'_{ij}\sigma'_{ij} &= 2k^2. \end{aligned}\right\} \quad (33)$$

By differentiating the yield criterion, we find $\sigma'_{ij}\, d\sigma'_{ij} = 0$. The incremental work of distortion $\sigma'_{ij}\, d\epsilon'_{ij}$ per unit volume is thus just the plastic work $dW_p = \sigma'_{ij}\, d\epsilon_{ij}^p$ (cf. Hencky's interpretation of von Mises' yield criterion, p. 20). Since the plastic work must be positive, that is,

$$dW_p = 2k^2\, d\lambda \geqslant 0,$$

a prescribed strain-increment $d\epsilon_{ij}$ produces loading if $\sigma'_{ij}\, d\epsilon'_{ij} > 0$; equations (33) then uniquely determine the stress-increment. When the work of distortion is negative, unloading takes place. If, on the other hand, a stress-increment satisfying $\sigma'_{ij}\, d\sigma'_{ij} = 0$ is prescribed (so that the element is still plastic), the elastic strain-increment is known, but $d\lambda$ and the magnitude of the plastic strain-increment are indeterminate from (33). In tension, for example, if the stress is maintained at the yield-point an arbitrary amount of extension may be produced. Other considerations determine the magnitude of the plastic strain, for example applied constraints or restrictions due to neighbouring elements of the material. The equations (32) and (33) are therefore to be contrasted in this respect.

The assumption that the hardening is zero does not diminish the practical value of the theory as much as one might suppose. For pre-strained metals the rate of hardening is comparatively small, and equations (33) are then a good approximation. If necessary, a semi-empirical correction factor can be applied to the calculated stresses; several examples will be described later. For an annealed metal the usual practice is to work with an arbitrarily chosen average value of the yield stress, though, as we shall see, it is sometimes possible, even in complicated problems, to allow for hardening in a more rational way.

(ii) *The geometrical representation of stress and strain in a plane diagram.* Returning to the equations (32), we now describe a geometrical representation of stress and strain in the special case when the principal axes of stress are fixed in the element.† The principal axes of the elastic and plastic strain-increments are then coincident. If the common principal axes are chosen as the fixed axes of reference, three of the six equations for the deviatoric strain components vanish identically, the remaining three being

$$\left.\begin{aligned} d\epsilon_1' &= \frac{3\sigma_1' \, d\bar{\sigma}}{2\bar{\sigma}H'} + \frac{d\sigma_1'}{2G}, \\ d\epsilon_2' &= \frac{3\sigma_2' \, d\bar{\sigma}}{2\bar{\sigma}H'} + \frac{d\sigma_2'}{2G}, \\ d\epsilon_3' &= \frac{3\sigma_3' \, d\bar{\sigma}}{2\bar{\sigma}H'} + \frac{d\sigma_3'}{2G}. \end{aligned}\right\} \quad (34)$$

Only two of these are independent, since their sum is identically zero. It should be carefully noted that $d\sigma_i$ ($i = 1, 2, 3$), defined in the first place as the normal components of the stress-increment referred to axes of reference fixed in the element, are also in this special case equal to the increments $d(\sigma_i)$ of the principal stress components. (When the principal axes rotate, it is not generally true, of course, that the principal components of the stress-increment are equal to the increments of the principal components of stress.) As there are only two non-vanishing

† Various geometrical representations have been proposed for certain special states of stress: for plane strain by L. Prandtl, *Proc. 1st Int. Cong. App. Mech.* Delft, (1924), 43; for torsion by A. Nadai, *Zeits. ang. Math. Mech.* **3** (1923), 442, and *Proc. Inst. Mech. Eng.* **157** (1947), 121; for combined torsion and tension by W. Prager, *Journ. App. Phys.* **15** (1944), 65. The construction described here is due to R. Hill, *Dissertation*, p. 66 (Cambridge 1948), issued by Ministry of Supply, Armament Research Establishment as Survey 1/48. It is a generalization of one due to A. Reuss, *Zeits. ang. Math. Mech.* **10** (1930), 266.

independent relations, the deviatoric stresses and strains can be represented in a plane diagram. When the principal axes of stress and stress-increment are not coincident, a five-dimensional space would normally be required, although the relations between the stress and the plastic strain-increment, or between the stress-increment and the elastic strain-increment, can still be *separately* represented in two dimensions.

In Fig. 7 **OP** and **OQ** represent the current deviatoric stress and strain in the plane Π, the coordinates of Q being $2G\epsilon'_i$ ($i = 1, 2, 3$), where ϵ_i

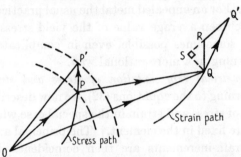

FIG. 7. Representation of the Reuss relations for a plastic element of work-hardening metal when the principal axes of stress are fixed in the element.

is defined in (18). **OP** also represents the deviatoric elastic strain recoverable by complete unloading. **PQ** therefore represents the total plastic strain. During continued deformation P and Q describe curves which respectively represent the stress- and strain-paths, or more exactly the projections on Π of the paths in $(\sigma_1, \sigma_2, \sigma_3)$ space. The hydrostatic components σ and ϵ are represented by vectors perpendicular to Π (see Sect. 2), and need not be considered further. Let **PP'** be an increment of stress and **QQ'** the corresponding increment of strain with components $d\epsilon'_i$. The elastic strain-increment $d\sigma'_i/2G$ is represented by the vector **QR** with components $d\sigma'_i$, and is therefore equal to the vector **PP'**. According to (34) **QQ'** is the sum of the vectors representing the elastic and plastic components of the strain, and so **RQ'** must be the plastic strain-increment. Let QS be the perpendicular from Q on to $Q'R$ produced. Since PP' is parallel to QR, and OP is parallel to RQ' by the Reuss equations, the acute angle between OP and PP' is equal to angle QRS. Now by (24) OP is of length $\sqrt{\tfrac{2}{3}}\,\bar{\sigma}$, and so

$$\sqrt{\tfrac{2}{3}}\, d\bar{\sigma} = PP' \cos \angle QRS = QR \cos \angle QRS = SR.$$

But from (24) we also have

$$d\bar{\sigma} = H'\, \overline{d\epsilon}^p = \frac{H'}{2G} \sqrt{\tfrac{2}{3}}\, RQ'.$$

On equating these expressions for $d\bar{\sigma}$ we obtain an equation for the position of R:

$$\frac{SR}{RQ'} = \frac{H'}{3G} \quad (0 \leqslant H' < \infty). \tag{35}$$

The construction for the strain-increment **QQ'** produced by a given stress-increment **PP'**, constituting loading, is then as follows. Through the strain-point Q draw QR equal to PP' and in the same direction. Let fall the perpendicular QS on to the line through R parallel to OP. The new strain-point Q' then lies on SR produced, the distance RQ' being given by (35). If **PP'** is such that P' lies within the yield circle through P, the element is unloaded and the strain-increment **QQ'**, being entirely elastic, is parallel and equal to **PP'**; the stress- and strain-points will then continue to trace identical paths until the former once more crosses the circle. If, on the other hand, **QQ'** is given, draw $Q'S$ parallel to OP, and let QS be the perpendicular to this line. The position of R is defined by (35), and **QR** is the elastic strain-increment. The stress-increment **PP'** is then determined since it is equal and parallel to **QR**. However, (35) does not fix the position of R uniquely since there are two points dividing SQ' in a given ratio, one inside SQ' and the other outside. Uniqueness is secured by the consideration that (i) P' must lie outside the yield circle through P and the plastic work **OP.RQ'** must at the same time be positive, or (ii) these two conditions cannot be satisfied simultaneously for either position of R, in which case unloading occurs and **PP'** is equal and parallel to **QQ'**. When the material does not work-harden, P' is constrained to move along the circle through P. **QR** is then always perpendicular to **RQ'**, and the work **OP.QR**$/2G$ expended in elastic distortion is zero, as already observed.

It is clear from (35) that when H' is of the order G the elastic component of the strain is at least of the same order of magnitude as the plastic component. This is so in the initial part of the stress-strain curve of an annealed metal, although after a logarithmic strain of about 0·05, H' is normally much less than G. However, the elastic component of strain may still be appreciable, and even predominant. This is so whenever **QQ'** makes a large angle with **OP**. In the limit when **QQ'** is perpendicular to **OP**, there is no plastic strain-increment at all; S and R coincide with Q', and the stress-point P' lies on the same yield circle. At the other extreme the elastic component is least when **QQ'** is parallel to **OP**. This may be realized in practice by increasing the stress components monotonically while holding their ratios constant. The stress-point then travels outwards along a fixed radius, corresponding to the assigned constant value of μ. Whatever the (μ, ν) relation the strain-point also moves in a fixed direction, according to the previous assumption about Γ. This type of experiment, which is often employed to determine the (μ, ν) relationship, possesses two great advantages: (i) the complicating effect of the elastic component of strain is reduced to a minimum, and is altogether negligible if the material is slightly pre-strained; (ii) since the expression (23) for ν is now equal to the analogous

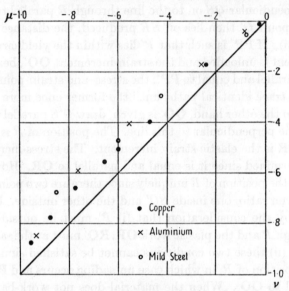

FIG. 8. Experimental results of Taylor and Quinney from combined torsion and tension tests on thin-walled tubes (μ and ν are Lode's variables).

expression in the finite strains (18), the measurement of ν may, for greater accuracy, be based on the latter.

(iii) *Experimental evidence*. The first experiments to investigate the validity of the Lévy–Mises relation were not made until 1926, when Lode[†] tested tubes of iron, copper, and nickel under combined tension and internal pressure. An approximately constant ratio of axial and circumferential stresses was maintained in each test. It was found that the relation was valid to a first approximation, but in spite of appreciable scatter in the data due to anisotropy[‡] in the drawn tubes the results indicated a probable deviation from the relation. Later tests by

[†] W. Lode, *Zeits. Phys.* 36 (1926), 913.
[‡] In many instances ν was not equal to -1 in simple tension, as it should be in an isotropic material from symmetry alone.

Lode† in 1929, with tubes machined from solid bars to minimize anisotropy, supported this view. The deviation was confirmed in 1931 by the classical experiments of Taylor and Quinney,‡ who stressed tubes of aluminium, copper, and pre-strained mild steel in combined tension and torsion. The axial load was held constant while the torque was increased, so that the stress ratios were not constant. The degree of anisotropy was kept within allowed limits by observations of the change in internal volume of the tubes during pure tension; this should be zero for a truly isotropic tube (elastic strains being neglected). The results of Taylor and Quinney are reproduced in Fig. 8. Prager§ found that the observations can be approximately fitted by taking

$$g(J'_2, J'_3) = J'_2\left(1 - 0.73\frac{J'^2_3}{J'^3_2}\right)$$

in place of $g = J'_2$, but the corresponding stress-strain relations (27) would be very cumbersome to use. On the other hand, Schmidt‖ (1932) tested mild steel and copper tubes in combined torsion and tension, the stress ratios being constant, and obtained results agreeing with the Lévy–Mises relation within experimental error. However, recent experiments by Davis†† (1943) on copper, and by Fraenkel‡‡ (1948) on mild steel, using tubes under internal pressure and axial load, have again shown deviations.

It seems fair to regard the Lévy–Mises relation as a reasonably good first approximation, though for applications requiring a high accuracy it will eventually be necessary to include the deviations in the theoretical framework. At present, however, the experimental data appears to be too slight to make the additional mathematical complexity worth while.

6. The Hencky stress-strain equations

Another system of stress-strain equations, due to Hencky§§ (1924), must be briefly referred to, as it has been frequently applied in special problems. This is unfortunate because the equations are, as we shall see, not entirely suitable for representing the observed behaviour of a metal.

---

† W. Lode, unpublished work, referred to by A. Nadai, *Trans. Am. Soc. Mech. Eng.* **55** (1933), 111.
‡ G. I. Taylor and H. Quinney, *Phil. Trans. Roy. Soc.* A, **230** (1931), 323.
§ W. Prager, *Mém. Sci. Math.* **87** (1937); *Journ. App. Phys.* **16** (1945), 837. See also G. H. Handelman, C. C. Lin, and W. Prager, op. cit., p. 33.
‖ R. Schmidt, *Ingenieur-Archiv*, **3** (1932), 215.
†† E. A. Davis, *Trans. Am. Soc. Mech. Eng.* **65** (1943), A–187.
‡‡ S. J. Fraenkel, *Journ. App. Mech.* **15** (1948), 193.
§§ H. Hencky, *Zeits. ang. Math. Mech.* **4** (1924), 323; *Proc. 1st Int. Cong. App. Mech.*, Delft (1924).

It is assumed that the strains are so small that their squares can be neglected, the tensor of total strain being defined as in elasticity by the equation
$$\epsilon_{ij} = \frac{1}{2}\left(\frac{\partial u_i}{\partial x_j}+\frac{\partial u_j}{\partial x_i}\right),$$
where u_i is the total displacement of a particle whose initial position was x_i. Hencky's equations are
$$\left.\begin{aligned}\epsilon'_{ij} &= \left(\phi+\frac{1}{2G}\right)\sigma'_{ij}, \\ \epsilon_{ii} &= \frac{(1-2\nu)}{E}\sigma_{ii},\end{aligned}\right\} \quad (36)$$
where ϕ is a scalar quantity, essentially positive during continued loading and zero during unloading. Equations of this type were previously used by Nadai† in the special problem of torsion. The plastic component of the strain is evidently
$$\epsilon^p_{ij} = \phi\sigma'_{ij}. \tag{37}$$
As a hypothesis of work-hardening ϕ is normally taken to be a function of $\bar{\sigma}$; this is clearly equivalent to the assumption that $\bar{\sigma}$ is a given function of $\bar{\epsilon}^p$. If $\bar{\sigma}$ is constant and the hardening zero (as in Hencky's original formulation), ϕ is an unspecified proportionality factor, analogous to $d\lambda$ in (33). According to (37), the components of *total* plastic strain are proportional to the corresponding deviatoric stress components, whereas in the Reuss equations (29) the components of the *increment* of plastic strain are proportional to the deviatoric stress components.‡ Again, in the Hencky equations the final state of strain is determined by the final state of stress, while in the Reuss equations there is no such unique correspondence. In general the two systems of equations lead to different conclusions.

The Reuss and Hencky theories can be directly contrasted by means of the previous geometrical representation, if we suppose for this purpose that the principal stress axes are fixed in direction. Equation (36) states that the strain-point Q lies on the prolongation of the stress vector **OP**. Thus, according to Hencky's equations, as Q describes some strain-path the stress vector rotates so that it is directed towards Q at every stage. According to the Reuss equations, on the other hand, this happens only when the strain-path is a straight line through the origin, that is, when

† A. Nadai, *Zeits. ang. Math. Mech.* 3 (1923), 442. See also *Plasticity* (McGraw-Hill Book Co., 1931).

‡ A. A. Ilyushin, *Prikladnaia Matematika i Mekhanika*, 9 (1945), 207, has designated theories of the Reuss and Hencky types as 'flow' and 'deformation' theories, respectively. Since 'flow' should be reserved to connôte deformation under constant stress, this terminology is not to be recommended. It is preferable to speak of 'incremental' and 'total' strain theories.

the stress and strain ratios are held constant.† Analytically, if $\sigma_{ij} = C\sigma^0_{ij}$, where σ^0_{ij} is constant and C is a monotonically increasing parameter, we have from the Reuss equations:

$$d\epsilon'_{ij} = C\sigma^{0\prime}_{ij} d\lambda + \frac{d\sigma'_{ij}}{2G}.$$

The strains being small, $\int d\epsilon_{ij}$ is equal to ϵ_{ij}, where this has the meaning defined above. Hence

$$\epsilon'_{ij} = \sigma^{0\prime}_{ij} \int C\, d\lambda + \frac{\sigma'_{ij}}{2G} = \left(\phi + \frac{1}{2G}\right)\sigma'_{ij},$$

if ϕ is written for $\left(\int C\, d\lambda\right)/C$. These are the Hencky equations. By a slight extension of the argument leading to (20), to allow for elastic strains, it may be verified that the respective work-hardening hypotheses are also identical. When the strain-path is curved, different stress states are predicted by the Reuss and Hencky theories;‡ possible experiments have been suggested by Drucker.§

It is very easy to show that the Hencky equations are unsuitable to describe the *complete* plastic behaviour of a metal. Suppose that after a certain plastic deformation the element is unloaded, partially or completely, and then reloaded to a different stress state on the same yield locus. While the stress-point lies inside the yield locus only elastic changes of strain can occur, and the total plastic strain is unchanged. According to (37), however, the plastic strain ratios are now entirely different since the state of stress has changed. This means that the plastic strain itself has altered during the unloading and reloading, which is absurd. In principle the same objection has been made, though less incisively, with regard to neutral changes of stress, by Handelman, Lin, and Prager,|| who have shown that it is impossible to satisfy the necessary continuity conditions between the elastic and plastic stress-strain equations even with very general total strain relations. None the less, in situations where the loading is continuous, the Hencky equations may lead to results in approximate agreement with observation. In some problems, too, total strain theories have certain advantages of mathematical convenience. It is apparently for this reason that the Hencky, or more general total

† A. A. Ilyushin, *Prikladnaia Matematika i Mekhanika*, 11 (1947), 293. See also W. Prager, *Journ. App. Phys.* 19 (1948), 540.

‡ It is a common fallacy to suppose that, since the strains are assumed small, it is legitimate to replace ϵ_{ij} by $d\epsilon_{ij}$ in (37). This would be equivalent to the statement that dy/dx is equal to y/x, which is only true, of course, when y/x is constant.

§ D. C. Drucker, *Proc. 1st Annual Symposium for App. Math.*, Am. Math. Soc. 1947. See also Sect. 3 (vi), Chap. V.

|| G. H. Handelman, C. C. Lin, and W. Prager, op. cit., p. 33.

strain, relations have often been used in applications where the strains are small,† particularly by Russian writers. Of course, when the strains are large the Hencky equations cannot be used without modification.‡

Throughout this book the Reuss equations will be universally employed for the sake of mathematical consistency and physical appropriateness.

7. Other theories

(i) *Theory of Swainger*. Another system of stress-strain equations, also establishing a one-one relation between the total (small) strain and the current stress, has been propounded by Swainger.§ In their simplest form Swainger's equations are

$$\epsilon'_{ij} = \frac{\sigma'_{ij}}{2G} + \frac{3}{2}\left(\frac{1}{P} - \frac{1}{E}\right)(\sigma'_{ij} - \sigma^{0'}_{ij}), \qquad \epsilon_{ii} = \frac{(1-2\nu)}{E}\sigma_{ii}.$$

The stress-strain curve in tension is supposed to have a sharp yield-point and a constant rate P of work-hardening. $\sigma^{0'}_{ij}$ is the stress in the element when it first yields; this is supposed to be governed by the criterion of von Mises. Swainger's system is inadmissible, not only because it is subject to the objection raised against Hencky's equations, but because it also conflicts with observation when the rate of hardening is very small or zero. For if P is very small so also is the difference between σ'_{ij} and $\sigma^{0'}_{ij}$ ($i, j = 1, 2, 3$). This means that no matter how the plastic strain is varied the deviatoric stress only changes by an infinitesimal amount, and in the limit not at all. This is true only for simple strain-paths where the strain ratios are held constant. If the stress ratios are held constant so that $\sigma_{ij} = C\sigma^0_{ij}$, then

$$\epsilon'_{ij} = \frac{\sigma'_{ij}}{2G} + \frac{3}{2}\left(\frac{1}{P} - \frac{1}{E}\right)\left(1 - \frac{1}{C}\right)\sigma^{0'}_{ij}.$$

The strain ratios are therefore also constant. By comparison with (36) it will be seen that for this special type of strain-path Swainger's equations are equivalent to Hencky's (with the appropriate work-hardening hypothesis), and so also to equations (32).

† A. A. Ilyushin, *Prikladnaia Matematika i Mekhanika*, **7** (1943), 245; **8** (1944), 337; **9** (1945), 101; **10** (1946), 623. Bending of plates and shells. W. W. Sokolovsky, *Prikladnaia Matematika i Mekhanika*, **8** (1944), 141. Bending of plates and shells. A. Winzer and W. Prager, *Journ. App. Mech.* **14** (1947), A-281. Expansion of a circular hole in a thin plate. A. Gleyzal, ibid. **68** (1946), A-261; **15** (1948), 288. Clamped circular diaphragm under pressure.

‡ No extension of the theory to large strains appears to have been suggested. Many definitions of finite strains are possible; perhaps the most natural in this context is $\int d\epsilon_{ij}$, where $d\epsilon_{ij}$ has the meaning of (10).

§ K. H. Swainger, *Phil. Mag.* **36** (1945), 443; **38** (1947), 422; *Proc. 7th Int. Cong. App. Mech.*, London (1948).

(ii) *Theory of Prager.* Of greater interest is a system of stress-strain equations proposed by Prager† to avoid the mathematical difficulties encountered in solving problems with the Reuss equations (33). These difficulties (about which more will be said later) arise from the different sets of equations in the elastic and plastic regions when the material is non-hardening. Prager proposed to avoid these difficulties by using one set of equations only, giving a gradual transition from the elastic to the plastic region. It may be shown that Prager's equations, initially expressed in his paper in a rather unusual form, are a special case of (32), the tensile stress-strain curve being

$$\sigma = Y \tanh\left(\frac{E\epsilon}{Y}\right), \qquad (38)$$

where σ is the stress and ϵ is the logarithmic strain. There is no sharp yield-point, the curve initially having the slope E and bending over to approach the stress Y asymptotically. The approach is extremely rapid; for example, when ϵ is only $4Y/E$, σ is already as much as $0{\cdot}999Y$. Prager's equations thus replace the abrupt yield-point of the Reuss material by a smooth but rapid bend in which the stress Y is very nearly attained in a strain of elastic order. The corresponding relation between $\bar\sigma$ and $\int \overline{d\epsilon^p}$ is

$$H' = \frac{d\bar\sigma}{\overline{d\epsilon^p}} = E\left(\frac{Y^2}{\bar\sigma^2}-1\right).$$

Substituting this in (32) we obtain

$$2G\,d\epsilon'_{ij} = d\sigma'_{ij} + \frac{\bar\sigma\,d\bar\sigma}{(Y^2-\bar\sigma^2)}\sigma'_{ij}, \quad \text{if } \nu = \tfrac{1}{2}, \qquad (39)$$

which, apart from notation, is one of the forms of his equations given by Prager. This system has received little application, and it seems that the desired simplification has not in fact been achieved.

† W. Prager, *Proc. 5th Int. Cong. App. Mech.*, Cambridge, Mass. (1938), 234; *Duke Math. Journ.* **9** (1942), 228; *Rev. Fac. Sci. Univ. Istanbul*, **5** (1941), 215.

III
GENERAL THEOREMS

1. The plastic potential

THE function $g(\sigma_{ij})$ defining the ratios of the components of the plastic strain-increment (equation (22) of Chap. II) is known as the plastic potential.

Now, if a polycrystalline element is strained, one of a number of possible systems of five independent glide directions must be active in each crystal grain, conceived as deforming uniformly. Such a glide-system demands for its operation a state of stress in which the critical shear stress is attained in these five directions and is not exceeded in any others. The glide-system determines both the ratios of the components of the strain-increment and also the ratios which the components of the applied stress must bear to the critical shear stress. It is likely, therefore, that there is a relation, from a statistical average over possible orientations of the grains in a polycrystal, between the plastic potential and the function $f(\sigma_{ij})$ defining the yield locus. It is not yet known what this should be, theoretically, for any particular metal.

It seems, however, that the simple relation $g = f$ has an especial significance in the mathematical theory of plasticity, for, as will be shown later, certain variational principles and uniqueness theorems can then be formulated. When $g = f$ (that is, when the curve Γ and the yield locus are similar in shape), the plastic stress-strain relations are

$$d\epsilon_{ij}^p = h \frac{\partial f}{\partial \sigma_{ij}} df. \qquad (1)$$

Notice that f must be independent of hydrostatic pressure, that is,

$$\partial f/\partial \sigma_{ii} = 0,$$

if the plastic volume change is to be zero. Also, if f is an even function, so that there is no Bauschinger effect, (1) implies that a reversal of the sign of the stress merely reverses the sign of the strain-increment; these properties are mutually consistent since each is true only in the absence of the same kind of internal stress (see pp. 8 and 35). We have already encountered an example of (1) in the Lévy–Mises or Reuss equations where $f = g = J_2' = \tfrac{1}{2}\sigma_{ij}'\sigma_{ij}'$, $\partial f/\partial \sigma_{ij} = \sigma_{ij}'$.

Suppose, now, that the plastic strain-increment $d\epsilon_{ij}^p$ is prescribed, and that the corresponding stress determined by (1) and the yield criterion is σ_{ij}. Let σ_{ij}^* be any other plastic state of stress, so that

$$f(\sigma_{ij}^*) = f(\sigma_{ij}) = c.$$

The work that would be done by σ_{ij}^* in the strain $d\epsilon_{ij}^p$ is $dW_p^* = \sigma_{ij}^* d\epsilon_{ij}^p$. This has a stationary value for varying plastic states σ_{ij}^* when

$$\frac{\partial}{\partial \sigma_{ij}^*}\{\sigma_{ij}^* d\epsilon_{ij}^p - f(\sigma_{ij}^*)\, d\lambda\} = 0,$$

where, following the method of Lagrange, we have introduced the constant multiplier $d\lambda$. Hence,

$$d\epsilon_{ij}^p = d\lambda \frac{\partial f(\sigma_{ij}^*)}{\partial \sigma_{ij}^*}.$$

This equation is satisfied when σ_{ij}^* is the actual stress σ_{ij} ($d\lambda$ being equal to $h\,df$). Thus, when the relations (1) hold, the plastic work done in a given plastic strain-increment has a stationary value in the actual state, with respect to varying stress systems satisfying the yield criterion. This theorem is due to von Mises† (1928). It becomes obvious when stated in terms of the geometrical representation of Fig. 6. The curve Γ is, by the hypothesis (1), similar to the yield locus C, and the vector **RQ**′ representing the plastic strain-increment is parallel to the normal at P, where **OP** is the stress vector σ_{ij}'. The plastic work done by any stress σ_{ij}^* is proportional to the projection of the corresponding stress vector **OP*** on **RQ**′ (cf. equation (25) of Chap. II). This evidently has a stationary value among neighbouring states when P^* coincides with P. Furthermore, when C is concave to the origin at P, the work done is a *maximum*. Now C must be concave to the origin at all points if the stress corresponding to a given strain-increment is to be unique, and in this case the work done is an absolute maximum, that is, a maximum for all plastic states of stress and not merely for infinitesimally near states. This is true, in particular, for the Reuss equations where the yield locus is a circle.‡ The stationary work theorem does not apply, of course, when Tresca's criterion is used in conjunction with the Lévy–Mises relation $\mu = \nu$.

If it were true that the plastic behaviour of a metal is such that the yield criterion and plastic potential are identical, it would be possible to derive the yield locus from an experimental determination of the (μ, ν) relation (or vice versa). This procedure has been suggested by

† R. von Mises, *Zeits. ang. Math. Mech.* **8** (1928), 161.
‡ R. Hill, E. H. Lee, and S. J. Tupper, *Proc. Roy. Soc.* A, **191** (1947), 278.

Taylor.† In the notation of Fig. 6, let θ and ψ be the angles made with the radius $\mu = 0$ by **OP** and **RQ'**. From equations (7) and (23) of Chapter II we have $\mu = -\sqrt{3}\tan\theta$, $\nu = -\sqrt{3}\tan\psi$. Suppose that ψ is a known function of θ, calculated from an experimentally determined (μ, ν) relation. If C and Γ are similar curves, the angle between the normal to the yield locus and the radius $\mu = 0$ is equal to ψ. The equation of the yield locus is therefore

$$\frac{1}{r}\frac{dr}{d\theta} = \tan(\theta - \psi), \quad \text{or} \quad r = r_0 \exp\left\{\int_0^\theta \tan(\theta - \psi)\, d\theta\right\}, \qquad (2)$$

where r is the length of the radius of inclination θ, and r_0 is the length of the radius $\theta = 0$. Taylor has applied this method to generate the yield locus for copper from the (μ, ν) relation measured by Taylor and Quinney (Fig. 8). The calculated locus lies between the Tresca and Mises criteria, though nearer the latter ($0 \leqslant \theta - \psi < 4°$). On the other hand, measurements of the yield locus agree closely with the criterion of von Mises (Fig. 4). In view of the scatter in the data due to variations in the material, and other causes, a definite decision as to the validity of equation (1) is hardly possible. Since large departures from the relation $\mu = \nu$ can be produced by a plastic potential differing comparatively little from a circle, it may be that the true yield criterion for copper is in fact slightly different from that of von Mises, and that the equation is correct.

If a metal yields in accordance with Tresca's criterion the yield locus is a regular hexagon (Fig. 2). If (1) holds, the plastic potential is represented by a similar hexagon and the strain-increment vector is directed normally to one or other of the sides; this direction corresponds, as we have seen, to a pure shear. It follows that the plastic strain-increment is the same for all stress-points on any one side of the hexagon, and that it is a shear in the direction of the maximum shear stress. The work done in a prescribed shear strain is the same for all possible corresponding states of stress (since the projection of the stress vector on the normal to the side is constant), and it is greater than the work done by any other plastic stress-system. Now, in the one metal which is observed to yield most closely in accordance with Tresca's criterion, namely annealed mild steel, the ensuing strain-increment in a Lüders band is a shear in the direction of the maximum shear stress. Hence, in this instance too, the plastic potential and the yield locus are at least approximately similar.

† G. I. Taylor, *Proc. Roy. Soc.* A, **191** (1947), 441.

2. Uniqueness of a stress distribution under given boundary conditions†

(i) *Work-hardening material.* Consider a mass of ideal material which has been deformed in any manner under applied surface forces or prescribed surface displacements. In general, a part of the mass will be plastic, a part will have unloaded from a previous plastic state, and the remainder will have been strained elastically throughout. The strain-history of each element will usually have been different, and the parameter c (Chap. II, equation (21)), specifying the final state of work-hardening, will vary from point to point. The current values of c and the stress σ_{ij} are regarded as known. Suppose, now, that further *infinitesimal* increments of stress or displacement are applied at the surface. We wish to find the circumstances in which the corresponding small changes in stress and strain throughout the mass are uniquely determined by the given boundary conditions and the stress-strain relations formulated for the ideal material. We shall assume that the function $f(\sigma_{ij})$, defining the yield locus, is also the plastic potential. The relations between infinitesimal changes in the stress and strain in any element are then taken in the form

$$\left. \begin{aligned} d\epsilon'_{ij} &= \frac{d\sigma'_{ij}}{2G} + h\frac{\partial f}{\partial \sigma_{ij}} df, \quad \text{wherever} \quad f(\sigma_{ij}) = c \text{ and } df \geqslant 0; \\ d\epsilon'_{ij} &= \frac{d\sigma'_{ij}}{2G}, \quad \begin{aligned} &\text{wherever} \quad f(\sigma_{ij}) < c, \\ &\text{or where} \quad f(\sigma_{ij}) = c \text{ and } df \leqslant 0; \end{aligned} \\ d\epsilon &= \frac{(1-2\nu)}{E} d\sigma \quad \text{everywhere} \quad (d\epsilon = \tfrac{1}{3}d\epsilon_{ii},\ d\sigma = \tfrac{1}{3}d\sigma_{ii}). \end{aligned} \right\} \quad (3)$$

The function h, apart from being positive, is not restricted in any way; we also do not need to retain the limitation that f is a homogeneous function. The components of the strain-increment must be derivable from a displacement-increment du_i such that

$$d\epsilon_{ij} = \frac{1}{2}\left\{\frac{\partial}{\partial x_j}(du_i) + \frac{\partial}{\partial x_i}(du_j)\right\}. \quad (4)$$

Elements at present stressed to the yield-point ($f(\sigma_{ij}) = c$) will be said to constitute the *plastic region*. In these elements continued plastic deformation is enforced if the stress-increment is such that $df > 0$; if $df = 0$ the element remains on the point of yielding, while if $df < 0$ it begins to unload elastically. Wherever $f(\sigma_{ij}) < c$, any further small change in stress produces only an elastic change of strain; these elements will be said to constitute the *elastic region*, even though some may

† The remainder of this chapter may be omitted on a first reading of the book.

previously have been plastically deformed. Finally, both the original stress and the change in stress must satisfy the equations of equilibrium

$$\frac{\partial}{\partial x_j}(\sigma_{ij}) = 0, \qquad \frac{\partial}{\partial x_j}(\delta\sigma_{ij}) = 0,$$

where $\delta\sigma_{ij}$ is the change of stress *at a fixed point*. This differs from the change $d\sigma_{ij}$ *in a given element* by the amount

$$d\sigma_{ij} - \delta\sigma_{ij} = du_k \frac{\partial \sigma_{ij}}{\partial x_k}, \tag{5}$$

which represents the change in stress due to the movement of the element. In terms of $d\sigma_{ij}$ the equations of equilibrium are thus

$$\frac{\partial}{\partial x_j}(d\sigma_{ij}) = \frac{\partial}{\partial x_j}\left(du_k \frac{\partial \sigma_{ij}}{\partial x_k}\right) = \frac{\partial}{\partial x_j}(du_k)\frac{\partial \sigma_{ij}}{\partial x_k}, \tag{6}$$

since

$$du_k \frac{\partial}{\partial x_j}\left(\frac{\partial \sigma_{ij}}{\partial x_k}\right) = du_k \frac{\partial}{\partial x_k}\left(\frac{\partial \sigma_{ij}}{\partial x_j}\right) = 0.$$

In the linear theory of elasticity the right-hand side of (6) is neglected since the change in strain is of order $1/E \times$ the change in stress; where plastic strains are concerned, however, the right-hand side is not necessarily negligible if the rate of work-hardening is small compared with E.

Suppose, now, that $(d\sigma_{ij}, du_i)$ and $(d\sigma_{ij}^*, du_i^*)$ could be two distinct distributions of stress- and displacement-increments satisfying (3), (4), and (6), and corresponding to the same changes of surface stress or surface displacement. Consider the integral

$$\int (d\sigma_{ij}^* - d\sigma_{ij})(d\epsilon_{ij}^* - d\epsilon_{ij}) \, dV$$

taken through the volume of the mass. By (4) and the symmetry of the stress tensor, the integral may be written as

$$\int (d\sigma_{ij}^* - d\sigma_{ij}) \frac{\partial}{\partial x_j}(du_i^* - du_i) \, dV.$$

This is evidently equal to

$$\int \frac{\partial}{\partial x_j}\{(d\sigma_{ij}^* - d\sigma_{ij})(du_i^* - du_i)\} \, dV - \int (du_i^* - du_i) \frac{\partial}{\partial x_j}(d\sigma_{ij}^* - d\sigma_{ij}) \, dV.$$

Since the small changes in stress and displacement are necessarily continuous in a plastic-elastic material, we may use Green's theorem to transform the first volume integral into an integral over the surface of the mass:

$$\int (d\sigma_{ij}^* - d\sigma_{ij})(du_i^* - du_i) l_j \, dS - \int (du_i^* - du_i) \frac{\partial}{\partial x_j}(d\sigma_{ij}^* - d\sigma_{ij}) \, dV,$$

where l_j is the unit outward normal to the surface. Now since (3), (4), and (6) are strictly satisfied only when the changes $d\sigma_{ij}$ and $d\sigma_{ij}^*$ are allowed to become infinitesimally small, having the meaning of differentials in the limit, the possibility that the *differentials* $d\sigma_{ij}$ and $d\sigma_{ij}^*$ are discontinuous must be reckoned with. This is frequently so, for instance, across the interface of the elastic and plastic regions, since the rate of change of the stress with respect to the external loads or displacements may be different in the two regions; the discontinuity in any element is, of course, only momentary as the element passes from one region to the other. In the limit, then, $\partial(d\sigma_{ij}^*-d\sigma_{ij})/\partial x_j$ becomes infinitely great at such an interface of discontinuity, and the interface makes a finite contribution to the volume integral in the last equation. Provided this interpretation is given to the volume integral the equation is true when $d\sigma_{ij}$, etc., are differentials. With the use of (6) we obtain finally

$$\int (d\sigma_{ij}^*-d\sigma_{ij})(d\epsilon_{ij}^*-d\epsilon_{ij})\,dV + \int (du_i^*-du_i)\frac{\partial}{\partial x_j}\left\{(du_k^*-du_k)\frac{\partial \sigma_{ij}}{\partial x_k}\right\}dV$$

$$= \int (d\sigma_{ij}^*-d\sigma_{ij})(du_i^*-du_i)l_j\,dS = 0 \quad (7)$$

if the boundary conditions are such that

$$l_j(d\sigma_{ij}^*-d\sigma_{ij})(du_i^*-du_i) = 0 \quad (8)$$

on the surface. At any interface Σ where the stress gradient is discontinuous, there is a contribution to the volume integral in (7) of the amount

$$\int (du_i^*-du_i)(du_k^*-du_k)\left[\frac{\partial \sigma_{ij}}{\partial x_k}\right]\lambda_j\,d\Sigma,$$

where the square brackets denote the jump of $\partial \sigma_{ij}/\partial x_k$ in the direction of the unit normal λ_j to Σ. Equation (8) is satisfied, in particular, when the external displacements are prescribed, so that $du_i^* = du_i$ on S. If the external stress $F_i = l_j\sigma_{ij}$ is prescribed, the boundary condition is that $d(l_j\sigma_{ij})$ is given. This does not imply that $l_j\,d\sigma_{ij}^*$ is equal to $l_j\,d\sigma_{ij}$, unless it is permissible to neglect the variation dl_j of the surface normal during the deformation. Thus, when the external stress is prescribed, the expression in (7) differs from zero by an amount depending on the change in shape of the surface.†

† Alternatively, one may take as the basis for an examination of uniqueness the equation
$$\int (\delta\sigma_{ij}^* - \delta\sigma_{ij})(d\epsilon_{ij}^* - d\epsilon_{ij})\,dV = \int (\delta\sigma_{ij}^* - \delta\sigma_{ij})(du_i^* - du_i)l_j\,dS$$
in combination with (5). In the proof of this equation it must be noted that for reasons of equilibrium $\lambda_j\,\delta\sigma_{ij}$ is continuous across a discontinuity interface.

From (3) we have

$$(d\sigma_{ij}^* - d\sigma_{ij})(d\epsilon_{ij}^* - d\epsilon_{ij})$$
$$= (d\sigma_{ij}^* - d\sigma_{ij})(d\epsilon_{ij}^{*\prime} - d\epsilon_{ij}^{\prime}) + 3(d\sigma^* - d\sigma)(d\epsilon^* - d\epsilon)$$
$$= h\frac{\partial f}{\partial \sigma_{ij}}(d\sigma_{ij}^* - d\sigma_{ij})(\alpha^* df^* - \alpha\, df) + \frac{1}{2G}(d\sigma_{ij}^{*\prime} - d\sigma_{ij}^{\prime})(d\sigma_{ij}^{*\prime} - d\sigma_{ij}^{\prime}) +$$
$$+ \frac{3(1-2\nu)}{E}(d\sigma^* - d\sigma)^2,$$

where

$\alpha = 1$ where $f(\sigma_{ij}) = c$ and $df \geq 0$;

$\alpha = 0$ where $f(\sigma_{ij}) < c$, or where $f(\sigma_{ij}) = c$ and $df < 0$;

$\alpha^* = 1$ where $f(\sigma_{ij}) = c$ and $df^* \geq 0$;

$\alpha^* = 0$ where $f(\sigma_{ij}) < c$, or where $f(\sigma_{ij}) = c$ and $df^* < 0$.

Since
$$df = \frac{\partial f}{\partial \sigma_{ij}} d\sigma_{ij}, \qquad df^* = \frac{\partial f}{\partial \sigma_{ij}} d\sigma_{ij}^*,$$

the above expression is equal to

$$h(df^* - df)(\alpha^* df^* - \alpha\, df) + \frac{1}{2G}(d\sigma_{ij}^{*\prime} - d\sigma_{ij}^{\prime})(d\sigma_{ij}^{*\prime} - d\sigma_{ij}^{\prime}) +$$
$$+ \frac{3(1-2\nu)}{E}(d\sigma^* - d\sigma)^2.$$

By considering the four possible combinations of signs which df and df^* may have at any one point, it may be shown immediately that the first term is never negative.† Since the last two terms together are always positive unless $d\sigma_{ij}^* = d\sigma_{ij}$, the whole expression is positive except when $d\sigma_{ij}^* = d\sigma_{ij}$. On the other hand, if the volume integral in (7) is negligible,

$$\int (d\sigma_{ij}^* - d\sigma_{ij})(d\epsilon_{ij}^* - d\epsilon_{ij})\, dV = 0,$$

provided the boundary condition (8) is satisfied. It follows that, *to the approximation involved in neglecting the volume integral*, there cannot be two distinct solutions. The volume integral is strictly zero when

$$\partial \sigma_{ij}/\partial x_k = 0;$$

that is, when the stress distribution is uniform. If the mass is entirely elastic, or if the existing state of stress is such that the elastic region

† The term is zero in the elastic region, or where both $d\sigma_{ij}$ and $d\sigma_{ij}^*$ produce unloading in the plastic region, since $\alpha = \alpha^* = 0$. Where both produce loading ($\alpha = \alpha^* = 1$) the term is equal to $h(df^* - df)^2$, which is never negative (h being essentially positive). Where $d\sigma_{ij}$ produces loading ($\alpha = 1$) and $d\sigma_{ij}^*$ unloading ($\alpha^* = 0$), the term is equal to $h(df^* - df)(-df)$; this is never negative since $df \geq 0$, $df^* \leq 0$. Where $d\sigma_{ij}$ produces unloading ($\alpha = 0$) and $d\sigma_{ij}^*$ loading ($\alpha^* = 1$), the term is equal to $h(df^* - df)\, df^*$; this again is never negative since $df \leq 0$, $df^* \geq 0$.

constrains strain-increments in the plastic region to be of order $1/E \times$ the stress-increment, then, to the usual approximation in the linear theory of elasticity, the solution is unique; this theorem is substantially due to Melan† (1938). When, however, large plastic strains are possible the above analysis fails to give a definite answer. In fact, it is easy to construct instances where two solutions are to be expected physically as well as mathematically. For example, if a hollow spherical shell is expanded sufficiently far by internal pressure, a stage is reached where a further plastic expansion does not require so large a pressure. After this stage there are clearly two solutions corresponding to a given small drop in pressure: one is the plastic expansion mentioned, the other is the incipient elastic recovery of the whole shell. On the other hand, in this example and also more generally, one might suspect that there is a unique solution corresponding to prescribed external *displacements*. This is a problem awaiting further research.

(ii) *Non-hardening material.* The stress-strain relations for a non-hardening material (c constant) are

$$\left. \begin{aligned} d\epsilon'_{ij} &= \frac{d\sigma'_{ij}}{2G} + \frac{\partial f}{\partial \sigma_{ij}} d\lambda, \quad \text{wherever} \quad f(\sigma_{ij}) = c \text{ and } df = 0; \\ d\epsilon'_{ij} &= \frac{d\sigma'_{ij}}{2G}, \quad \begin{aligned} &\text{wherever} \quad f(\sigma_{ij}) < c, \\ &\text{or where} \quad f(\sigma_{ij}) = c \text{ and } df < 0; \end{aligned} \\ d\epsilon &= \frac{(1-2\nu)}{E} d\sigma \quad \text{everywhere.} \end{aligned} \right\} \quad (9)$$

When the stress-increment maintains a plastic element at the yield-point ($df = 0$) the associated strain-increment must be such that the plastic work is positive. Now

$$dW_p = \sigma_{ij} d\epsilon^p_{ij} = \sigma_{ij} \frac{\partial f}{\partial \sigma_{ij}} d\lambda.$$

The scalar product of σ_{ij} and $\partial f/\partial \sigma_{ij}$ is positive since the vector representing $\partial f/\partial \sigma_{ij}$ in the plane Π is directed along the outward normal to the yield locus. The requirement is satisfied, therefore, if

$$d\lambda \geqslant 0. \qquad (10)$$

With the notation of the previous section, it follows from (9) that

$$(d\sigma^*_{ij} - d\sigma_{ij})(d\epsilon^*_{ij} - d\epsilon_{ij}) = (df^* - df)(\alpha^* d\lambda^* - \alpha d\lambda) + \\ + \frac{1}{2G}(d\sigma^{*\prime}_{ij} - d\sigma'_{ij})(d\sigma^{*\prime}_{ij} - d\sigma'_{ij}) + \frac{3(1-2\nu)}{E}(d\sigma^* - d\sigma)^2.$$

† E. Melan, *Ingenieur-Archiv*, **9** (1938), 116.

The first term on the right-hand side is always positive or zero, while the last two together are positive unless $d\sigma_{ij}^* = d\sigma_{ij}$. The argument continues as before, with the conclusion that the stress distribution is unique when only small strains are possible; this was proved by Greenberg[†] for the Reuss material, and by Bauer[‡] for a material with a general function f. However, we cannot deduce, without further investigation, that the displacements are also unique. Whereas in a work-hardening material the strain-increment in any element is determined by the stress-increment, the magnitude of the plastic component of the strain-increment in a non-hardening material is indeterminate from equations (9) *alone*, since $d\lambda$ can have any positive value. It is not yet known, in general, for what boundary conditions and existing stress states the displacements are unique; proofs of uniqueness can, however, be given in special cases (see, for example, Chap. IX, Sect. 2 (iii)).

(iii) *Plastic-rigid material.* A material is described as plastic-rigid when Young's modulus is assigned an indefinitely large value. The elastic component of strain is zero in the limit (it follows that there is no change of volume), and an element is therefore rigid when stressed below the yield-point. In many problems where the plastic strains are large the stress distribution calculated for a plastic-rigid material closely approximates that for a plastic-elastic material. Since the solution of problems is often much simpler for the former, it is worth while to investigate the corresponding requirements for uniqueness.

In the first place, the uniqueness theorems of the previous section are naturally still valid for a material in which E has a very large, though finite, value. If the existing state of stress is such that the non-plastic material constrains the remainder, the whole mass is rigid in the limit, and there is only one distribution of stress-increments corresponding to prescribed increments of the external loads. On the other hand, if all or part of the plastic region is free to deform there is not necessarily a unique solution.

When, however, increments in the surface displacements are prescribed, we have to distinguish between the behaviour of plastic-rigid material and that of plastic-elastic material with a finite Young's modulus. It is helpful to refer to the geometrical representation of stress and strain (Chap. II, Sect. 5 (ii)). Consider a plastic-elastic element to which is applied a given strain-increment producing no change in volume. If the element is not already plastic, or if it is caused to unload

[†] H. J. Greenberg, *Quart. App. Math.* 7 (1949), 85.
[‡] F. B. Bauer, *Brown Univ. Rep.* A11–27 (1949).

from a plastic state, the strain-increment is entirely elastic. If the element remains in a plastic state the vector representing the elastic part of the strain-increment has a constant component perpendicular to the corresponding normal to the yield locus. It follows that the stress-increment, being equal to $2G \times$ the elastic strain-increment, always increases indefinitely as E tends to infinity. In the limit, then, the stress in a plastic-rigid element changes discontinuously. The only exception is for a plastic state in which the corresponding normal to the yield locus is parallel to the vector representing the prescribed strain-increment (and in which the principal axes of the stress and the strain-increment coincide); the strain-increment will then be said to be consistent with the existing stress. If it is not consistent, the stress vector rotates so that it becomes so. If the element is now taken along a certain strain-path the stress vector rotates so that it is always in the direction of the tangent to the strain-path; provided the strain-path has a continuously turning tangent the stress also varies continuously.† In general, then, when increments of displacement (producing no change in volume) are applied at the surface of a mass of plastic-rigid material, the stress changes discontinuously so that it becomes consistent with the strain-increment at all points. Thenceforward, provided the stress distribution changes smoothly with the surface displacements, the theorems established in the previous sections are applicable.

It is natural to inquire whether the consistent distribution of stress produced by applied surface displacements is unique, or whether it depends on the state of stress existing beforehand. In partial answer to this question, we now prove that there is not more than one consistent distribution for which the whole mass deforms plastically.‡ This theorem has a rather restricted application since the mass is brought to a completely plastic state only by special distributions of surface displacements. Suppose that (σ_{ij}, du_i) and (σ_{ij}^*, du_i^*) could be two consistent solutions corresponding to the same boundary conditions. It may be shown by the use of Green's theorem, and the equations of equilibrium, that

$$\int (\sigma_{ij}^* - \sigma_{ij})(d\epsilon_{ij}^* - d\epsilon_{ij}) \, dV = \int (\sigma_{ij}^* - \sigma_{ij})(du_i^* - du_i) l_j \, dS = 0 \quad (11)$$

if
$$(F_i^* - F_i)(du_i^* - du_i) = 0 \quad (12)$$

on the surface, where $F_i = l_j \sigma_{ij}$ denotes the external stress. This

† Note that in a non-hardening plastic-rigid material the stress-increment does not depend on the change in length of the strain-path, but on the change in direction of the strain-path.
‡ R. Hill, *Quart. Journ. Mech. App. Math.* 1 (1948), 18.

condition is satisfied, in particular, when the surface displacements are prescribed. Now, if the two solutions are such that the whole mass is plastic,
$$f(\sigma_{ij}) = f(\sigma_{ij}^*) = c$$
at each point, where c specifies the local state of hardening immediately before the application of the incremental displacements. Hence

$$(\sigma_{ij}^*-\sigma_{ij})(d\epsilon_{ij}^*-d\epsilon_{ij}) = (\sigma_{ij}^*-\sigma_{ij})\left(\frac{\partial f}{\partial \sigma_{ij}^*}d\lambda^* - \frac{\partial f}{\partial \sigma_{ij}}d\lambda\right), \qquad (13)$$

where $d\lambda$ and $d\lambda^*$ are positive or zero, and where $\partial f/\partial \sigma_{ij}^*$ denotes
$$\partial f(\sigma_{ij}^*)/\partial \sigma_{ij}^*$$
(for the present purpose we do not need to write $d\lambda$ as $h\,df$ when the material work-hardens). Now $(\sigma_{ij}^*-\sigma_{ij})\partial f/\partial \sigma_{ij}^*$ is proportional to the scalar product of the outward normal to the yield locus at the point $\sigma_{ij}^{*\prime}$ and the chord joining σ_{ij}' to $\sigma_{ij}^{*\prime}$. This product is obviously positive since f, by hypothesis, is never convex to the origin. Since $(\sigma_{ij}-\sigma_{ij}^*)\partial f/\partial \sigma_{ij}$ is similarly positive, the quantity (13) is also positive unless $\sigma_{ij}' = \sigma_{ij}^{*\prime}$. Hence, from (11), σ_{ij} and σ_{ij}^* can only differ at each point by a hydrostatic stress which, for equilibrium, must be uniform.

The boundary condition (12) is also satisfied if the external stresses are given ($F_i^* = F_i$), or if the normal component of the displacement and the resultant tangential stress are given (since the vector $du_i^*-du_i$ lies in the tangent plane and the vector $F_i^*-F_i$ is directed along the normal, their scalar product vanishes). Thus, in a plastic-rigid material, there cannot be two distinct plastic states of stress satisfying the same boundary conditions. On the other hand, the increments of strain are not necessarily unique.

3. Extremum and variational principles

(i) *Elastic material.* For the purposes of later comparison, it is helpful to retrace the proofs of some well-known extremum principles in the theory of elasticity. Let (σ_{ij}, e_{ij}) be the stress and strain in an elastic mass which has been loaded from a stress-free state by prescribed external stresses F_i over a part S_F of its surface, and by prescribed displacements over the remainder S_U. If x_j is the initial position of an element, the equations of equilibrium are

$$\frac{\partial \sigma_{ij}}{\partial x_j} = 0,$$

… EXTREMUM AND VARIATIONAL PRINCIPLES …

where, as usual in the theory of elasticity, the displacement of the element is neglected. The stress-strain relations are

$$e'_{ij} = \frac{\sigma'_{ij}}{2G}, \qquad e = \frac{(1-2\nu)}{E}\sigma.$$

e_{ij} must be derivable from a continuous displacement function u_i such that

$$e_{ij} = \frac{1}{2}\left(\frac{\partial u_i}{\partial x_j} + \frac{\partial u_j}{\partial x_i}\right).$$

Let $(\sigma^*_{ij}, e^*_{ij})$ be any other distribution of stress and strain satisfying the equilibrium equations, the stress-strain relations and the stress boundary conditions, but not necessarily derivable from a displacement function.

Consider the integral

$$\int (\sigma^*_{ij} - \sigma_{ij}) e_{ij}\, dV$$

taken through the volume of the mass in the unstrained configuration. By the use successively of the equilibrium equations and Green's theorem, the integral is equal to

$$\int (\sigma^*_{ij} - \sigma_{ij})\frac{\partial u_i}{\partial x_j} dV = \int \frac{\partial}{\partial x_j}\{(\sigma^*_{ij} - \sigma_{ij})u_i\}\, dV = \int (\sigma^*_{ij} - \sigma_{ij}) u_i l_j\, dS$$

$$= \int_{S_U} (F^*_i - F_i) u_i\, dS$$

since $F^*_i = F_i$ over S_F. Now

$$2(\sigma^*_{ij} - \sigma_{ij}) e_{ij} \equiv (\sigma^*_{ij} e^*_{ij} - \sigma_{ij} e_{ij}) - [\sigma^*_{ij}(e^*_{ij} - e_{ij}) + e_{ij}(\sigma_{ij} - \sigma^*_{ij})]$$
$$= (\sigma^*_{ij} e^*_{ij} - \sigma_{ij} e_{ij}) - (\sigma^*_{ij} - \sigma_{ij})(e^*_{ij} - e_{ij}),$$

by virtue of the reciprocity relation

$$\sigma^*_{ij} e_{ij} = \frac{\sigma'_{ij} \sigma^{*\prime}_{ij}}{2G} + \frac{3(1-2\nu)\sigma\sigma^*}{E} = \sigma_{ij} e^*_{ij}.$$

But

$$(\sigma^*_{ij} - \sigma_{ij})(e^*_{ij} - e_{ij}) = \frac{1}{2G}(\sigma^{*\prime}_{ij} - \sigma'_{ij})(\sigma^{*\prime}_{ij} - \sigma'_{ij}) + \frac{3(1-2\nu)}{E}(\sigma^* - \sigma)^2 \geqslant 0,$$

and therefore $\quad \frac{1}{2}(\sigma^*_{ij} e^*_{ij} - \sigma_{ij} e_{ij}) \geqslant (\sigma^*_{ij} - \sigma_{ij}) e_{ij}$

for all stresses σ_{ij} and σ^*_{ij}. The equality holds if, and only if, $\sigma^*_{ij} = \sigma_{ij}$. Hence, unless σ^*_{ij} is equal to σ_{ij} at all points,

$$\tfrac{1}{2}\int (\sigma^*_{ij} e^*_{ij} - \sigma_{ij} e_{ij})\, dV > \int (\sigma^*_{ij} - \sigma_{ij}) e_{ij}\, dV = \int_{S_U} (F^*_i - F_i) u_i\, dS.$$

Rearranging this:

$$\tfrac{1}{2}\int \sigma_{ij}^* e_{ij}^* \, dV - \int_{S_U} F_i^* u_i \, dS > \tfrac{1}{2}\int \sigma_{ij} e_{ij} \, dV - \int_{S_U} F_i u_i \, dS$$

$$= \tfrac{1}{2}\int_{S_F} F_i u_i \, dS - \tfrac{1}{2}\int_{S_U} F_i u_i \, dS, \qquad (14)$$

with the use of Green's theorem and the equations of equilibrium. This is the well-known extremum principle, to the effect that the expression on the left-hand side of (14) takes an absolute minimum value in the actual state. (It follows as an immediate corollary that the boundary conditions define a unique solution.) By using the stress-strain relations to express the left-hand side solely in terms of σ_{ij}^* or of e_{ij}^* we obtain two complementary forms of the principle; either of these is applied, according to convenience, in the approximate solution of special problems.

A weaker statement is sometimes given in the form of a variational principle for distributions σ_{ij}^* differing *infinitesimally* from the actual one. This is proved as follows. Let E^* denote the potential energy $\tfrac{1}{2}\sigma_{ij}^* e_{ij}^*$ of an element in which the stress is σ_{ij}^*. We wish to find when the expression

$$\int E^* \, dV - \int_{S_U} F_i^* u_i \, dS$$

assumes a stationary value with respect to varying σ_{ij}^* satisfying the equations of equilibrium. By the usual transformation the expression is equal to

$$\int (E^* - \sigma_{ij}^* e_{ij}) \, dV + \int_{S_F} F_i^* u_i \, dS.$$

Since $F_i^* = F_i$ on S_F this has a stationary value when

$$\frac{\partial}{\partial \sigma_{ij}^*}(E^* - \sigma_{ij}^* e_{ij}) = 0, \quad \text{or} \quad \frac{\partial E^*}{\partial \sigma_{ij}^*} = e_{ij},$$

at every point. Since σ_{ij}^* and e_{ij}^* are related by the stress-strain equations,

$$e_{ij}^* = \frac{\partial E^*}{\partial \sigma_{ij}^*}.$$

The expression has, therefore, a stationary value when $e_{ij}^* = e_{ij}$; that is, in an actual state.

A second extremum principle is concerned with distributions $(\sigma_{ij}^*, e_{ij}^*)$ where e_{ij}^* is derivable from a displacement u_i^* satisfying the boundary conditions but σ_{ij}^* does not satisfy the equations of equilibrium. We begin with the integral

$$\int (e_{ij}^* - e_{ij}) \sigma_{ij} \, dV,$$

and transform it into $\int_{S_F} (u_i^* - u_i) F_i \, dS$,

where use has been made of the fact that $u_i^* = u_i$ over S_U. Now, by the reciprocity relation,
$$(e_{ij}^* - e_{ij})\sigma_{ij} = (\sigma_{ij}^* - \sigma_{ij})e_{ij}.$$
Hence, by the inequality proved above,
$$\tfrac{1}{2}(\sigma_{ij}^* e_{ij}^* - \sigma_{ij} e_{ij}) > (e_{ij}^* - e_{ij})\sigma_{ij}$$
unless $\sigma_{ij}^* = \sigma_{ij}$. Integrating this through the volume, and rearranging the result, we find
$$\int_{S_F} u_i^* F_i \, dS - \tfrac{1}{2} \int \sigma_{ij}^* e_{ij}^* \, dV < \int_{S_F} u_i F_i \, dS - \tfrac{1}{2} \int \sigma_{ij} e_{ij} \, dV$$
$$= \tfrac{1}{2} \int_{S_F} F_i u_i \, dS - \tfrac{1}{2} \int_{S_U} F_i u_i \, dS. \qquad (15)$$

On comparing this with (14), it will be observed that the expression on the right-hand side is the same in both equations, and that it is the absolute minimum of the function of $(\sigma_{ij}^*, e_{ij}^*)$ in (14) and the absolute maximum of the function of $(\sigma_{ij}^*, e_{ij}^*)$ in (15). These extremum principles furnish the means of obtaining upper and lower bounds in the approximate solution of special problems.

It is noted, finally, that similar extremum principles may be proved in identical fashion for *increments* of stress and strain.

(ii) *Plastic-elastic material.* We restrict our attention to existing states of stress in which increments of plastic strain are constrained to be of order $1/E \times$ the stress-increments. Let $(d\sigma_{ij}, d\epsilon_{ij})$ be the actual increments of stress and strain produced by given external stress-increments dF_i over a part S_F of the surface, and by given displacement-increments over the remainder S_U. Let $(d\sigma_{ij}^*, d\epsilon_{ij}^*)$ be any increments which satisfy the stress-strain relations, the stress boundary conditions, and the equations of equilibrium, but such that $d\epsilon_{ij}^*$ is not necessarily derivable from a continuous displacement. In other words $(d\sigma_{ij}^*, d\epsilon_{ij}^*)$ are physically possible increments for an element if it were free, but not necessarily when it is subject to the constraint of surrounding material. It may be shown, in the usual way, that
$$\int \{(d\sigma_{ij}^* - d\sigma_{ij}) \, d\epsilon_{ij}\} \, dV = \int_{S_U} \{(dF_i^* - dF_i) \, du_i\} \, dS.$$
Now,
$$2(d\sigma_{ij}^* - d\sigma_{ij}) \, d\epsilon_{ij}$$
$$\equiv (d\sigma_{ij}^* d\epsilon_{ij}^* - d\sigma_{ij} d\epsilon_{ij}) - [d\sigma_{ij}^*(d\epsilon_{ij}^* - d\epsilon_{ij}) + d\epsilon_{ij}(d\sigma_{ij} - d\sigma_{ij}^*)].$$

But, in a work-hardening material,

$$d\sigma_{ij}^*(d\epsilon_{ij}^* - d\epsilon_{ij}) + d\epsilon_{ij}(d\sigma_{ij} - d\sigma_{ij}^*)$$

$$= \frac{1}{2G}(d\sigma_{ij}^{*\prime} - d\sigma_{ij}')(d\sigma_{ij}^{*\prime} - d\sigma_{ij}') + \frac{3(1-2\nu)}{E}(d\sigma^* - d\sigma)^2 +$$

$$+ [h\,df^*(\alpha^*\,df^* - \alpha\,df) + \alpha h\,df(df - df^*)]$$

and, in a non-hardening material,

$$= \frac{1}{2G}(d\sigma_{ij}^{*\prime} - d\sigma_{ij}')(d\sigma_{ij}^{*\prime} - d\sigma_{ij}') + \frac{3(1-2\nu)}{E}(d\sigma^* - d\sigma)^2 +$$

$$+ [df^*(\alpha^*\,d\lambda^* - \alpha\,d\lambda) + \alpha\,d\lambda(df - df^*)].$$

In both cases, the first two terms on the right-hand side are together positive except when $d\sigma_{ij}^* = d\sigma_{ij}$, while the term in square brackets is never negative.† It follows that

$$\tfrac{1}{2}(d\sigma_{ij}^*\,d\epsilon_{ij}^* - d\sigma_{ij}\,d\epsilon_{ij}) > (d\sigma_{ij}^* - d\sigma_{ij})\,d\epsilon_{ij}$$

except when $d\sigma_{ij}^* = d\sigma_{ij}$. Thus, unless $d\sigma_{ij}^* = d\sigma_{ij}$ at every point,

$$\tfrac{1}{2}\int (d\sigma_{ij}^*\,d\epsilon_{ij}^* - d\sigma_{ij}\,d\epsilon_{ij})\,dV > \int_{S_U}\{(dF_i^* - dF_i)\,du_i\}\,dS,$$

or

$$\tfrac{1}{2}\int (d\sigma_{ij}^*\,d\epsilon_{ij}^*)\,dV - \int_{S_U}(dF_i^*\,du_i)\,dS > \tfrac{1}{2}\int (d\sigma_{ij}\,d\epsilon_{ij})\,dV - \int_{S_U}(dF_i\,du_i)\,dS$$

$$= \tfrac{1}{2}\int_{S_F}(dF_i\,du_i)\,dS - \tfrac{1}{2}\int_{S_U}(dF_i\,du_i)\,dS. \quad (16)$$

This extremum principle is substantially due to Hodge and Prager for a work-hardening material, to Greenberg for a non-hardening Reuss material, and to Bauer for a non-hardening material with a general function f.‡ From (16) we may immediately deduce the uniqueness of the stress-increment distribution, already proved more directly in Section 2 (i) and (ii).

† For both materials the term is zero in the elastic region, or where both $d\sigma_{ij}$ and $d\sigma_{ij}^*$ produce unloading in the plastic region, since $\alpha = \alpha^* = 0$. Where both produce loading ($\alpha = \alpha^* = 1$), the term is equal to $h(df^* - df)^2$ or to zero, respectively; it is therefore never negative. Where $d\sigma_{ij}$ produces loading ($\alpha = 1$) and $d\sigma_{ij}^*$ unloading ($\alpha^* = 0$), the term is equal to $h(df^2 - 2df\,df^*)$ and $-2df^*\,d\lambda$, respectively; both are non-negative since $df \geq 0$, $df^* < 0$, $d\lambda \geq 0$. When $d\sigma_{ij}$ produces unloading ($\alpha = 0$) and $d\sigma_{ij}^*$ loading ($\alpha^* = 1$), the term is equal to $h\,df^{*2}$ and zero, respectively; again it is non-negative.

‡ P. Hodge and W. Prager, *Journ. Math. and Phys.* **27** (1948), 1; H. J. Greenberg, op. cit., p. 58; F. B. Bauer, op. cit., p. 58. The first-named writers proved the theorem under the restriction that no part of the plastic region should unload; the present proof was supplied later by the author; letter to Prof. W. Prager, 10 June 1949.

It is instructive to compare the above proof with that of the weaker variational principle, asserting the existence of a stationary value of

$$\tfrac{1}{2}\int (d\sigma_{ij}^* \, d\epsilon_{ij}^*) \, dV - \int_{S_U} (dF_i^* \, du_i) \, dS$$

in the actual state.† By transformation of the surface integral, there results

$$\int (\tfrac{1}{2} d\sigma_{ij}^* d\epsilon_{ij}^* - d\sigma_{ij}^* d\epsilon_{ij}) \, dV + \int_{S_F} (dF_i^* \, du_i) \, dS.$$

Since $dF_i^* = dF_i$ on S_F, the last integral is a constant and the expression has a stationary value for varying $d\sigma_{ij}^*$ when

$$\delta(\tfrac{1}{2} d\sigma_{ij}^* d\epsilon_{ij}^* - d\sigma_{ij}^* d\epsilon_{ij}) = 0$$

at every point. Now, for a work-hardening material, it follows from (3) that

$$d\sigma_{ij}^* d\epsilon_{ij}^* = \frac{3(1-2\nu)}{E} d\sigma^{*2} + \frac{d\sigma_{ij}^{*\prime} d\sigma_{ij}^{*\prime}}{2G} + \alpha^{*} h \, df^{*2}.$$

Therefore,

$$\delta(\tfrac{1}{2} d\sigma_{ij}^* d\epsilon_{ij}^*) = \frac{3(1-2\nu)}{E} d\sigma^{*} \delta(d\sigma^{*}) + \left(\frac{d\sigma_{ij}^{*}}{2G} + \alpha^{*} h \frac{\partial f}{\partial \sigma_{ij}} df^{*}\right) \delta(d\sigma_{ij}^{*\prime})$$

since

$$\delta(df^*) = \delta\left(\frac{\partial f}{\partial \sigma_{ij}} d\sigma_{ij}^*\right) = \frac{\partial f}{\partial \sigma_{ij}} \delta(d\sigma_{ij}^*).$$

Hence, $\qquad \delta(\tfrac{1}{2} d\sigma_{ij}^* d\epsilon_{ij}^*) = d\epsilon_{ij}^* \delta(d\sigma_{ij}^*).$

Also, $\qquad \delta(d\sigma_{ij}^* d\epsilon_{ij}) = d\epsilon_{ij} \delta(d\sigma_{ij}^*).$

Thus, the condition for a stationary value is $d\epsilon_{ij}^* = d\epsilon_{ij}$; that is, the increments must be the actual ones.

Just as with a purely elastic material there is a second extremum principle for distributions $(d\sigma_{ij}^*, d\epsilon_{ij}^*)$ where $d\epsilon_{ij}^*$ is derivable from a displacement du_i^* satisfying the boundary conditions but $d\sigma_{ij}^*$ is not in equilibrium. Following the usual method, we begin with the equation

$$\int \{(d\epsilon_{ij}^* - d\epsilon_{ij}) \, d\sigma_{ij}\} \, dV = \int_{S_F} \{(du_i^* - du_i) \, dF_i\} \, dS$$

and the identity

$$2(d\epsilon_{ij}^* - d\epsilon_{ij}) \, d\sigma_{ij}$$
$$\equiv (d\sigma_{ij}^* d\epsilon_{ij}^* - d\sigma_{ij} d\epsilon_{ij}) - [d\epsilon_{ij}^*(d\sigma_{ij}^* - d\sigma_{ij}) + d\sigma_{ij}(d\epsilon_{ij} - d\epsilon_{ij}^*)].$$

By the argument employed previously the term in square brackets can be shown to be positive for both a non-hardening and a work-hardening material, unless $d\sigma_{ij}^* = d\sigma_{ij}$. Thus,

$$\tfrac{1}{2}(d\sigma_{ij}^* d\epsilon_{ij}^* - d\sigma_{ij} d\epsilon_{ij}) > (d\epsilon_{ij}^* - d\epsilon_{ij}) d\sigma_{ij}$$

† This was proved for the case $f \equiv J_2$ (Reuss material) by W. Prager, *Proc. 6th Int. Cong. App. Mech.*, Paris (1946).

except when $d\sigma_{ij}^* = d\sigma_{ij}$. Hence, unless $d\sigma_{ij}^* = d\sigma_{ij}$ at every point,

$$\int_{S_F} (du_i^* \, dF_i) \, dS - \tfrac{1}{2} \int (d\sigma_{ij}^* \, d\epsilon_{ij}^*) \, dV < \int_{S_F} (du_i \, dF_i) \, dS - \tfrac{1}{2} \int (d\sigma_{ij} \, d\epsilon_{ij}) \, dV$$

$$= \tfrac{1}{2} \int_{S_F} (dF_i \, du_i) \, dS - \tfrac{1}{2} \int_{S_U} (dF_i \, du_i) \, dS. \qquad (17)$$

Comparing this with (16) we observe that in the first principle the right-hand side is the minimum of one function of $(d\sigma_{ij}^*, d\epsilon_{ij}^*)$, and in the second the maximum of another. This theorem is substantially due to Greenberg for a non-hardening Reuss material, and to Bauer for a general non-hardening material.†

(iii) *Plastic-rigid material.* In Section 2 (iii) a uniqueness theorem was proved for the consistent distribution of stress under boundary conditions such that the whole mass is deforming plastically. By analogy with the previous results it is natural to expect two related extremum principles pertaining to a consistent state. In one we consider plastic states $(\sigma_{ij}^*, d\epsilon_{ij}^*)$ such that σ_{ij}^* satisfies the equations of equilibrium and stress boundary-conditions but $d\epsilon_{ij}^*$ is not derivable from a displacement; in the other, $d\epsilon_{ij}^*$ *is* derivable from du_i^* satisfying the displacement boundary-conditions, but the distribution σ_{ij}^* is not in equilibrium.

In the proof of the first principle the natural starting-point is the equation

$$\int \{(\sigma_{ij}^* - \sigma_{ij}) d\epsilon_{ij}\} dV = \int_{S_U} \{(F_i^* - F_i) du_i\} dS.$$

Now, in any element, the vectors representing $\sigma_{ij}^{*\prime}$ and σ_{ij}' lie, by hypothesis, on the same yield locus, while $d\epsilon_{ij}$ is parallel to the outward normal to the locus at the point σ_{ij}'. Hence, $(\sigma_{ij}^* - \sigma_{ij}) d\epsilon_{ij}$ is proportional to the scalar product of the outward normal to the yield locus at the point σ_{ij}' with the chord joining σ_{ij}' to $\sigma_{ij}^{*\prime}$; it is therefore negative, since f is concave to the origin. Thus,

$$(\sigma_{ij}^* - \sigma_{ij}) \, d\epsilon_{ij} < 0$$

unless $\sigma_{ij}^{*\prime} = \sigma_{ij}'$, and so

$$\int_{S_U} (F_i^* \, du_i) dS < \int_{S_U} (F_i \, du_i) dS, \qquad (18)$$

unless σ_{ij}^* and σ_{ij} differ only by a uniform hydrostatic stress. Thus, the work done by the actual surface forces in the prescribed displacements is greater than that done by the surface forces corresponding to any other

† H. J. Greenberg, op. cit., p. 58; F. B. Bauer, op. cit., p. 58. These writers proved the theorem for boundary conditions such that no plastic element unloaded; the present proof is due to the author, op. cit., p. 64.

equilibrium state of plastic stress.† This is known as the principle of maximum plastic work.

For the second principle, we begin with the equation

$$\int \{(d\epsilon_{ij}^*-d\epsilon_{ij})\sigma_{ij}\}dV = \int_{S_F} \{(du_i^*-du_i)F_i\}dS,$$

and consider the special case of the Reuss material ($f \equiv \sigma'_{ij}\sigma'_{ij}$). For this, the vectors representing σ'_{ij} and $d\epsilon_{ij}$ are parallel, and so their scalar product $\sigma'_{ij}d\epsilon_{ij}$ (or $\sigma_{ij}d\epsilon_{ij}$) is equal to the product of their lengths or moduli, $|\sigma'_{ij}|$ and $|d\epsilon_{ij}|$, where

$$|\sigma'_{ij}| = \sqrt{(\sigma'_{ij}\sigma'_{ij})}, \qquad |d\epsilon_{ij}| = \sqrt{(d\epsilon_{ij}d\epsilon_{ij})}.$$

In passing, it may be recalled that the moduli are proportional to the equivalent stress and strain-increment; in fact, according to equations (14) and (16) of Chapter II, $\bar{\sigma} = \sqrt{\tfrac{3}{2}}|\sigma'_{ij}|$, $\overline{d\epsilon} = \sqrt{\tfrac{2}{3}}|d\epsilon_{ij}|$. Now, the scalar product of the two non-parallel vectors representing σ'_{ij} and $d\epsilon_{ij}^*$ is, of course, less than the product of their moduli. Hence,

$$\sigma_{ij}(d\epsilon_{ij}^*-d\epsilon_{ij}) < |\sigma'_{ij}|(|d\epsilon_{ij}^*|-|d\epsilon_{ij}|)$$

unless $\sigma_{ij}^{*\prime} = \sigma'_{ij}$. It follows‡ that, unless σ_{ij}^* and σ_{ij} differ only by a uniform hydrostatic stress,

$$\int (|\sigma'_{ij}|\,|d\epsilon_{ij}^*|)dV - \int_{S_F}(F_i\,du_i^*)dS > \int(|\sigma'_{ij}|\,|d\epsilon_{ij}|)dV - \int_{S_F}(F_i\,du_i)dS$$
$$= \int_{S_U}(F_i\,du_i)dS. \qquad (19)$$

The right-hand sides of (18) and (19) are identical. In this equation the quantity $|\sigma'_{ij}|$, specifying the initial state of hardening, is to be regarded as known. If the state of hardening is uniform ($|\sigma'_{ij}| = $ constant) and if, further, the displacements are given at all points of the surface ($S_F = 0$) then (19) reduces to

$$\int |d\epsilon_{ij}^*|dV > \int |d\epsilon_{ij}|dV.$$

† R. Hill, op. cit., p. 59. The same principle applies also for the Hencky stress-strain relations when the material is incompressible and non-hardening; see A. H. Philippidis, *Journ. App. Mech.* **15** (1948), 241, and the discussion by H. J. Greenberg, ibid. **16** (1949), 103. When an elastic region is present, the Hencky relations for a non-hardening material are associated with a variational principle suggested by A. Haar and Th. von Karman, *Göttinger Nachrichten, math.-phys. Klasse*, (1909), 204; a proof of the Haar-Karman principle under these conditions has been given by H. J. Greenberg, Brown University, Division of App. Math., *Rep.* A11–S4 (1949).

A heuristic principle of maximum plastic 'resistance' was stated by M. A. Sadowsky, *Journ. App. Mech.* **10** (1943), A–65, on the basis of its success in a few instances; the circumstances in which it happens to furnish the correct solution have been stated in the paper by Hill.

‡ R. Hill, ibid. **17** (1950), 64.

In this restricted form the principle was first stated by Markov.†

(iv) *Inversion of the stress-strain relations.* For the application of (16) and (17) to special problems it must be possible to express $d\sigma_{ij}^* d\epsilon_{ij}^*$ solely in terms of the independently varied quantity, $d\sigma_{ij}^*$ or $d\epsilon_{ij}^*$, respectively. We show now that this may always be done.

To eliminate $d\epsilon_{ij}^*$ we multiply

$$d\epsilon_{ij}^{*\prime} = \frac{d\sigma_{ij}^{*\prime}}{2G} + \alpha^* h \frac{\partial f}{\partial \sigma_{ij}} df^*$$

by $d\sigma_{ij}^{*\prime}$, and sum. We obtain

$$d\sigma_{ij}^{*\prime} d\epsilon_{ij}^{*\prime} = \frac{d\sigma_{ij}^{*\prime} d\sigma_{ij}^{*\prime}}{2G} + \alpha^* h \, df^{*2}. \tag{20}$$

For a non-hardening material, $h \, df^*$ is replaced by $d\lambda^*$. Now $\alpha^* = 0$ where $f(\sigma_{ij}) < c$, or where $f(\sigma_{ij}) = c$ and $df^* < 0$; elsewhere the $d\sigma_{ij}^*$ are restricted so that $df^* = 0$. Thus

$$d\sigma_{ij}^{*\prime} d\epsilon_{ij}^{*\prime} = \frac{d\sigma_{ij}^{*\prime} d\sigma_{ij}^{*\prime}}{2G} \tag{21}$$

everywhere, for a non-hardening material.

To eliminate $d\sigma_{ij}^*$ for a plastic element undergoing continued loading ($\alpha^* = 1$) we begin by forming the product

$$\frac{\partial f}{\partial \sigma_{ij}} d\epsilon_{ij}^* = \frac{\partial f}{\partial \sigma_{ij}} d\epsilon_{ij}^{*\prime} = \left(\frac{1}{2G} + h \frac{\partial f}{\partial \sigma_{ij}} \frac{\partial f}{\partial \sigma_{ij}}\right) df^* \quad (f(\sigma_{ij}) = c, df^* \geqslant 0) \tag{22}$$

for a work-hardening material, and

$$\frac{\partial f}{\partial \sigma_{ij}} d\epsilon_{ij}^* = \frac{\partial f}{\partial \sigma_{ij}} \frac{\partial f}{\partial \sigma_{ij}} d\lambda^* \quad (f(\sigma_{ij}) = c, df^* = 0, d\lambda^* \geqslant 0) \tag{23}$$

for a non-hardening material. Since $(\partial f/\partial \sigma_{ij})(\partial f/\partial \sigma_{ij})$ is a positive quantity, being a sum of squares, a given strain-increment produces continued loading of a plastic element if

$$\frac{\partial f}{\partial \sigma_{ij}} d\epsilon_{ij}^* \geqslant 0. \tag{24}$$

If the strain-increment is such that the inequality is reversed, the

† A. A. Markov, *Prikladnaia Matematika i Mekhanika*, **11** (1947), 339.

element unloads. Using (22) and (23) to eliminate df^* and $d\lambda^*$ from the respective stress-strain relations, we obtain

$$\frac{d\sigma_{ij}^{*'}}{2G} = d\epsilon_{ij}^{*'} - \frac{\alpha^*\left(\dfrac{\partial f}{\partial \sigma_{kl}} d\epsilon_{kl}^*\right)}{\left(\dfrac{1}{2Gh} + \dfrac{\partial f}{\partial \sigma_{kl}}\dfrac{\partial f}{\partial \sigma_{kl}}\right)} \frac{\partial f}{\partial \sigma_{ij}} \quad (25)$$

for a work-hardening material, and

$$\frac{d\sigma_{ij}^{*'}}{2G} = d\epsilon_{ij}^{*'} - \frac{\alpha^*\left(\dfrac{\partial f}{\partial \sigma_{kl}} d\epsilon_{kl}^*\right)}{\left(\dfrac{\partial f}{\partial \sigma_{kl}}\dfrac{\partial f}{\partial \sigma_{kl}}\right)} \frac{\partial f}{\partial \sigma_{ij}} \quad (26)$$

for a non-hardening material, where

$\alpha^* = 0$ when $f(\sigma_{ij}) < c$, or when $f(\sigma_{ij}) = c$ and $\dfrac{\partial f}{\partial \sigma_{ij}} d\epsilon_{ij}^* < 0$,

$\alpha^* = 1$ when $f(\sigma_{ij}) = c$ and $\dfrac{\partial f}{\partial \sigma_{ij}} d\epsilon_{ij}^* \geqslant 0$.

Hence, $\quad d\sigma_{ij}^{*'} d\epsilon_{ij}^{*'} = 2G\left[d\epsilon_{ij}^{*'} d\epsilon_{ij}^{*'} - \dfrac{\alpha^*\left(\dfrac{\partial f}{\partial \sigma_{ij}} d\epsilon_{ij}^*\right)^2}{\left(\dfrac{1}{2Gh} + \dfrac{\partial f}{\partial \sigma_{ij}}\dfrac{\partial f}{\partial \sigma_{ij}}\right)}\right] \quad (27)$

for a work-hardening material, and

$$d\sigma_{ij}^{*'} d\epsilon_{ij}^{*'} = 2G\left[d\epsilon_{ij}^{*'} d\epsilon_{ij}^{*'} - \frac{\alpha^*\left(\dfrac{\partial f}{\partial \sigma_{ij}} d\epsilon_{ij}^*\right)^2}{\left(\dfrac{\partial f}{\partial \sigma_{ij}}\dfrac{\partial f}{\partial \sigma_{ij}}\right)}\right] \quad (28)$$

for a non-hardening material.

If equations (20), (21), (27), or (28), are combined with the identities

$$d\sigma_{ij}^* d\epsilon_{ij}^* - d\sigma_{ij}^{*'} d\epsilon_{ij}^{*'} = 3\, d\sigma^* d\epsilon^* = \frac{3(1-2\nu)}{E} d\sigma^{*2} = \frac{3E}{(1-2\nu)} d\epsilon^{*2}, \quad (29)$$

the quantity $d\sigma_{ij}^* d\epsilon_{ij}^*$ is expressed in terms of $d\sigma_{ij}^*$ or $d\epsilon_{ij}^*$ alone.

IV
THE SOLUTION OF PLASTIC-ELASTIC PROBLEMS. I

1. Introduction

THE complete solution of a general problem in plasticity involves a calculation of the stress and the deformation in both the elastic and plastic regions. In the former the stress is directly connected with the total strain by means of the elastic equations. In the latter there is, as we have seen, no such unique correspondence, and the stress-strain differential relations have to be integrated by following the history of the deformation from the initiation of plasticity at some point of the body. A process of plastic deformation has to be considered mathematically as a succession of small increments of strain, even where the overall distortion is so small that the change in external surfaces can be neglected. When the strains are large the determination of the changing shape of free plastic surfaces necessitates, in itself, the following of the deformation from moment to moment.

The solutions in the elastic and plastic regions are interrelated by certain continuity conditions in the stresses and displacements which must be satisfied along the plastic-elastic boundary. This boundary is itself one of the unknowns, and is usually of such an awkward shape that even the stress distribution in the elastic region can only be obtained by laborious numerical methods. The complete and accurate solution of a plastic problem is practicable in relatively few cases, and can normally only be expected when the problem has some special symmetry or other simplifying feature. This chapter and the following one are concerned with problems for which analytic solutions are known, either in an explicit form or such that the solution can be completed by direct numerical integration. From them it is possible to obtain a general insight into the interrelation of the states of stress and strain in the elastic and plastic regions. We shall then be in a position to formulate reasonable approximations which assist solutions of the more complicated problems of technical importance.

The problems described in the present chapter are particularly simple, and fall broadly into two groups. In the first group there is no plastic-elastic boundary to be calculated, since all elements are simultaneously stressed to the yield limit. In the second group plastic and elastic regions exist side by side, but the strains are restricted to be small.

2. Theory of Hohenemser's experiment

(i) When Reuss formulated his stress-strain equations he also suggested an experiment by which their validity might be tested. This was necessary because although the relation $\mu = \nu$ had been shown by Lode to be approximately true when the stress-increment and the elastic component of the strain were negligible, this might not be so when the elastic and plastic components of the strain were comparable. In other words the ratios of the components of the plastic strain-increment might not depend only on the stress ratios, but on the stress-increment as well. The suggested experiment consisted in twisting a hollow cylindrical tube to the point of yielding, and then extending it longitudinally while holding the twist constant; the wall of the tube was sufficiently thin to obtain an approximately uniform distribution of stress, and the metal pre-strained to secure a sharp yield-point and reduce the rate of hardening to a value small compared with the elastic modulus. If σ and ϵ are the axial stress and logarithmic strain, and τ and γ are the shear stress and strain, the Reuss equations ((33) of Chap. II) give

$$d\gamma = 0 = \tau\,d\lambda + \frac{d\tau}{2G},$$

$$d\epsilon = \tfrac{2}{3}\sigma\,d\lambda + \frac{d\sigma}{E},$$

$$\sigma^2 + 3\tau^2 = Y^2.$$

There are other equations from which the transverse contraction of the tube wall can be found if required. Eliminating $d\lambda$:

$$d\epsilon = -\frac{\sigma}{3\tau}\frac{d\tau}{G} + \frac{d\sigma}{E}.$$

Substituting for τ from the yield criterion, and integrating:

$$\frac{6G}{Y}\epsilon = \ln\!\left(\frac{1+s}{1-s}\right) + \frac{1-2\nu}{1+\nu}s, \quad \text{where } s = \sigma/Y. \tag{1}$$

As ϵ is increased, σ rises from zero (initially with rate E) and approaches Y asymptotically and so rapidly that σ is already within 1 per cent. of Y when ϵ is only $3Y/E$. The shear stress τ correspondingly diminishes from $Y/\sqrt{3}$ towards zero. Because of this initial rapid change of stress, the elastic strain-increment is comparable with the plastic strain-increment, as desired. After a tensile strain of only a few times the yield-point strain Y/E the elastic strain-increment is entirely negligible and the stress is virtually simple tension. The prescribed twist has by then been rendered permanent.

The experiment was carried out by Hohenemser,[†] using pre-strained mild steel. Agreement with the predictions of Reuss's equations was probably as good as could be expected in view of various uncertainties, for example the hysteresis loop and rounding of the yield-point, elastic after-effects, and the probable deviations from the relation $\mu = \nu$ indicated by Lode's experiments. Secondary factors of this kind, not embodied in the theory, tend to prevent any significant test of the influence of the stress-increment when the strain is small. There appear to be no other published data from experiments designed to check the Reuss equations.[‡] It seems that this is another reason for the use of the Hencky equations by some writers in small-strain problems where the loading is continuous, notwithstanding the objection mentioned in the last chapter. It happens, too, that the variation of the torque in Hohenemser's experiment agrees rather better with Hencky's theory than with that of Reuss.[§] The agreement is fortuitous, and no fundamental significance should be attached to it.

(ii) When the work-hardening is not so small as we have assumed, the elastic strain-increment continues to be comparable with the plastic strain-increment so long as the rate of hardening is comparable with E, quite apart from the influence of the prescribed strain-path. In detail, we have from (32) of Chapter II:

$$0 = \frac{3\tau\, d\bar{\sigma}}{2H'\bar{\sigma}} + \frac{d\tau}{2G},$$

$$d\epsilon = \frac{\sigma\, d\bar{\sigma}}{H'\bar{\sigma}} + \frac{d\sigma}{E},$$

[†] K. Hohenemser, *Proc. 3rd Int. Cong. App. Mech.*, Stockholm, 2 (1930); *Zeits. ang. Math. Mech.* 11 (1931), 15. See also K. Hohenemser and W. Prager, ibid. 12 (1932), 1.

[‡] Since the book went to press, J. L. M. Morrison and W. M. Shepherd, *Proc. Inst. Mech. Eng.*, (1950), have reported experiments on the stressing of tubes of steel and aluminium by various combinations of torsion and tension. Plastic and elastic strains were comparable, and the measured variations of length and twist were in substantial agreement with the predictions of the Reuss equations (work-hardening being allowed for by the equivalent stress-strain curve). In particular, the initial torsional rigidity of the tube when twisted under an axial load producing a permanent extension was exactly equal to G (see (iii) below). The Hencky equations, on the other hand, were frequently seriously in error.

[§] W. Prager, *Mém. Sci. Math.* 87 (1937); *Journ. App. Phys.* 15 (1944), 65.

where $\bar{\sigma} = (\sigma^2 + 3\tau^2)^{\frac{1}{2}}$ and H' is the rate of hardening (a known function of $\bar{\sigma}$). From the first equation:

$$\int_Y^{\bar{\sigma}} \frac{d\bar{\sigma}}{H'\bar{\sigma}} = \int_\tau^{Y/\sqrt{3}} \frac{d\tau}{3G\tau} = \frac{1}{3G} \ln\left(\frac{Y}{\sqrt{3}\tau}\right).$$

This is a relation between σ and τ, in general requiring numerical integration. Then, by eliminating σ from the leading term on the right-hand side of the second equation, a differential relation is obtained from which the fall of torque with increasing strain may be calculated.

(iii) An allied problem of some interest is the twisting of a tube already stressed in tension to the point of yielding. Suppose the tube has been extended to the point where its yield stress is Y, and that an increasing torque is then applied, the tensile stress being held constant and equal to Y. We then have

$$\bar{\sigma}^2 = 3\tau^2 + Y^2, \qquad \bar{\sigma}\,d\bar{\sigma} = 3\tau\,d\tau,$$

and the equations of Reuss take the form

$$d\epsilon = \frac{3Y\tau\,d\tau}{H'(3\tau^2 + Y^2)},$$

$$d\gamma = \frac{9\tau^2\,d\tau}{2H'(3\tau^2 + Y^2)} + \frac{d\tau}{2G}.$$

From the latter equation we see that initially, when $\tau = 0$, the effective torsional rigidity is still G, even though the tube is plastic. The general relation between γ and τ can be determined explicitly when H' is constant. On performing the integrations we obtain

$$\epsilon = \frac{Y}{2H'} \ln\left(1 + \frac{3\tau^2}{Y^2}\right),$$

$$\gamma = \frac{3}{2H'}\left[\tau - \frac{Y}{\sqrt{3}} \tan^{-1}\left(\frac{\sqrt{3}\tau}{Y}\right)\right] + \frac{\tau}{2G}.$$

The (τ, γ) relation has a tangent of slope $2G$ at the origin, and gradually bends over until the slope of the tangent becomes $2H'/3$. The meaning of γ in this context should be carefully noted. It is equal to

$$\int_0^\theta \frac{r\,d\theta}{2l},$$

where θ is the relative twist of the ends of the tube, and l and r are the current values of the length and radius of the tube, which can be calculated by using the Reuss equations for the radial and circumferential strain-increments. γ has no such simple geometrical interpretation as in pure torsion, where it is equal to $\tfrac{1}{2}\tan\phi$, ϕ being the slope of the helix into which an original generator of the tube is distorted.

3. Combined torsion and tension of a thin-walled tube

We now investigate the stresses in a thin-walled tube subjected to an arbitrary combination of twist and extension. The strain-path and the rate of hardening are supposed such that the elastic component of the strain can be neglected. Let l_0, r_0, and t_0 be the initial length, mean radius, and thickness of the tube, and let l, r, and t be their current values. From (32) of Chapter II expressed in cylindrical coordinates (r,θ,z):

$$\left. \begin{aligned} d\epsilon_z &= \frac{dl}{l} = \frac{\sigma\,d\bar\sigma}{H'\bar\sigma}, & d\gamma_{\theta z} &= \frac{r\,d\theta}{2l} = \frac{3\tau\,d\bar\sigma}{2H'\bar\sigma}, \\ d\epsilon_\theta &= \frac{dr}{r} = -\frac{\sigma\,d\bar\sigma}{2H'\bar\sigma}, & d\epsilon_r &= \frac{dt}{t} = -\frac{\sigma\,d\bar\sigma}{2H'\bar\sigma}, \end{aligned} \right\} \quad (2)$$

where θ is the relative angular twist of the extreme sections of the tube. It follows that $dr/r = -dl/2l$ and hence, that

$$r^2 l = \text{constant} = r_0^2 l_0. \qquad (3)$$

The internal volume of the tube therefore remains constant during the distortion. This is necessarily true for an isotropic tube under pure tension as a simple consequence of the equality of the radial and circumferential strain-increments. It is only true under combined torsion and tension, however, when the relation $\mu = \nu$ is satisfied. Taylor and Quinney made these two facts the basis of their experimental determination[†] of the (μ,ν) relation, and measured the internal volume by filling the tubes with water and observing the movement of the meniscus in a connected capillary. By first straining each specimen in tension they were able both to pre-strain the material by any desired amount and also to detect anisotropy in the material.

From the first and second equations of (2):

$$r\frac{d\theta}{dl} = r_0\sqrt{\left(\frac{l_0}{l}\right)}\frac{d\theta}{dl} = \frac{3\tau}{\sigma}. \qquad (4)$$

If the strain-path is prescribed $d\theta/dl$ may be regarded as a given function of the length of the tube. Accordingly (4) determines the ratio of the tensile and shear stresses at each stage. Now

$$\bar\sigma = (\sigma^2 + 3\tau^2)^{\tfrac{1}{2}}$$

[†] G. I. Taylor and H. Quinney, *Phil. Trans. Roy. Soc.* A, **230** (1931), 323.

and so, eliminating τ between this and (4), we obtain

$$\bar{\sigma} = \sigma\left[1 + \frac{r_0^2 l_0}{3l}\left(\frac{d\theta}{dl}\right)^2\right]^{\frac{1}{2}}. \tag{5}$$

Substituting for σ in the first equation of (2) and integrating:

$$\int_Y^{\bar{\sigma}} \frac{d\bar{\sigma}}{H'} = \int_{l_0}^l \left[1 + \frac{r_0^2 l_0}{3l}\left(\frac{d\theta}{dl}\right)^2\right]^{\frac{1}{2}} \frac{dl}{l}, \tag{6}$$

where Y is the initial yield stress of the tube. This is the relation between $\bar{\sigma}$ and the length of the tube (the left-hand side is evidently $\int d\bar{\epsilon}$). When the integral has been evaluated by numerical integration, or otherwise, σ and τ are directly obtainable from (4) and (5). The tensile load L is $2\pi r t \sigma$, and the torque G is $2\pi r^2 t \tau$. Since the volume $2\pi r l t$ of the tube material is constant we have $rlt = r_0 l_0 t_0$. Hence from (3):

$$\frac{r}{r_0} = \frac{t}{t_0} = \sqrt{\frac{l_0}{l}}. \tag{7}$$

Therefore
$$L = 2\pi r_0 t_0 \left(\frac{l_0}{l}\right)\sigma, \qquad G = 2\pi r_0^2 t_0 \left(\frac{l_0}{l}\right)^{\frac{3}{2}}\tau. \tag{8}$$

Conversely, if some restriction is placed on the stresses applied to the tube (for example, the load or the torque may be held constant), the elimination of $\bar{\sigma}$, σ, and τ between (4), (5), (6), and (8) gives a relation between θ and l from which the resulting strain-path can be calculated.

4. Combined torsion and tension of a cylindrical bar

It is supposed that the elastic component of the strain is negligible, and that the bar is everywhere plastic. With these assumptions the analysis for the thin-walled tube leads us to expect that a cylindrical bar behaves as if it were an assembly of concentric shells, in each of which the states of stress and strain develop independently. Since the internal volume of each shell remains constant the material inside any shell can be accommodated without producing other components of stress, and transverse sections will remain plane. These expectations are confirmed by the strict analysis. For this it is convenient to introduce the components u, v, w of the velocity of an element referred to cylindrical coordinates (r, θ, z). Assuming that transverse sections remain plane, and that the local rate of strain is a function only of r, we try the following expressions for the velocities:

$$u = -\frac{\dot{l}r}{2l}, \quad v = \frac{\dot{\theta}rz}{l}, \quad w = \frac{\dot{l}z}{l}, \tag{9}$$

where the origin of coordinates is taken at the centre of one end, which is held fixed. l is the length of the bar, and θ is the relative twist of the ends; a dot denotes the time derivative. Assuming further that the only non-vanishing components of stress are $\sigma_z = \sigma$ and $\tau_{\theta z} = \tau$, equations (32) of Chapter II reduce† to

$$\dot\epsilon_r = -\frac{\dot l}{2l} = -\frac{\sigma\dot{\bar\sigma}}{2H'\bar\sigma}; \qquad \dot\gamma_{\theta z} = \frac{r\dot\theta}{2l} = \frac{3\tau\dot{\bar\sigma}}{2H'\bar\sigma};$$

$$\dot\epsilon_\theta = -\frac{\dot l}{2l} = -\frac{\sigma\dot{\bar\sigma}}{2H'\bar\sigma}; \qquad \dot\gamma_{rz} = 0;$$

$$\dot\epsilon_z = \frac{\dot l}{l} = \frac{\sigma\dot{\bar\sigma}}{H'\bar\sigma}; \qquad \dot\gamma_{r\theta} = 0.$$

These are all satisfied provided that

$$\frac{\dot l}{l} = \frac{\sigma\dot{\bar\sigma}}{H'\bar\sigma}, \quad \text{and} \quad \frac{r\dot\theta}{l} = \frac{3\dot\tau}{\sigma}. \tag{4'}$$

The stresses are functions of r only and so the equations of equilibrium are automatically satisfied. (9) and (4') therefore constitute a compatible system of stresses and velocities, and the conception of the solid bar as an assembly of independent shells is justified. This would not be true in general for any other (μ,ν) relation, since the internal volume of a shell is not constant (the radial and circumferential strain-increments being unequal). Transverse sections must necessarily become warped, except when the bar is loaded either by a pure tension ($\mu = \nu = -1$) or by a pure torque ($\mu = \nu = 0$).

When there is no hardening the stress distribution is immediately found from the equations

$$\sigma^2 + 3\tau^2 = Y^2, \qquad \frac{\sigma}{\tau} = \frac{3}{r}\frac{dl}{d\theta} = \frac{\alpha}{r}, \quad \text{say,}$$

where α is independent of r but varies from moment to moment. Then

$$\sigma = \frac{\alpha Y}{\sqrt{(\alpha^2+3r^2)}}, \qquad \tau = \frac{rY}{\sqrt{(\alpha^2+3r^2)}}. \tag{10}$$

These formulae appear to have been first stated by Nadai.‡ The shear stress is zero on the axis of the bar, and rises to a maximum at the surface. If the strain-path is prescribed, α must be regarded as a known function of the length of the bar. Conversely, if some condition is placed on the

† See Appendix II for the expressions for the rate-of-strain components in cylindrical coordinates.

‡ A. Nadai, *Trans. Am. Soc. Mech. Eng.* **52** (1930), 193. The parameter α has a different meaning in Nadai's formula since he used the Hencky stress-strain relations.

relative values of the load and torque, the parameter α and the strain-path can be calculated.

5. Compression under conditions of plane strain

A rectangular block of metal with edges parallel to the axes of x, y, and z is compressed in the x direction between overlapping rigid plates. Expansion is allowed only in the y direction, and is prevented in the z direction by rigid dies. All surfaces of contact are supposed to be perfectly lubricated, so that the deformation is plane and uniform. This experimental arrangement has been employed by Bridgman.† The solution‡ of the problem illustrates features common to all problems of plane strain. Work-hardening is supposed zero, and the metal is assumed to yield in accordance with Tresca's law. While the block is still stressed elastically $\sigma_z = \nu\sigma_x$. It will appear *a posteriori* that σ_z is always the intermediate principal stress; consequently $\sigma_x = -Y$ at yielding and throughout subsequent plastic distortion. The Reuss equations are

$$\left. \begin{aligned} d\epsilon_x &= \tfrac{1}{3}(2\sigma_x-\sigma_y-\sigma_z)\,d\lambda+\frac{1}{E}(d\sigma_x-\nu\,d\sigma_y-\nu\,d\sigma_z), \\ d\epsilon_y &= \tfrac{1}{3}(2\sigma_y-\sigma_z-\sigma_x)\,d\lambda+\frac{1}{E}(d\sigma_y-\nu\,d\sigma_z-\nu\,d\sigma_x), \\ 0 = d\epsilon_z &= \tfrac{1}{3}(2\sigma_z-\sigma_x-\sigma_y)\,d\lambda+\frac{1}{E}(d\sigma_z-\nu\,d\sigma_x-\nu\,d\sigma_y). \end{aligned} \right\} \quad (11)$$

Substituting $\sigma_x = -Y$, $\sigma_y = 0$, and eliminating $d\lambda$ from the first and third equations, and integrating:

$$E(\epsilon_x - \epsilon_{x0}) = -\frac{3Y}{4}\ln\left(\frac{1-2\nu}{1+2\sigma_z/Y}\right) + (\tfrac{1}{2}-\nu)(\sigma_z+\nu Y),$$

where $\epsilon_{x0} = -(1-\nu^2)Y/E$ is the strain when the block first yields. ϵ_y can be determined from the second Reuss equation if required. As the compression proceeds the last term becomes unimportant and σ_z rises rapidly from its value $-\nu Y$ at yielding to the limiting value $-\tfrac{1}{2}Y$. For example, if $\nu = 0{\cdot}3$ (a typical value for metals), σ_z is already equal to $-0{\cdot}498Y$ after a plastic strain of only five times the yield-point strain. Owing to this rapid initial change of σ_z the elastic strain-increment is comparable with the plastic strain-increment up to total strains of three or four times the yield-point strain. Thereafter the elastic strain-increment quickly diminishes in relative value, and σ_z can be taken equal to

† P. W. Bridgman, *Journ. App. Phys.* **17** (1946), 225.
‡ R. Hill, *Journ. App. Mech.* **16** (1949), 295.

$-\tfrac{1}{2}Y$, or $\tfrac{1}{2}\sigma_x$, to an extremely close approximation. It is instructive to consider this by means of the geometrical representation described on p. 42.

Similar conclusions are reached if the metal yields according to the law of von Mises:
$$\sigma_x^2 - \sigma_x \sigma_z + \sigma_z^2 = Y^2. \qquad (12)$$

Since $\sigma_z = \nu \sigma_x$ in the elastic state, yielding begins when
$$\sigma_{x0} = \frac{-Y}{\sqrt{(1-\nu+\nu^2)}}, \quad \epsilon_{x0} = \frac{-(1-\nu^2)Y}{\sqrt{(1-\nu+\nu^2)}E}.$$

For the usual values of ν, the yield stress is numerically a little less than $2Y/\sqrt{3}$; thus $\sigma_{x0} = -1{\cdot}127Y$ when $\nu = 0{\cdot}3$. On putting $\sigma_y = 0$ in (11), and eliminating $d\lambda$ from the first and third equations,
$$E\,d\epsilon_x = \left[1 - 2\nu + \frac{3Y^2}{(\sigma_x - 2\sigma_z)(2\sigma_x - \sigma_z)}\right] d\sigma_x$$
after using (12) and its differential form
$$(2\sigma_x - \sigma_z)\,d\sigma_x + (2\sigma_z - \sigma_x)\,d\sigma_z = 0.$$
Hence,
$$E(\epsilon_x - \epsilon_{x0}) = (1-2\nu)\sigma_x - \frac{\sqrt{3}\,Y}{4}\ln\!\left(\frac{2Y/\sqrt{3} - \sigma_x}{2Y/\sqrt{3} + \sigma_x}\right) + \text{const.}, \qquad (13)$$
where the constant can be obtained by substituting the conditions at the first yielding. As ϵ_x is increased, σ_x approaches $-2Y/\sqrt{3}$, the centre term quickly becoming negligible. σ_z correspondingly decreases rapidly towards $\tfrac{1}{2}\sigma_x$, virtually attaining this limiting value in a total strain of elastic order.

Experimental verification of these conclusions would be valuable, but no dies are absolutely rigid and some allowance would need to be made for their elasticity. Bridgman's experiments, already mentioned, were not sufficiently refined for this purpose, but confirmed that when elastic strains are negligible the plane-strain yield stress of medium-carbon alloy steel is about 15 per cent. higher than the uniaxial yield stress (until anisotropy develops). This has also been roughly confirmed for other metals by Baranski,[†] who extended bars of brass and mild steel under approximate conditions of plane strain (the corresponding analysis for tension is obtained by changing the sign of Y in the above equations).

The foregoing analysis suggests the general conclusion that in plane strain the stress σ_z perpendicular to the planes of flow may be taken equal to the mean $\tfrac{1}{2}(\sigma_x + \sigma_y)$ of the other two normal stresses to a very good

---

† G. Baranski, *Zeits. Metallkunde*, 26 (1934), 173. See also H. Ford, *Proc. Inst. Mech. Eng.* 159 (1948), 115, and the discussion by the writer, ibid. 157.

approximation after a plastic strain of a few times the yield-point strain. The elastic component of strain is correspondingly negligible, and the Lévy–Mises relation ((28) of Chap. II) may be used instead of the Reuss equations. This helpful simplification continues to be valid during the ensuing plastic deformation so long as there is no sharp 'bend' in the strain-path, where the stress-increment is of order E times the strain-increment. The conclusion is also unaffected by the presence of work-hardening, provided the rate is small compared with E, as it will normally be after a plastic strain of a few per cent. On the other hand, whenever

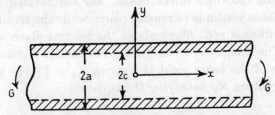

Fig. 9. Coordinate system for analysis of the bending of a rectangular beam in plane strain, with plastic zones shown shaded.

the stress-increment is of order E times the strain-increment, considerable inaccuracies would be introduced by neglecting the elastic strain-increment (no matter what the previous plastic strain may have been). This will be further illustrated in later examples.

6. Bending under conditions of plane strain

As our first example of a problem in which plastic and elastic regions exist side by side, consider the bending of a uniform rectangular beam or sheet of non-hardening material under conditions of plane strain. We choose coordinate axes so that the z-axis is in the direction in which strain is prevented (normal to the paper in Fig. 9), and the xz plane coincides with the central plane of the beam. The axis of y is taken positive towards the convex side of the beam, which is bent under terminal couples G in the sense shown. It is supposed that the radius of curvature is so large in relation to the thickness that the induced transverse stresses in the y direction can be neglected; the effect of these will be considered in a later chapter when the problem is analysed for strains of any magnitude. The neutral (unstrained) fibre Ox is bent into a circular arc of radius R. While the beam is stressed below the yield-point, it is known from elastic theory that the longitudinal stress σ_x is distributed linearly

across a transverse section according to the equation
$$\sigma_x = \frac{Ey}{(1-\nu^2)R}.$$
σ_x is tensile above Ox and compressive below it. Apart from σ_z, which is equal to $\nu\sigma_x$, all other stress components are zero. The required couple per unit width is
$$G = \int_{-a}^{a} \sigma_x y \, dy = \frac{2Ea^3}{3(1-\nu^2)R}, \tag{14}$$
where $2a$ is the thickness of the beam. The deformation is such that transverse planes continue to remain plane, while the tensile strain in a longitudinal fibre is y/R; fibres above the neutral plane are extended and those below compressed. According to von Mises' criterion (12) the surface fibres of the beam yield at a stress $\sigma_{x0} = Y/\sqrt{(1-\nu+\nu^2)}$, and a radius of curvature R_Y satisfying the equation
$$\frac{Ea}{(1-\nu^2)R_Y} = \frac{Y}{\sqrt{(1-\nu+\nu^2)}}.$$
As the couple is further increased, plastic regions spread inwards from both surfaces. If $2c$ is the width of the remaining elastic zone at any moment, the corresponding radius of curvature R is evidently given by
$$\frac{R}{R_Y} = \frac{c}{a}; \qquad R_Y = \frac{\sqrt{(1-\nu+\nu^2)}Ea}{(1-\nu^2)Y}. \tag{15}$$
This equation ceases to hold, even approximately, when the strains are no longer small. The natural expectation that transverse planes continue to remain plane while part of the beam is plastic, and that the state of stress is still a longitudinal tension or compression, may be justified by substitution in the Reuss equations and the equations of equilibrium. It is evident that the analysis is in fact identical with that of the last section, and that the state of stress in any one fibre develops independently of the rest,† the strain ϵ_x being equal to y/R. The transverse contraction or expansion in any fibre is automatically accommodated, since all are free to move transversely. At each stage of the bending σ_x and σ_z can be calculated from equations (12) and (13), with the appropriate choice of the sign of Y in the regions of compression and tension. The longitudinal stress in a fibre increases from the value σ_{x0}, when the fibre first becomes plastic, towards the limiting value $2Y/\sqrt{3}$. However,

---

† This would not be so if the beam were bent while under a constant tension. The position of the neutral axis, and hence the strain in any fibre, would have to be determined at each stage by the condition that the resultant of the longitudinal stresses is equal to the applied tension.

the approach to the limiting value is not particularly close, except in the extreme fibres, in the range of strain for which the present analysis is accurate. The integral (14) for the bending couple must be evaluated numerically.

The treatment of the problem is appreciably simplified if Tresca's yield criterion is adopted, since the longitudinal stress in the plastic region is always $\pm Y$. In this case

$$R = \frac{c}{a}R_Y = \frac{Ec}{(1-\nu^2)Y}.$$

The couple G per unit width is

$$2\int_0^c \frac{Ey^2\,dy}{(1-\nu^2)R} + 2\int_c^a Yy\,dy = \frac{2Ec^3}{3(1-\nu^2)R} + Y(a^2-c^2) = Y(a^2-\tfrac{1}{3}c^2). \quad (16)$$

This result was first obtained by Saint-Venant. Even though the material has a sharp yield-point, the relation between the couple G and the curvature $1/R$ is a smooth curve with a continuously turning tangent.

In the conventional engineering treatment of this problem no account is taken of the elastic component of the strain in the plastic region. A discontinuity in the calculated stress σ_z across the plastic-elastic boundary can then only be avoided by assuming the fictitious value 0·5 for Poisson's ratio. For in the elastic region $\sigma_z = \nu\sigma_x$, while in the plastic region $\sigma_z = \tfrac{1}{2}\sigma_x$; consequently if the plastic boundary is determined by the condition that σ_x is continuous across it, σ_z must be discontinuous. This does not affect the relation between the couple and the curvature when Tresca's criterion is used, but it obviously does in the case of von Mises' criterion. The longitudinal stress σ_x is then also discontinuous, since it is equal to $2Y/\sqrt{3}$ in the plastic region but to $Y/\sqrt{(1-\nu+\nu^2)}$ on the elastic side of the plastic boundary. In the conventional treatment this is avoided at the expense of an inconsistency by formally continuing the 'elastic' stress distribution $\sigma_x = Ey/(1-\nu^2)R$ to the fibre in which $\sigma_x = 2Y/\sqrt{3}$.

7. Bending of a prismatic beam

A uniform prismatic beam of non-hardening material is bent by equal and opposite couples G applied at its ends. For simplicity it is supposed that a transverse section of the beam (the yz plane in Fig. 10) has an axis of symmetry Oy, and that the axes of the couples are parallel to Oz. The origin O is taken on the neutral fibre in the xy plane. It is required to calculate the relation between the curvature of the beam and the applied couple when part of the beam has yielded plastically. The strain is restricted to values sufficiently small for the curvature to be neglected in so far as the equilibrium equations are concerned. It is normally assumed that the plastic-elastic boundaries are parallel to the neutral plane, and that all stress components apart from σ_x are zero. That neither of these assumptions is in fact correct can be seen without difficulty by considering the deformation of elements situated momentarily just on the elastic or plastic side of the common interface. Let the longitudinal curvature of the beam be $1/R$, and consider a further small change $d(1/R)$. If the stress were in fact a simple tension, the change in the anticlastic (or transverse) curvature of elastic elements would be $\nu\, d(1/R)$. On the other hand, the change in the anticlastic curvature of plastic elements would be $\frac{1}{2}d(1/R)$, since the stress after yielding remains equal to Y and there is no elastic strain-increment.

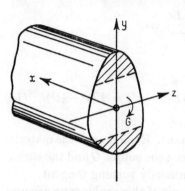

FIG. 10. Coordinate system for analysis of the bending of a prismatic beam, with plastic zones shown shaded (Poisson's ratio = 0·5).

Elements on opposite sides of the plastic-elastic interface would therefore not fit together. The necessary condition of continuity can only be satisfied by introducing transverse stresses, and rejecting the original over-simplified assumptions. The boundary of the plastic zone can no longer be assumed plane, and the problem becomes extremely difficult; the general solution is not yet known. By a slight extension of this argument the same conclusion may be proved for a work-hardening material without a sharp yield-point.

When Poisson's ratio is $\frac{1}{2}$ the difficulty does not arise, and the simple treatment is correct. The position of the neutral plane xz is determined by the condition that the longitudinal stresses should have zero resultant, namely,

$$\iint \sigma_x\, dy\, dz = 0$$

taken over a transverse section. While the beam is still elastic the stress σ_x is distributed linearly according to the formula $\sigma_x = Ey/R$, where R is the radius of curvature of the fibre coinciding with the x-axis. Hence O must be the centroid of the section. The fibre lying farthest from the centroid yields first, and, if the section is not symmetrical about Oz, the plastic zone penetrates some way from this side of the beam before the other side yields. The stress in a plastic fibre is a constant tension or compression of amount Y. When the couple has increased to an extent such that there are two plastic zones, the plastic-elastic interfaces are at equal distances $c = RY/E$ from the neutral plane (Fig. 10). In general when the beam is partly plastic the neutral (unstressed) plane varies from moment to moment, and does not necessarily contain the line of centroids. So long as the fibres in the unstressed plane have only been strained elastically during the prior bending (as they will have for all usual shapes of section), the resultant strain in the neutral plane is zero, and the components u, v, and w of the total displacement are

$$u = \frac{xy}{R}, \quad v = -\frac{1}{4R}(2x^2+y^2-z^2), \quad w = -\frac{yz}{2R}, \qquad (17)$$

in both elastic and plastic zones. These are the well-known expressions for purely elastic bending.† It may be verified that they are compatible with the assumed stress distribution in the elastic region, and that the distortion is such that transverse planes remain plane during the bending. In the plastic region we have to show that the strain-increment produced by a small change $d(1/R)$ in the curvature is compatible with the assumed stress distribution on the basis of the Reuss equations. The incremental components of displacement in a plastic element are

$$du = xy\, d\!\left(\frac{1}{R}\right) + \frac{x}{R}dy,$$

$$dv = -\tfrac{1}{4}(2x^2+y^2-z^2)\, d\!\left(\frac{1}{R}\right) - \frac{y}{2R}dy,$$

$$dw = -\tfrac{1}{2}yz\, d\!\left(\frac{1}{R}\right) - \frac{z}{2R}dy,$$

since the position of the element does not change (to the order of approximation involved in the assumption of small strains), but the y coordinate varies due to the relative movement of the neutral plane. The terms in dy represent a uniform longitudinal extension of amount dy/R; the remaining terms represent the strain-increment which would result if the neutral plane were fixed. The increments of strain are then

---

† See, for example, S. Timoshenko, *Theory of Elasticity*, p. 224 (McGraw-Hill Book Co., 1934); R. V. Southwell, *Introduction to the Theory of Elasticity*, p. 327 (Clarendon Press, 1936); I. S. Sokolnikoff, *Mathematical Theory of Elasticity*, p. 114 (McGraw-Hill Book Co., 1946).

(remembering that dy is the same for all elements)

$$d\epsilon_x = y\,d\left(\frac{1}{R}\right) + \frac{dy}{R} = d\left(\frac{y}{R}\right),$$

$$d\epsilon_y = d\epsilon_z = -\tfrac{1}{2}y\,d\left(\frac{1}{R}\right) - \frac{1}{2}\frac{dy}{R} = -\tfrac{1}{2}d\left(\frac{y}{R}\right),$$

$$d\gamma_{xy} = d\gamma_{yz} = d\gamma_{zx} = 0.$$

The strain-increment in every fibre is a uniform extension with equal lateral contractions of half its amount, and this is consistent with the assumption that the fibres are stressed in simple tension (irrespective of whether the material work-hardens or not).

The calculation of the bending moment $\iint \sigma_x y\,dydz$ in a beam of given section is straightforward; examples can be found in the books of Nadai[†] and Sokolovsky.[‡] The same theory is also commonly applied to beams bent under shearing forces, where the stress distribution and the position of the plastic-elastic interface vary along the length of the beam. It is evident that this involves further approximations of uncertain accuracy.

8. Torsion of a prismatic bar

(i) *Elastic Torsion.* In all the problems so far investigated the stress has been a function of at most one geometrical variable. We now consider a problem of greater difficulty where the stress depends on two independent coordinates. A uniform, long, prismatic bar of arbitrary section (enclosed by a single curve) is twisted about an axis parallel to the generators by equal and opposite couples. Cartesian coordinates x, y, and z are taken with the z-axis parallel to the generators; the origin and orientation of the x- and y-axes is immaterial in a general analysis.

In the elastic range it is found§ that, if the only non-vanishing stress components are τ_{xz} and τ_{yz}, the deformation consists of a rotation of originally plane sections about a line parallel to the z-axis, together with a longitudinal warping of constant amount. Apart from an arbitrary rigid-body movement, the components u, v, and w of the total displacement are

$$u = -\theta yz, \qquad v = \theta xz, \qquad w = w(x,y,\theta),$$

[†] A. Nadai, *Plasticity*, chap. 22 (McGraw-Hill Book Co., 1931).
[‡] W. W. Sokolovsky, *Theory of Plasticity*, chap. 11 (Moscow, 1946). See also J. W. Roderick, *Phil. Mag.* 39 (1948), 529; H. A. Williams, *Journ. Aero. Sci.* 14 (1947), 457.
§ S. Timoshenko, *Theory of Elasticity*, p. 232 (McGraw-Hill Book Co., 1934); R. V. Southwell, *Introduction to the Theory of Elasticity*, p. 324 (Clarendon Press, 1936); I. S. Sokolnikoff, *Mathematical Theory of Elasticity*, p. 122 (McGraw-Hill Book Co., 1946).

where w is the warping function and θ is the twist per unit length, right-handed about the positive z-axis. The angular rotation of a transverse plane is evidently proportional to its distance from the fixed origin. The warping is, of course, proportional to the twist in the elastic range, but w is written in the more general form as the proportionality cannot be expected to continue when the bar becomes plastic. The components of strain are

$$\epsilon_x = \epsilon_y = \epsilon_z = \gamma_{xy} = 0,$$

$$2\gamma_{xz} = \frac{\partial w}{\partial x} - \theta y,$$

$$2\gamma_{yz} = \frac{\partial w}{\partial y} + \theta x.$$

The shear-stress components are

$$\tau_{xz} = 2G\gamma_{xz} = G\left(\frac{\partial w}{\partial x} - \theta y\right),$$

$$\tau_{yz} = 2G\gamma_{yz} = G\left(\frac{\partial w}{\partial y} + \theta x\right).$$

By eliminating w we obtain the compatibility relation

$$\frac{\partial \tau_{yz}}{\partial x} - \frac{\partial \tau_{xz}}{\partial y} = 2G\theta. \tag{18}$$

The equation of equilibrium is

$$\frac{\partial \tau_{xz}}{\partial x} + \frac{\partial \tau_{yz}}{\partial y} = 0,$$

and this is satisfied by introducing the stress function $\phi(x, y)$, where

$$\tau_{xz} = \frac{\partial \phi}{\partial y}, \quad \tau_{yz} = -\frac{\partial \phi}{\partial x}. \tag{19}$$

The resultant shear stress acting over a transverse plane is directed at each point along the contour curves of constant ϕ, which are therefore called *lines of shearing stress*. The magnitude of the shear stress is $|\text{grad}\,\phi|$ or $|\partial\phi/\partial n|$, where the derivative is taken normal to a contour line. The resultant shear stress acting over a transverse plane must be directed parallel to the boundary in a surface element, since the lateral surface of the bar is free from stress. The boundary must therefore be a contour of ϕ, and we may conveniently take the constant value to be zero. The couple M required to produce a twist θ is

$$M = \iint (x\tau_{yz} - y\tau_{xz})\,dxdy = -\iint \left(x\frac{\partial \phi}{\partial x} + y\frac{\partial \phi}{\partial y}\right) dxdy = 2\iint \phi\,dxdy \tag{20}$$

on integrating by parts and using the condition $\phi = 0$ on the boundary. M is therefore equal to twice the volume under the surface representing the stress function. Substituting the expressions (19) in the compatibility relation (18) we obtain finally

$$\nabla^2 \phi \equiv \frac{\partial^2 \phi}{\partial x^2} + \frac{\partial^2 \phi}{\partial y^2} = -2G\theta. \tag{21}$$

For a given cross-section and a prescribed twist θ this equation, together with the boundary condition $\phi = 0$, is sufficient to determine ϕ, and hence the stresses, over the whole section. Prandtl[†] (1903) first pointed out the analogy with the deflexion of a uniformly tensioned membrane, fixed at its edge to a plane support of the same shape as the section of the bar, and stressed laterally on one side by a uniform pressure. The deflexion satisfies the same equation and boundary condition as the stress function ϕ, the constant on the right-hand side depending on the applied tension and pressure. Thus the surface assumed by the membrane in its equilibrium position is similar to the ϕ surface, and the contours of equal deflexion correspond to the shearing lines, while the slope of the membrane at any point is proportional to the resultant shear stress at the corresponding point of the bar. The membrane analogy furnishes an experimental means of calculating the stresses for shapes of section difficult to treat by mathematical analysis.

(ii) *Stress distribution in a partly plastic bar.* The bar yields when the resultant shear stress in some element reaches a critical value k, where k is equal to $Y/2$ according to Tresca's criterion or to $Y/\sqrt{3}$ according to von Mises' criterion. It is a property of equation (21) that $|\text{grad}\,\phi|$ attains its greatest value on the boundary;[‡] this is physically self-evident from the membrane analogy, since the membrane would naturally be expected to assume a shape everywhere concave to the applied pressure. Thus plastic yielding first takes place somewhere on the boundary. In a bar of elliptical section, for example, the weakest points are the extremities of the minor axis; in a bar whose section is a square or an equilateral triangle the centres of the sides yield first (the shear stress is zero at projecting corners). As the applied couple is increased plastic zones spread from these points. Assuming for the moment that τ_{xz} and τ_{yz} are still the only non-vanishing stress components, it follows that the resultant shear stress remains equal to k if

[†] L. Prandtl, *Zeits. Phys.* **4** (1903).
[‡] See, for example, I. S. Sokolnikoff, *Mathematical Theory of Elasticity* (McGraw-Hill Book Co., Inc.), 1946, p. 130.

there is no work-hardening. Hence, in the plastic region the stress function satisfies the equation

$$\tau_{xz}^2 + \tau_{yz}^2 = \left(\frac{\partial \phi}{\partial x}\right)^2 + \left(\frac{\partial \phi}{\partial y}\right)^2 = k^2, \qquad (22)$$

while in the elastic region ϕ continues to satisfy (21). The plastic boundary is determined by the conditions (i) that the shear stress is continuous across it, and (ii) that the shear stress is not greater than k inside the elastic region. Otherwise expressed, the slope of the ϕ surface must be equal to k in the plastic region, and must not be greater than k in the elastic region, while the height and slope of the surface must be continuous across the plastic boundary. As was first pointed out by Nadai,[†] this suggests an extension of the membrane analogy to simulate the stress function for a partially plastic bar when the twist is so small that changes in the external contour can be neglected. It is only necessary to erect a fixed roof of constant slope over the membrane with the outer boundary of the section as its base. The area under the part of the roof where the membrane is pressed against it by the applied pressure represents the plastic region corresponding to a certain torque.[‡]

It is evident from this analogy that the stress distribution for a given torque is determined solely by the shape of the boundary (the strains being small), and that the equations (21) and (22), together with the boundary condition $\phi = 0$, are sufficient for calculating the stresses without reference to the deformation. For this reason the torsion problem is said to be *statically determined*. It is worth observing that the stress distribution and plastic region corresponding to a given torque are independent of the value of the shear modulus G, since this only occurs in conjunction with θ; if the torque is prescribed, equation (20) supplies a condition to determine the constant parameter $2G\theta$ in (21). The twist for a given torque is therefore inversely proportional to G.

It is easy to see from an analytic expression of the roof analogy that the stress at a point in the plastic region is uniquely determined by the shape of the external boundary. The stress function at a point in the plastic region is equal to k times its distance from the boundary, measured along the normal through the point. The shearing stress lines of constant ϕ are orthogonal to the normals (the lines of greatest slope), and are evidently spaced parallel to the boundary at constant distances from it. They are shown by the full curves in Fig. 11 a, the broken curve

[†] A. Nadai, *Zeits. ang. Math. Mech.* 3 (1923), 442.
[‡] For further examples, see A. Nadai, *Plasticity*, chap. 19 (McGraw-Hill Book Co., 1931).

representing the plastic-elastic boundary. The shearing-stress lines may also be thought of as the family of involutes of the evolute of the boundary. If the contour contains a re-entrant angle (Fig. 11 b) the vertex is a singularity for the stress, and the shearing lines within a certain angular span are circular arcs. On the other hand, the shear stress is always zero at a projecting corner, which therefore never becomes plastic.

(iii) *Calculation of the displacements.* Now since the strains are assumed so small that changes of the external contour and displacements of elements can be neglected (as in elasticity), the stress in an

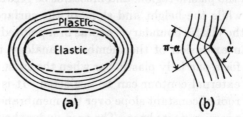

Fig. 11. Shear-stress trajectories in the cross-section of a twisted prismatic bar with (a) a smooth contour, and (b) a re-entrant angle.

element remains constant in magnitude and direction once the element has become plastic. The elastic strain-increments are then zero to the same order of approximation, and the Reuss equations reduce to

$$d\epsilon_x = d\epsilon_y = d\epsilon_z = d\gamma_{xy} = 0; \qquad \frac{d\gamma_{xz}}{\tau_{xz}} = \frac{d\gamma_{yz}}{\tau_{yz}}.$$

Hence, since the shear-strain components γ_{xz} and γ_{yz} are proportional to the respective stress components τ_{xz} and τ_{yz} while the element is elastic, they must continue so when it is plastic. In the geometrical representation of Fig. 7 the strain-path is a straight line through the origin, the stress-point remaining fixed. We may therefore write

$$\frac{\gamma_{xz}}{\gamma_{yz}} = \frac{(\partial w/\partial x) - \theta y}{(\partial w/\partial y) + \theta x} = \frac{\tau_{xz}}{\tau_{yz}} \qquad (23)$$

in the plastic region.† The original assumption that τ_{xz} and τ_{yz} are the only non-vanishing stress components has now been justified, since the system of displacements assumed for the elastic region is also compatible with the plastic state of stress. (23) is an equation determining the warping function, in which the shear stresses are to be regarded as

† This is identical with the corresponding equations of the Hencky theory ((36) of Chap. II) which thus leads to the same warping function, as remarked by H. Geiringer and W. Prager, *Ergebnisse d. exakten Naturwiss.* 13 (1934), 310.

previously calculated functions of x and y. The equation is hyperbolic and its characteristics are the normals to the external contour.† If dn denotes a small increment of distance measured outward along a normal, we have
$$\frac{dw}{dn} = \cos\psi \frac{\partial w}{\partial x} + \sin\psi \frac{\partial w}{\partial y},$$
where ψ is the angular orientation of a normal to the positive x-axis. Now
$$\tau_{xz} = -k\sin\psi, \qquad \tau_{yz} = k\cos\psi,$$
and so
$$\frac{dw}{dn} = \frac{\tau_{yz}}{k}\frac{\partial w}{\partial x} - \frac{\tau_{xz}}{k}\frac{\partial w}{\partial y}$$
$$= \frac{\theta}{k}(x\tau_{xz} + y\tau_{yz}) \quad \text{from (23)}.$$
$$= \theta(-x\sin\psi + y\cos\psi).$$
Hence, if p denotes the length of the perpendicular from the origin on to a normal,
$$\frac{dw}{dn} = \pm\theta p. \tag{24}$$

The warping therefore varies linearly along any normal;‡ in particular, if the section has two axes of symmetry w is zero along them in both the elastic and plastic regions when the centroid is taken as the origin. Since, for a given twist, w is known on the plastic-elastic boundary from the solution for the warping function in the elastic region, w can be immediately calculated everywhere in the plastic region. In view of the varying shape of the plastic-elastic boundary as it moves inward, there is evidently no reason to expect that w is proportional to the twist.

(iv) *Bar of circular section.* When the section is circular the warping is zero and the shearing-stress contours are concentric circles. The solution was originally obtained by Saint-Venant. If c is the radius of the plastic-elastic boundary corresponding to a twist $\theta = k/Gc$, the distribution of the shearing stress is
$$\tau = kr/c \quad (0 \leqslant r \leqslant c),$$
$$\tau = k \quad (c \leqslant r \leqslant a).$$
Hence the torque is
$$M = 2\pi \int_0^a r^2 \tau\, dr = \tfrac{2}{3}\pi k(a^3 - \tfrac{1}{4}c^3), \qquad c = k/G\theta, \tag{25}$$

† With the notation of Appendix III, $P = \tau_{yz}$, $Q = -\tau_{xz}$, and the characteristics are the family of normals $dx/\tau_{yz} = -dy/\tau_{xz}$.
‡ This equation has been applied to the calculation of the warping in an I-beam by P. G. Hodge, *Journ. App. Mech.* **16** (1949), 399. It is shown that the contribution of the elastic core to the warping becomes negligible at large angles of twist; hence w can be assigned the value zero on the curve representing the evanescent elastic core.

while the torque when the bar first yields is $M_Y = \tfrac{1}{2}\pi ka^3$. The value of the torque required to cause complete yielding of the bar is theoretically $\tfrac{2}{3}\pi ka^3$, or $\tfrac{4}{3}M_Y$. This condition cannot, of course, be obtained in practice since no matter how large the twist there is always a core of material which is still elastic.

(v) *Sokolovsky's solution for a bar of oval section.* An ingenious inverse method of solution for a certain oval contour has been formulated by Sokolovsky.† His method consists in assuming a mathematically convenient shape for the plastic-elastic boundary, and determining from this the appropriate contour of the section. If the plastic-elastic boundary is the ellipse
$$\frac{x^2}{a^2}+\frac{y^2}{b^2} = 1,$$
the stresses in the elastic region are
$$\tau_{xz} = -\left(\frac{2G\theta}{a+b}\right)ay, \qquad \tau_{yz} = \left(\frac{2G\theta}{a+b}\right)bx, \qquad (26)$$
where θ, as usual, is the twist per unit length. These obviously satisfy the equation of equilibrium and the compatibility relation (18). On the ellipse:
$$\tau_{xz}^2+\tau_{yz}^2 = \left(\frac{2G\theta}{a+b}\right)^2(a^2y^2+b^2x^2) = \left(\frac{2G\theta ab}{a+b}\right)^2.$$
Yielding therefore occurs all along the ellipse (but not inside it) when the twist is
$$\theta = \frac{k(a+b)}{2Gab}. \qquad (27)$$
Setting $\qquad \tau_{xz} = -k\sin\psi, \qquad \tau_{yz} = k\cos\psi,$
in the plastic region as before, we have for the value of ψ at a point (ξ, η) on the ellipse:
$$\tan\psi = -\frac{\tau_{xz}}{\tau_{yz}} = \frac{a\eta}{b\xi}; \qquad \frac{\xi^2}{a^2}+\frac{\eta^2}{b^2} = 1.$$
Hence $\qquad \xi = a\cos\psi, \qquad \eta = b\sin\psi.$

Since the stress must be continuous across the ellipse, the normals to the shearing-stress contours in the plastic region are obtained by constructing a set of lines outward from each point of the ellipse with the appropriate inclination ψ (Fig. 12). The shearing lines are the orthogonal trajectories of this set, and any one of the shearing lines is a suitable external contour. The particular value of the twist when the plastic-elastic boundary coincides with the ellipse is given by (27). Clearly the

† W. W. Sokolovsky, *Prikladnaia Matematika i Mekhanika*, 6 (1942), 241.

method will only be of value if, for the chosen contour, the plastic-elastic boundary is an ellipse for a *range* of values of θ. We could always choose some elastic stress distribution, take the plastic-elastic boundary as the curve where the shear stress is k, and hence construct an external contour and its normals. In general, however, we should only have solved the problem for that contour for one value of θ, and should be no nearer solving it for other values. The success of Sokolovsky's method in this instance is due to the circumstance that the plastic-elastic boundary happens to be an ellipse, with a similar elastic stress distribution, for a range of angles of twist. Now the equation of a normal to the shearing stress lines is

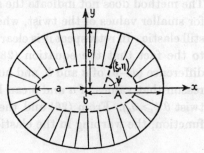

Fig. 12. Plastic and elastic zones in a twisted prismatic bar of oval section (diagrammatic, after Sokolovsky).

$$y-\eta = (x-\xi)\tan\psi.$$

Substituting for ξ and η:

$$y = x\tan\psi - (a-b)\sin\psi.$$

The orthogonal trajectories of this family of lines have the parametric equations

$$\left.\begin{aligned}x &= \cos\psi[A+(A-B)\sin^2\psi],\\ y &= \sin\psi[B-(A-B)\cos^2\psi],\end{aligned}\right\} \qquad (28)$$

where $A-B = \tfrac{1}{2}(a-b)$.

When $2B \geqslant A$, the contour is a closed oval curve with semi-axes A and B; it differs very little from an ellipse. Changing our point of view, let us regard the contour (28) as specified by certain values of A and B. When the twist is θ, the plastic-elastic boundary is an ellipse with semi-axes a and b given by

$$\frac{a+b}{ab} = \frac{2G\theta}{k}, \qquad a-b = 2(A-B).$$

Hence

$$\left.\begin{aligned}a &= \frac{k}{2G\theta}+A-B+\left[\frac{k^2}{4G^2\theta^2}+(A-B)^2\right]^{\tfrac{1}{2}},\\ b &= \frac{k}{2G\theta}+B-A+\left[\frac{k^2}{4G^2\theta^2}+(A-B)^2\right]^{\tfrac{1}{2}}.\end{aligned}\right\} \qquad (29)$$

The solution is obviously only valid when this ellipse lies entirely within the contour, that is, when $a \leqslant A$. Hence θ must be greater than

$$\theta_0 = \frac{Bk}{A(2B-A)G}. \tag{30}$$

The method does not indicate the shape of the plastic-elastic boundary for smaller values of the twist, when points of the external contour are still elastic. In retrospect it is clear that the success of the method is due to the fact that the equation (28) for the contour only involves the difference $(a-b)$ of a and b, and not their separate values. It is for this reason that values of a and b can be found corresponding to any given twist θ ($\geqslant \theta_0$). From (26) and the original equations for the warping function, the warping in the elastic region is

$$w = -\frac{(a-b)}{(a+b)}xy\theta.$$

The warping in the plastic region can now be computed from (24), using the fact that $p = (a-b)\sin\psi\cos\psi$. A numerical example can be found in Sokolovsky's original paper, where a formula for the torque is also given.

(vi) *Other solutions.* The only other accurate solution known at present is Trefftz's conformal transformation method[†] for determining the plastic region at the re-entrant corner of an L-beam, where there is a concentration of stress. Trefftz has also discussed the stress distribution round a small circular hole after the yield-point has been reached. The mathematical difficulties in dealing with a section of general shape are so great that a numerical treatment is practically unavoidable. Relaxation methods have recently been applied to bars with triangular and I-sections by Christopherson,[‡] and to bars of hollow section by Shaw, and also by Southwell.[§] A fair approximation to the maximum torque that the bar can withstand can easily be obtained by supposing that the bar is rigid up to the yield-point. This means that G is infinite and that no twist is possible until the bar is completely plastic. The maximum torque is thus twice the volume under the plastic roof covering the whole section. For example, when the section is a square of side $2a$ the roof is a tetrahedron of height ka, and the torque is therefore

---

[†] E. Trefftz, *Zeits. ang. Math. Mech.* **5** (1925), 64.

[‡] D. G. Christopherson, *Journ. App. Mech.* **7** (1940), A-1. D. G. Christopherson and R. V. Southwell, *Proc. Roy. Soc.* A, **168** (1938), 317.

[§] F. S. Shaw, Australian Council for Aeronautics, Rep. ACA-11 (1944). R. V. Southwell, *Quart. Journ. Mech. App. Math.* **2** (1949), 385. See also R. V. Southwell, *Relaxation Methods in Theoretical Physics*, p. 193 (Clarendon Press, Oxford, 1946).

$8ka^3/3$. Other examples are given by Nadai[†] and Sokolovsky.[‡] However, such calculations are of limited significance since an actual bar work-hardens and becomes non-prismatic at strains for which the elastic core is negligible. It is worth noting that the stress is discontinuous across lines which are the projections of ridges on the roof. This is a general feature of the stress distribution in a twisted bar of rigid material. It is easy to see that the component of shear stress normal to such a line is the same on both sides; the discontinuity occurs in the component parallel to the line.

(vii) *Torsion of an annealed bar*. The torsion of a bar of annealed metal, with no well-defined yield-point, has not yet been thoroughly investigated. In one respect the problem is simpler than for a pre-strained metal, since there is no plastic-elastic boundary and the same equations hold all over the section. On the other hand, the final differential equation is more complicated. From (32) of Chapter II we have

$$d\epsilon_x = d\epsilon_y = d\epsilon_z = d\gamma_{xy} = 0;$$

$$2G\,d\gamma_{xz} = G\left(\frac{\partial^2 w}{\partial x \partial \theta} - y\right)d\theta = d\tau_{xz} + \frac{3G\tau_{xz}\,d\bar{\sigma}}{H'\bar{\sigma}};$$

$$2G\,d\gamma_{yz} = G\left(\frac{\partial^2 w}{\partial y \partial \theta} + x\right)d\theta = d\tau_{yz} + \frac{3G\tau_{yz}\,d\bar{\sigma}}{H'\bar{\sigma}}.$$

It is convenient to introduce a function $f(\bar{\sigma})$ defined by the equation

$$\ln f = \int\limits_0^{\bar{\sigma}} \frac{3G\,d\bar{\sigma}}{H'\bar{\sigma}}. \tag{31}$$

Then

$$G\left(\frac{\partial^2 w}{\partial x \partial \theta} - y\right) = \frac{1}{f}\frac{\partial}{\partial \theta}(f\tau_{xz}),$$

$$G\left(\frac{\partial^2 w}{\partial y \partial \theta} + x\right) = \frac{1}{f}\frac{\partial}{\partial \theta}(f\tau_{yz}).$$

Introducing the stress function ϕ, and eliminating w, we obtain an equation due to Prager:[§]

$$\frac{\partial}{\partial x}\left\{\frac{1}{f}\frac{\partial}{\partial \theta}\left(f\frac{\partial \phi}{\partial x}\right)\right\} + \frac{\partial}{\partial y}\left\{\frac{1}{f}\frac{\partial}{\partial \theta}\left(f\frac{\partial \phi}{\partial y}\right)\right\} = -2G. \tag{32}$$

Since $\bar{\sigma} = \sqrt{3}(\tau_{xz}^2 + \tau_{yz}^2)^{\frac{1}{2}} = \sqrt{3}|\text{grad}\,\phi|,$

f must be regarded as a given function of $|\text{grad}\,\phi|$ in (32). In general, integration can hardly be effected without recourse to numerical

---

[†] A. Nadai, *Plasticity*, chap. 19 (McGraw-Hill Book Co., 1931).
[‡] W. W. Sokolovsky, *Theory of Plasticity*, p. 70 (Moscow, 1946).
[§] W. Prager, *Journ. App. Phys.* 18 (1947), 375.

methods. Exceptionally, when the section is circular, the solution is trivial. As with the non-hardening bar there is no warping and the shearing-stress contours are circles. The engineering shear strain γ at a radius r is $r\theta$. The shear stress τ is a known function of γ, say $\tau(\gamma)$, obtained directly from a torsion test on a thin-walled tube or indirectly from a tension test. If the section is of radius a, the torque† is

$$M(\theta) = 2\pi \int_0^a r^2 \tau \, dr = \frac{2\pi}{\theta^3} \int_0^{a\theta} \gamma^2 \tau(\gamma) \, d\gamma.$$

Conversely, if the relation between the torque and the twist has been measured for the solid bar we can derive from it the stress-strain curve in pure shear. By differentiating the above equation:

$$\frac{d}{d\theta}(M\theta^3) = 2\pi a^3 \theta^2 \tau(a\theta),$$

i.e.
$$\tau(a\theta) = \frac{1}{2\pi a^3}\left(\theta \frac{dM}{d\theta} + 3M\right). \tag{33}$$

$\tau(a\theta)$ is the shear stress at the boundary of the section, where the shear strain is $a\theta$. Since the derivative $dM/d\theta$ must be computed by numerical differentiation of the measured torque-twist curve, this method (due to Nadai)‡ does not allow a particularly high accuracy in the initial part of the curve. However, the accuracy here can be improved by writing (33) as

$$\tau(a\theta) = \frac{1}{2\pi a^3}\left\{\theta^2 \frac{d}{d\theta}\left(\frac{M}{\theta}\right) + 4M\right\}.$$

The function M/θ is constant in the elastic range of strain and thereafter steadily decreases. Consequently the term involving the derivative makes no contribution in the elastic range and only a small one while the rate of work-hardening is high. In the later part of the stress-strain curve, where the rate of hardening is small and fairly constant, greater accuracy can be achieved with (33).§

9. Torsion of a bar of non-uniform section

When a bar with a non-uniform circular section is stressed in torsion below the yield limit, it is known from elastic theory‖ that transverse

---

† When work-hardening is present the Reuss and Hencky theories lead to identical results only for the circular section (cf. the footnote on p. 88).
‡ A. Nadai, *Plasticity*, chap. 18 (McGraw-Hill Book Co., 1931).
§ For examples of its practical application, and experimental verification, see J. L. M. Morrison, *Proc. Inst. Mech. Eng.* **159** (1948), 81; also the discussion by H. W. Swift on p. 110.
‖ S. Timoshenko, *Theory of Elasticity*, p. 276 et seq. (McGraw-Hill Book Co., 1934).

sections remain plane and circular, though radial lines become curved. The displacements, referred to cylindrical coordinates, are

$$u = 0, \quad v \equiv v(r, z), \quad w = 0,$$

while the stresses are

$$\sigma_r = \sigma_\theta = \sigma_z = \tau_{rz} = 0,$$

$$\tau_{r\theta} = G\left(\frac{\partial v}{\partial r} - \frac{v}{r}\right), \quad \tau_{\theta z} = G\frac{\partial v}{\partial z}.$$

If these are substituted in the single equation of equilibrium

$$\frac{\partial \tau_{\theta z}}{\partial z} + \frac{\partial \tau_{r\theta}}{\partial r} + \frac{2\tau_{r\theta}}{r} = 0,$$

there results

$$\frac{\partial^2 v}{\partial z^2} + \frac{\partial^2 v}{\partial r^2} + \frac{1}{r}\frac{\partial v}{\partial r} - \frac{v}{r^2} = 0. \tag{34}$$

Since the external surface is stress-free, the resultant shear stresses acting on a surface element, over both a meridian and a transverse section, must be tangential to the contour; the second condition is automatically satisfied, while the first requires that

$$\frac{\tau_{r\theta}}{\tau_{\theta z}} = \tan \alpha,$$

where α is the inclination of the tangent with respect to the z axis. The solution of (34), with this boundary condition, determines the elastic state of stress.

When part of the bar is plastic, the stresses in the plastic region satisfy

$$\tau_{r\theta}^2 + \tau_{\theta z}^2 = k^2.$$

It is convenient to introduce a parameter ψ, defining the inclination of the resultant shear stress over a meridian plane, such that

$$\tau_{r\theta} = k \sin \psi, \quad \tau_{\theta z} = k \cos \psi.$$

The equation of equilibrium is satisfied, provided that

$$\frac{\partial \psi}{\partial r} - \tan \psi \frac{\partial \psi}{\partial z} + \frac{2 \tan \psi}{r} = 0. \tag{35}$$

The characteristics of this equation are in the directions

$$\frac{dz}{dr} = -\tan \psi,$$

and are therefore orthogonal to the lines of shearing stress in a meridian plane. The variation of ψ along a characteristic is

$$d\psi = \frac{\partial \psi}{\partial r}dr + \frac{\partial \psi}{\partial z}dz = \left(\frac{\partial \psi}{\partial r} - \tan \psi \frac{\partial \psi}{\partial z}\right)dr = -2 \tan \psi \frac{dr}{r}.$$

This is immediately integrable, and we find that

$$r^2 \sin \psi = \text{constant} \tag{36}$$

along a characteristic, where the value of the constant is determined by the boundary condition $\psi = \alpha$. This result is due to Sokolovsky.† The characteristic field, and hence the shearing lines, can be constructed by integrating (36) inward from the surface. In particular, the characteristic is a straight line orthogonal to the surface wherever the tangent to the surface is either parallel or perpendicular to the z-axis. Since the plastic state of stress depends only on the shape of the contour, the stress in any plastic element remains effectively constant so long as the total distortion is negligibly small.

Cross-sections continue to remain plane when the bar is partly plastic, provided the relation $\mu = \nu$ holds good. Since the direction of the stress vector does not alter during the twisting,

$$\frac{\gamma_{r\theta}}{\gamma_{\theta z}} = \frac{(\partial v/\partial r)-(v/r)}{\partial v/\partial z} = \frac{\tau_{r\theta}}{\tau_{\theta z}} = \tan \psi,$$

or
$$\frac{\partial v}{\partial r} - \tan \psi \frac{\partial v}{\partial z} - \frac{v}{r} = 0. \tag{37}$$

This evidently has the same characteristics as equation (35), while the variation of v along them is

$$dv = v \, dr/r.$$

Therefore,
$$v \propto r \tag{38}$$

along each characteristic. Thus, in the plastic region, any surface of revolution formed by characteristics is simply rotated during the twisting, without any distortion. The amount of rotation is equal to the angular displacement of elements where the surface of revolution meets the plastic-elastic interface. The non-distorted surfaces are orthogonal to the shearing lines; this is true also in the elastic region.

There appears to be no specific contour for which the complete analytic solution is known. Sokolovsky investigated the plastic distribution of stress in a conical bar and in a stepped bar consisting of two cylinders of different radii, but did not calculate the plastic boundary or the displacements. Eddy and Shaw‡ have applied relaxation methods in an approximate computation of the stress distribution and plastic boundary in a shaft with a collar.

† W. W. Sokolovsky, *Prikladnaia Matematika i Mekhanika*, **9** (1945), 343.
‡ R. P. Eddy and F. S. Shaw, *Journ. App. Mech.* **16** (1949), 139.

V
THE SOLUTION OF PLASTIC-ELASTIC PROBLEMS. II

This chapter is concerned with problems of spherical and axial symmetry. The solution of spherically symmetric problems is straightforward in principle, and for the normal boundary conditions the equations can often be integrated explicitly. As a concrete example we shall analyse the expansion of a spherical thick-walled shell by internal pressure; in the extreme case when the outer radius is taken indefinitely large we have to do with the formation of a spherical cavity in an infinite solid medium. Problems with axial symmetry are more difficult, even when, as here, the stresses are supposed not to vary in the longitudinal direction. Even with this simplification, it is only in special circumstances that an explicit analytic solution can be derived, prolonged numerical calculations being generally unavoidable. We shall be concerned mostly with the expansion of a long cylindrical tube by uniform internal pressure. This is a long-standing problem because of its connexion with the autofrettage of pressure vessels and gun-barrels; the process has also been used to determine the influence of a stress gradient on the criterion of yielding. Although many investigators have examined the problem, it is only comparatively recently that an accurate method of solution has been formulated.

1. The expansion of a spherical shell

(i) *Calculation of the stresses.* Let the current internal and external radii of the shell be denoted by a and b, and their initial values by a_0 and b_0. No restriction is placed on the magnitude of the expansion, an essential task of the analysis being to determine the variation of the radii with the pressure p. This is distributed uniformly over the inner surface, while the external pressure is negligible compared with the yield stress of the material. As the pressure is gradually increased from zero the shell is first stressed elastically. According to the familiar Lamé solution† the stress components in spherical polar coordinates

---

† See, for example, S. Timoshenko, *Theory of Elasticity*, p. 325 (McGraw-Hill Book Co., 1934); R. V. Southwell, *Introduction to the Theory of Elasticity*, p. 337 (Clarendon Press, 1936).

are given by the equations

$$\sigma_r = -p\left(\frac{b_0^3}{r^3}-1\right)\bigg/\left(\frac{b_0^3}{a_0^3}-1\right),$$
$$\sigma_\theta = \sigma_\phi = p\left(\frac{b_0^3}{2r^3}+1\right)\bigg/\left(\frac{b_0^3}{a_0^3}-1\right). \qquad (1)$$

The radial displacement, measured outwards, is

$$u = \frac{p}{E}\left\{(1-2\nu)r + \frac{(1+\nu)b_0^3}{2r^2}\right\}\bigg/\left(\frac{b_0^3}{a_0^3}-1\right). \qquad (2)$$

Since, by virtue of the symmetry, the state of stress is everywhere just a hydrostatic tension $(\sigma_\theta, \sigma_\theta, \sigma_\theta)$ superposed on a uniaxial compressive stress $(\sigma_\theta-\sigma_r, 0, 0)$, the yield criterion is

$$\sigma_\theta - \sigma_r = Y. \qquad (3)$$

For the present it is supposed that the metal does not work-harden appreciably in the relevant range of strain, so that Y is a constant. Now from (1)

$$\sigma_\theta - \sigma_r = \frac{3pa_0^3}{2r^3}\bigg/\left(1-\frac{a_0^3}{b_0^3}\right).$$

Since this is greatest on the inner surface yielding begins there, the corresponding pressure being

$$p_0 = \frac{2Y}{3}\left(1-\frac{a_0^3}{b_0^3}\right). \qquad (4)$$

The displacements of the internal and external surfaces are

$$u_0(a_0) = \frac{Ya_0}{E}\left\{\frac{2(1-2\nu)a_0^3}{3b_0^3}+\frac{1+\nu}{3}\right\}, \quad u_0(b_0) = \frac{(1-\nu)Ya_0^3}{Eb_0^2}. \qquad (5)$$

When the shell is very thin, it yields at a pressure $p_0 = 2Yt_0/a_0$, approximately (t_0 = thickness of the shell). The circumferential tension is equal to $p_0 a_0/2t_0$, that is to Y, and is therefore the effective agent causing yielding. Even when the shell is very thick ($b_0 \to \infty$), the disruptive tendency of the circumferential tension is still appreciable, and the inner layer yields at a pressure equal to two-thirds of the yield stress.

With increasing pressure a plastic region spreads into the shell. For reasons of symmetry the plastic boundary in a homogeneous material must be a spherical surface; its radius at any moment is denoted by c (Fig. 13). In the elastic region the stresses are still of the form

$$\sigma_r = -A\left(\frac{b_0^3}{r^3}-1\right), \qquad \sigma_\theta = \sigma_\phi = A\left(\frac{b_0^3}{2r^3}+1\right),$$

where A is a parameter to be determined (we may naturally still neglect changes in the radius b_0 to the usual order of approximation). Now the material just on the elastic side of the plastic boundary must be on the point of yielding and so, by substitution in equation (3), A is found to be $2Yc^3/3b_0^3$. The elastic stress distribution is therefore

$$\left.\begin{aligned}\sigma_r &= -\frac{2Yc^3}{3b_0^3}\left(\frac{b_0^3}{r^3}-1\right), \\ \sigma_\theta &= \frac{2Yc^3}{3b_0^3}\left(\frac{b_0^3}{2r^3}+1\right)\end{aligned}\right\} \quad (c \leqslant r \leqslant b_0). \qquad (6)$$

The displacement is

$$u = \frac{2Yc^3}{3Eb_0^3}\left\{(1-2\nu)r + \frac{(1+\nu)b_0^3}{2r^2}\right\} \quad (c \leqslant r \leqslant b_0). \qquad (7)$$

Thus the solution in the elastic region is dependent only on the parameter c; this is a particularly simple form of interrelation between the solutions in the elastic and plastic regions.

In the latter we have the equilibrium equation

$$\frac{\partial \sigma_r}{\partial r} = \frac{2(\sigma_\theta - \sigma_r)}{r}$$

and the yield criterion (3) governing the two unknown stress components. We immediately derive

$$\sigma_r = 2Y \ln r + B,$$

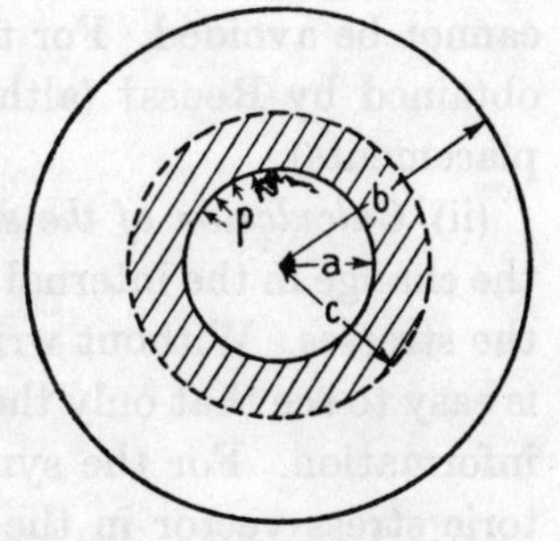

FIG. 13. Plastic region round a cylindrical or spherical cavity expanded by uniformly distributed pressure.

where B is another parameter. Since σ_r must be continuous across the plastic boundary, we find from (6) that

$$B = -2Y \ln c - \frac{2Y}{3}\left(1 - \frac{c^3}{b_0^3}\right).$$

Hence

$$\left.\begin{aligned}\sigma_r &= -2Y \ln\left(\frac{c}{r}\right) - \frac{2Y}{3}\left(1 - \frac{c^3}{b_0^3}\right), \\ \sigma_\theta &= Y - 2Y \ln\left(\frac{c}{r}\right) - \frac{2Y}{3}\left(1 - \frac{c^3}{b_0^3}\right)\end{aligned}\right\} \quad (a \leqslant r \leqslant c). \qquad (8)$$

The radial stress decreases in magnitude with increasing radius, while owing to the yield criterion the circumferential stress correspondingly increases. Both stress components decrease through the elastic region, so that σ_θ is greatest on the plastic boundary. By substituting $r = a$ in (8), the internal pressure needed to produce plastic flow to a radius c is found to be

$$p = 2Y \ln\left(\frac{c}{a}\right) + \frac{2Y}{3}\left(1 - \frac{c^3}{b_0^3}\right). \qquad (9)$$

If the ratio b_0/a_0 of the external and internal radii is not too large (say less than 4 or 5) the strains and the displacement of the inner surface are small so long as the shell has not yielded completely. As in elasticity, variations in a_0 can then be neglected so far as the stresses are concerned. For a given pressure the stress distribution is completely determined parametrically through c without the need to calculate the displacements in the plastic region. The problem is then statically determined, with the characteristic feature that a limited amount of information can be obtained without a full solution. On the other hand, if the internal displacement is regarded as the independent variable and the pressure-expansion relation is required, a complete solution cannot be avoided. For this simple case the first correct solution was obtained by Reuss† (although he did not explicitly evaluate the displacements).

(ii) *Calculation of the strains.* When the strains are large, however, the change in the internal radius has to be calculated‡ in order to obtain the stresses. Without writing down the Reuss stress-strain relations it is easy to see that only the compressibility equation provides additional information. For the symmetry of the problem constrains the deviatoric stress vector in the plane diagram (Fig. 7) to the position corresponding to uniaxial compression, and consequently the strain-path must be a straight line in the same direction (the deviatoric elastic strain-increments being zero). This is obviously so in a symmetric expansion since the two circumferential components of an increment of strain are equal. In calculating the displacement of any individual particle it is convenient to take the movement of the plastic boundary as the scale of 'time' or progress of the expansion, since the parameter c appears in the formulae for the stresses. We may speak of the velocity v of a particle, meaning that the particle is displaced by an amount $v\,dc$ when the plastic boundary moves outwards a further distance dc. v can be expressed directly in terms of the total displacement u. For the incremental displacement of a particle can also be written as

$$du = \frac{\partial u}{\partial c}dc + \frac{\partial u}{\partial r}dr = \left(\frac{\partial u}{\partial c} + v\frac{\partial u}{\partial r}\right)dc,$$

where r and c are taken as the independent variables. Equating this expression to $v\,dc$ we obtain

$$v = \frac{\partial u/\partial c}{1-(\partial u/\partial r)}. \qquad (10)$$

---

† A. Reuss, *Zeits. ang. Math. Mech.* **10** (1930), 266.
‡ R. Hill, *Journ. App. Mech.* **16** (1949), 295.

Now the compressibility equation is

$$d\epsilon_r + d\epsilon_\theta + d\epsilon_\phi = \frac{(1-2\nu)}{E}(d\sigma_r + d\sigma_\theta + d\sigma_\phi),$$

where we must take care to evaluate the increments of stress and strain following a given element, not at a given point in space. Thus

$$d\epsilon_r = \frac{\partial}{\partial r}(du) = \frac{\partial v}{\partial r}dc; \qquad d\epsilon_\theta = d\epsilon_\phi = \frac{du}{r} = \frac{v\,dc}{r};$$

$$d\sigma_r = \left(\frac{\partial \sigma_r}{\partial c} + v\frac{\partial \sigma_r}{\partial r}\right)dc; \qquad d\sigma_\theta = d\sigma_\phi = \left(\frac{\partial \sigma_\theta}{\partial c} + v\frac{\partial \sigma_\theta}{\partial r}\right)dc.$$

Hence
$$\frac{\partial v}{\partial r} + \frac{2v}{r} = \frac{(1-2\nu)}{E}\left(\frac{\partial}{\partial c} + v\frac{\partial}{\partial r}\right)(\sigma_r + 2\sigma_\theta). \qquad (11)$$

Substituting the expressions for σ_r and σ_θ from (8), we obtain a differential equation for the velocity v:

$$\frac{\partial v}{\partial r} + \frac{2v}{r} = 6(1-2\nu)\frac{Y}{E}\left[\frac{v}{r} - \frac{1}{c}\left(1 - \frac{c^3}{b_0^3}\right)\right].$$

v is known on the plastic boundary from the solution for the displacements in the elastic region; thus, from (7) and (10),

$$v_{r=c} = \frac{Y}{E}\left[2(1-2\nu)\frac{c^3}{b_0^3} + (1+\nu)\right].$$

On integrating the equation for v, and neglecting second and higher powers of Y/E, there results

$$v = \frac{3(1-\nu)Yc^2}{Er^2} - \frac{2(1-2\nu)Y}{E}\left(1 - \frac{c^3}{b_0^3}\right)\frac{r}{c}. \qquad (12)$$

Since $v = da/dc$ on the inner surface, the relation between c and a is

$$\frac{da}{dc} = \frac{3(1-\nu)Yc^2}{Ea^2} - \frac{2(1-2\nu)Y}{E}\left(1 - \frac{c^3}{b_0^3}\right)\frac{a}{c}. \qquad (13)$$

Integrating, and neglecting second order terms in Y/E, we find after some rearrangement that

$$\frac{a^3}{a_0^3} = 1 + \frac{3(1-\nu)Yc^3}{Ea_0^3} - \frac{2(1-2\nu)Y}{E}\left[3\ln\left(\frac{c}{a_0}\right) + 1 - \frac{c^3}{b_0^3}\right]. \qquad (14)$$

Care is needed in the approximation since for very thick shells c^3/a^3 may become of order E/Y.

When the thickness of the shell is not too great ($b_0^3/a_0^3 \ll E/Y$) and the expansion is small during the partly plastic state, the total displacement of the inner surface since the first application of pressure is found from (14) to be

$$u(a_0) = \frac{Ya_0}{E}\left[(1-\nu)\frac{c^3}{a_0^3} - \tfrac{2}{3}(1-2\nu)\left(1 - \frac{c^3}{b_0^3}\right) - 2(1-2\nu)\ln\left(\frac{c}{a_0}\right)\right]. \qquad (15)$$

This may alternatively be obtained by using the compressibility equation in the integrated form

$$\epsilon_r + 2\epsilon_\theta = \frac{\partial u}{\partial r} + \frac{2u}{r} = \frac{(1-2\nu)}{E}(\sigma_r + 2\sigma_\theta),$$

which is valid to the present order of approximation when the strains are small. The shell becomes completely plastic when the internal displacement is

$$u_1(a_0) = \frac{Ya_0}{E}\left[(1-\nu)\frac{b_0^3}{a_0^3} - 2(1-2\nu)\ln\left(\frac{b_0}{a_0}\right)\right].$$

From (7) the corresponding external displacement is

$$u_1(b_0) = \frac{(1-\nu)Yb_0}{E}.$$

The fractional *increase* in the volume of shell material at this stage is therefore

$$\frac{3[b_0^2 u_1(b_0) - a_0^2 u_1(a_0)]}{b_0^3 - a_0^3} = \frac{6(1-2\nu)Y\ln(b_0/a_0)}{E[(b_0^3/a_0^3)-1]}.$$

From (9) we find that

$$\frac{\partial p}{\partial c} = \frac{2Y}{c}\left(1 - \frac{c^3}{b_0^3}\right).$$

Thus, when $c = b_0$, $\partial p/\partial c$ is zero and the pressure reaches a true maximum. During the subsequent expansion the pressure is equal to $2Y\ln(b/a)$, and this evidently decreases steadily. In practice this stage of the expansion could not be followed since the equilibrium is unstable and any slight local weakness in the shell would induce necking and eventual bursting of the shell.

(iii) *Residual stresses.* If the shell is unloaded from a partly plastic state, the residual stresses are obtained (if a possible Bauschinger effect is neglected) by subtracting from (6) and (8) the elastic stress distribution (1):

$$\left.\begin{aligned}
\sigma_r &= -\frac{2Y}{3}\left(\frac{c^3}{a_0^3} - \frac{p}{p_0}\right)\left(\frac{a_0^3}{r^3} - \frac{a_0^3}{b_0^3}\right) \\
\sigma_\theta &= \frac{2Y}{3}\left(\frac{c^3}{a_0^3} - \frac{p}{p_0}\right)\left(\frac{a_0^3}{2r^3} + \frac{a_0^3}{b_0^3}\right)
\end{aligned}\right\} (c \leqslant r \leqslant b_0);$$

$$\left.\begin{aligned}
\sigma_r &= -\frac{2Y}{3}\left\{\frac{p}{p_0}\left(1 - \frac{a_0^3}{r^3}\right) - 3\ln\left(\frac{r}{a_0}\right)\right\} \\
\sigma_\theta &= \frac{2Y}{3}\left\{\frac{3}{2} + 3\ln\left(\frac{r}{a_0}\right) - \frac{p}{p_0}\left(1 + \frac{a_0^3}{2r^3}\right)\right\}
\end{aligned}\right\} (a_0 \leqslant r \leqslant c);$$

(16)

where
$$p = \frac{2Y}{3}\left\{3\ln\left(\frac{c}{a_0}\right) + \left(1 - \frac{c^3}{b_0^3}\right)\right\} \tag{9'}$$

and
$$p_0 = \frac{2Y}{3}\left(1 - \frac{a_0^3}{b_0^3}\right). \tag{4'}$$

The contraction of the outer layers of the shell compresses the inner layers, and leaves the internal surface in a state of tangential compression of amount $Y(p/p_0 - 1)$. A subsequent application of pressure less than the original maximum only strains the shell elastically. In this way the shell is strengthened by an initial overstrain (when the material work-hardens the strengthening is still greater). It is necessary, as in all residual stress calculations, to examine the tacit assumption that no element is unloaded and stressed to the yield-point in the reverse direction during the removal of the applied load. For this we need to know the magnitude of $\sigma_\theta - \sigma_r$, which, according to (3), is the quantity governing yielding. At any intermediate stage of the unloading when a fraction λp $(0 \leqslant \lambda \leqslant 1)$ of the pressure has been removed, it follows from (1), (3), and (6) that

$$\sigma_\theta - \sigma_r = Y\left(1 - \lambda\frac{pa_0^3}{p_0 r^3}\right) \quad (a_0 \leqslant r \leqslant c),$$

$$\sigma_\theta - \sigma_r = Y\left(\frac{c^3}{r^3} - \lambda\frac{pa_0^3}{p_0 r^3}\right) \quad (c \leqslant r \leqslant b_0),$$

where p_0 is given by equation (4). As the internal pressure is released, and λ increases from zero, $\sigma_\theta - \sigma_r$ steadily decreases, finally becoming negative within a radius $a_0(p/p_0)^{\frac{1}{3}}$ (which may easily be shown from (4) and (9) to be less than c). The numerical magnitude of $\sigma_\theta - \sigma_r$ is greatest on the internal surface, and the condition that yielding does not restart here during the unloading is evidently

$$p \leqslant 2p_0. \tag{17}$$

Thus a single application of pressure cannot strengthen a shell by more than a factor of two; this optimum cannot be achieved if the shell is too thin ($b_0/a_0 < 4.92$), the critical thickness being given by the equation

$$ln\left(\frac{b_0}{a_0}\right) = \frac{2}{3}\left(1 - \frac{a_0^3}{b_0^3}\right).$$

(iv) *Expansion of a spherical cavity in an infinite medium.* At the other extreme, when a spherical cavity is expanded from zero radius in an infinite medium, the stresses are functions of r/a only, and the ratio c/a

remains constant. We find from (14), or directly from (13), that

$$\frac{c}{a} = \left\{\frac{E}{3(1-\nu)Y}\right\}^{\frac{1}{3}} \tag{18}$$

to a sufficient approximation. From (9) the constant internal pressure is

$$p = \frac{2Y}{3}\left\{1+\ln\left(\frac{E}{3(1-\nu)Y}\right)\right\}. \tag{19}$$

This may also be thought of as the work needed to make unit volume of the cavity. For the common pre-strained metals E/Y is of order 300 to 400, while ν is usually in the range 0·25 to 0·35. Hence c/a normally lies between 5 and 6, while p is about $4Y$. The expressions (18) and (19) are to be regarded as upper limits to which c/a and p approach if a hole is expanded from some finite radius in a very thick shell. It may be shown that the plastic and elastic strains are of comparable size throughout a large part of the plastic region. This means that it would not be a particularly good approximation to neglect changes of volume in the plastic region, even though the strains become infinitely large near the inner surface. Thus Bishop, Hill, and Mott[†] derived the value $\{E/(1+\nu)Y\}^{\frac{1}{3}}$ for c/a on this assumption, a formula which overestimates c/a by some 20 per cent. A closer approximation can be obtained by neglecting volume changes throughout the entire shell; on putting $\nu = \frac{1}{2}$ in (18) we obtain $c/a = (2E/3Y)^{\frac{1}{3}}$, which is an overestimate of about 12 per cent. On the other hand, considerable errors would result if volume changes were neglected in the expansion of a thin shell, since plastic and elastic strains are then comparable everywhere in the plastic region.

(v) *Inclusion of work-hardening in the analysis.* We have now to consider how to extend the analysis to include work-hardening. The problem is no longer statically determined even if the strains are small since the solutions for the stresses and strains are linked by the law of work-hardening and have to be carried through together. For convenience the stress-strain curve of the material in compression is taken in the form
$$\sigma = Y + H(\epsilon),$$
where Y is the yield stress, and H is the amount of hardening expressed as a function of the logarithmic total strain ϵ. It has already been remarked that each element of the shell is stressed, apart from a superposed hydrostatic tension of amount σ_θ, under a uniaxial compression $\sigma_\theta - \sigma_r$ in the radial direction. Consequently we can directly apply the

[†] R. F. Bishop, R. Hill, and N. F. Mott, *Proc. Phys. Soc.* **57** (1945), 147.

data obtained in simple compression provided we allow for the additional elastic strain produced by the hydrostatic tension.

Making this allowance we have

$$\sigma_\theta - \sigma_r = Y + H\left\{-\epsilon_r + (1-2\nu)\frac{\sigma_\theta}{E}\right\}.$$

Now
$$\epsilon_r + 2\epsilon_\theta = \frac{(1-2\nu)}{E}(\sigma_r + 2\sigma_\theta),$$

and
$$\epsilon_\theta = \ln(r/r_0),$$

where r_0 is the initial radius to the element. Hence

$$\sigma_\theta - \sigma_r = Y + H\left\{2\ln\left(\frac{r}{r_0}\right) - \frac{(1-2\nu)}{E}(\sigma_r + \sigma_\theta)\right\}. \quad (20)$$

This replaces the yield criterion (3), and, together with the equation of equilibrium and the boundary condition on $r = c$, determines the stresses at each stage of the expansion. The inclusion of the second term in the argument of H is essential when the plastic and elastic components of the strain are comparable. If H is a general function, it is evident that the integration can only be effected by a small-arc process; no investigation of this appears to have been published. However, a great simplification results if the material is treated as incompressible, so that $\nu = \frac{1}{2}$. We then have

$$\frac{d\sigma_r}{dr} = \frac{2(\sigma_\theta - \sigma_r)}{r} = \frac{2Y}{r} + \frac{2H\{2\ln(r/r_0)\}}{r},$$

and so
$$\sigma_r = -2Y\ln\left(\frac{c}{r}\right) - 2\int_r^c H\left\{2\ln\left(\frac{r}{r_0}\right)\right\}\frac{dr}{r} - \frac{2Y}{3}\left(1 - \frac{c^3}{b_0^3}\right).$$

The internal pressure is thus

$$p = 2Y\ln\left(\frac{c}{a}\right) + 2\int_a^c H\left\{2\ln\left(\frac{r}{r_0}\right)\right\}\frac{dr}{r} + \frac{2Y}{3}\left(1 - \frac{c^3}{b_0^3}\right). \quad (21)$$

Since the material is assumed incompressible

$$r^3 - r_0^3 = a^3 - a_0^3.$$

But in the elastic region

$$r^3 - r_0^3 = (r_0 + u)^3 - r_0^3 \simeq 3r_0^2 u = 3Yc^3/2E$$

to the usual order of approximation. Hence

$$c^3 = \frac{2E}{3Y}(a^3 - a_0^3). \quad (22)$$

p can be calculated as a function of the internal displacement from (21) and (22). The overall error of the analysis should be least for a real material when the plastic strains are large, since the effect of elastic compressibility is then smallest. In the most favourable case, when a cavity is expanded from zero radius in an infinite medium,

$$\frac{c}{a} = \left(\frac{2E}{3Y}\right)^{\frac{1}{3}}, \qquad r_0^3 = r^3 - a^3,$$

and
$$p = \frac{2Y}{3}\left\{1 + \ln\left(\frac{2E}{3Y}\right)\right\} + 2\int_1^{c/a} H\left\{\frac{2}{3}\ln\left(\frac{t^3}{t^3-1}\right)\right\}\frac{dt}{t}; \qquad t = \frac{r}{a}. \quad (23)$$

On the inner surface $t = 1$ and the strain becomes infinite. Nevertheless, the integral converges to a finite limit if we suppose the rate of hardening to approach a constant or zero value at very large strains. In particular, if $H \equiv H'\epsilon$, where H' is a constant rate of hardening (so that the stress-strain curve is approximated by two straight lines), Bishop, Hill, and Mott have shown that

$$p = \frac{2Y}{3}\left[1 + \ln\left(\frac{2E}{3Y}\right)\right] + \frac{2\pi^2 H'}{27}.$$

For cold-worked copper they took $Y = 17 \cdot 5$ tn./in.², $H' = 6 \cdot 5$ tn./in.², and $E = 8{,}000$ tn./in.² With these values the formula gives $p = 84 \cdot 5$ tn./in.², of which $4 \cdot 8$ tn./in.² is due to strain-hardening. The theoretical value of p provides an estimate of the steady maximum resistive pressure in deep penetration by a punch into a quasi-infinite medium. The work needed to make unit volume of a cavity *deep* in a medium should not depend greatly on the shape of the indenter or, within broad limits, on the way the cavity is produced, provided friction is negligible. Using a rotated 40° conical punch with a cut-back shank, Bishop, Hill, and Mott observed a maximum pressure of about 85 tn./in.² in copper with the above properties; this was attained after a penetration of some five punch-diameters, and is close to the calculated value. There appears to be no other reliable published data on deep punching.

2. The expansion of a cylindrical tube

(i) *Elastic expansion and plastic yielding.* In the spherically symmetrical problem both the shape of the plastic boundary and the strain-path are determined by symmetry alone. In the special problems of cylindrical symmetry to be investigated here only the shape of the plastic boundary is known beforehand. The strain-path for every individual particle has to be calculated, along with the stresses, step by step as the plastic region develops. This greatly adds to the difficulties of a solution.

Consider a long cylindrical tube simultaneously expanded by internal pressure and loaded at both ends by equal and opposite forces directed along the axis. The length of the tube is supposed so great in relation to its mean width that the distribution of stress and strain sufficiently far from the ends does not vary along the tube, to any desired approximation. It then follows by symmetry that any originally plane transverse section remains plane, and hence that the longitudinal strain ϵ_z is the same for all elements at each stage of the expansion. Let a and b be the current internal and external radii (initial values a_0 and b_0), and let c be the radius of the plastic boundary (Fig. 13). At present, no restriction is placed on the amount of expansion. When the tube is stressed elastically the most general radial displacement is known to be of the form

$$u = Ar + B/r.$$

The elastic equations in cylindrical coordinates (r, θ, z) are

$$E\epsilon_r = E\frac{\partial u}{\partial r} = E\left(A - \frac{B}{r^2}\right) = \sigma_r - \nu(\sigma_\theta + \sigma_z),$$

$$E\epsilon_\theta = E\frac{u}{r} = E\left(A + \frac{B}{r^2}\right) = \sigma_\theta - \nu(\sigma_z + \sigma_r),$$

$$E\epsilon_z = \sigma_z - \nu(\sigma_r + \sigma_\theta),$$

where σ_r, σ_θ, and σ_z, are the principal stresses. Solving these equations:

$$(1+\nu)(1-2\nu)\frac{\sigma_r}{E} = A - (1-2\nu)\frac{B}{r^2} + \nu\epsilon_z,$$

$$(1+\nu)(1-2\nu)\frac{\sigma_\theta}{E} = A + (1-2\nu)\frac{B}{r^2} + \nu\epsilon_z,$$

$$(1+\nu)(1-2\nu)\frac{\sigma_z}{E} = 2\nu A + (1-\nu)\epsilon_z.$$

A noteworthy property of the stress distribution is that the longitudinal stress σ_z and the hydrostatic component of the stress $\frac{1}{3}(\sigma_r + \sigma_\theta + \sigma_z)$ are both constant over the section. A and B are determined by the conditions that $\sigma_r = -p$ on $r = a_0$, and $\sigma_r = 0$ on $r = b_0$:

$$A = -\nu\epsilon_z + \frac{(1+\nu)(1-2\nu)p}{E(b_0^2/a_0^2 - 1)}; \qquad B = \frac{(1+\nu)b_0^2 p}{E(b_0^2/a_0^2 - 1)}.$$

The final expressions for the stresses are then

$$\left.\begin{aligned}\sigma_r &= -p\left(\frac{b_0^2}{r^2} - 1\right)\bigg/\left(\frac{b_0^2}{a_0^2} - 1\right), \\ \sigma_\theta &= p\left(\frac{b_0^2}{r^2} + 1\right)\bigg/\left(\frac{b_0^2}{a_0^2} - 1\right), \\ \sigma_z &= E\epsilon_z + 2\nu p\bigg/\left(\frac{b_0^2}{a_0^2} - 1\right).\end{aligned}\right\} \quad (24)$$

The displacement is

$$u = -\nu\epsilon_z r + \frac{(1+\nu)p}{E(b_0^2/a_0^2-1)}\left\{(1-2\nu)r+\frac{b_0^2}{r}\right\}. \qquad (25)$$

The resultant longitudinal force applied to the ends of the tube is

$$L = \pi(b_0^2-a_0^2)\sigma_z = \pi a_0^2\left\{2\nu p+E\epsilon_z\left(\frac{b_0^2}{a_0^2}-1\right)\right\}.$$

Thus the longitudinal strain corresponding to a given axial load and internal pressure is

$$\epsilon_z = \left(\frac{L}{\pi a_0^2}-2\nu p\right)\bigg/E\left(\frac{b_0^2}{a_0^2}-1\right). \qquad (26)$$

In the autofrettage process the tube is either closed at both ends by plugs firmly attached to the tube, or by floating pistons which allow it to expand freely. The first method is described as autofrettage under closed-end conditions, and the second as under open-end conditions. For a tube with closed ends the resultant longitudinal force acting on the central section is a tension equal to the force exerted by the internal pressure on each plug, viz. $\pi a_0^2 p$. Hence, from (26),

$$\left.\begin{aligned}\epsilon_z &= (1-2\nu)p\bigg/E\left(\frac{b_0^2}{a_0^2}-1\right),\\ \sigma_z &= p\bigg/\left(\frac{b_0^2}{a_0^2}-1\right) = \tfrac{1}{2}(\sigma_r+\sigma_\theta)\end{aligned}\right\} \text{ (closed end).} \qquad (27)$$

When the ends of the tube are open, $L = 0$, and

$$\left.\begin{aligned}\epsilon_z &= -2\nu p\bigg/E\left(\frac{b_0^2}{a_0^2}-1\right),\\ \sigma_z &= 0\end{aligned}\right\} \text{ (open end).} \qquad (28)$$

The longitudinal strain is a contraction. A third end condition, sometimes considered owing to its relative mathematical simplicity, is that of plane strain, or zero extension, in which

$$\left.\begin{aligned}\epsilon_z &= 0,\\ \sigma_z &= \nu(\sigma_r+\sigma_\theta) = 2\nu p\bigg/\left(\frac{b_0^2}{a_0^2}-1\right)\end{aligned}\right\} \text{ (plane strain).} \qquad (29)$$

This case is intermediate to the first two, though nearer to the closed-end condition, to which it is strictly equivalent when the material is incompressible ($\nu = \tfrac{1}{2}$). In all three cases the radial and circumferential stresses are the same, and only depend on p. The longitudinal stress is different, but it will be observed that it is always intermediate to the other two principal stresses.

If, then, Tresca's criterion of yielding is adopted, we have to consider the magnitude of
$$\sigma_\theta - \sigma_r = \frac{2pb_0^2}{r^2}\bigg/\bigg(\frac{b_0^2}{a_0^2}-1\bigg).$$
Since this is greatest when $r = a_0$ yielding begins on the internal surface at a pressure
$$p_0 = \tfrac{1}{2}Y\bigg(1-\frac{a_0^2}{b_0^2}\bigg). \tag{30}$$
This is the same for all end conditions.

On the other hand, if von Mises' criterion is adopted, the end condition affects the pressure at which yielding begins. From (24),
$$(\sigma_r-\sigma_\theta)^2+(\sigma_\theta-\sigma_z)^2+(\sigma_z-\sigma_r)^2$$
$$= \frac{6b_0^4 p^2}{(b_0^2/a_0^2-1)^2 r^4} + 2\bigg(\frac{p}{b_0^2/a_0^2-1}-\sigma_z\bigg)^2.$$
Since this is greatest when $r = a_0$, the pressure p_0 which causes yielding at the internal surface satisfies
$$\frac{3p_0^2}{(1-a_0^2/b_0^2)^2}+\bigg(\frac{p_0}{b_0^2/a_0^2-1}-\sigma_z\bigg)^2 = Y^2. \tag{31}$$
Inserting the values of σ_z for the three end conditions (27), (28), and (29), the respective values of p_0 are

$$\left.\begin{aligned}p_0 &= \frac{Y}{\sqrt{3}}\bigg(1-\frac{a_0^2}{b_0^2}\bigg) & \text{(closed end);} \\ p_0 &= \frac{Y}{\sqrt{3}}\bigg(1-\frac{a_0^2}{b_0^2}\bigg)\bigg/\sqrt{\bigg(1+\frac{a_0^4}{3b_0^4}\bigg)} & \text{(open end);} \\ p_0 &= \frac{Y}{\sqrt{3}}\bigg(1-\frac{a_0^2}{b_0^2}\bigg)\bigg/\sqrt{\bigg\{1+(1-2\nu)^2\frac{a_0^4}{3b_0^4}\bigg\}} & \text{(plane strain).}\end{aligned}\right\} \tag{32}$$

We see that the tube with open ends yields at the lowest pressure. The yield pressures for the closed-end and plane-strain conditions differ by less than 3 per cent. for the usual values of ν, the former being the higher; when b_0/a_0 is 2 (a typical dimension for a gun-barrel) the difference is about 0·2 per cent. The differences between the three pressures become vanishingly small as $b_0 \to \infty$, but even for $b_0/a_0 = 2$ they are less than 1 per cent.

(ii) *Partly plastic tube with strains of any magnitude.* When the tube is partly plastic (Fig. 13) the stresses in the elastic region are still of the form
$$\sigma_r = -C\bigg(\frac{b_0^2}{r^2}-1\bigg), \qquad \sigma_\theta = C\bigg(\frac{b_0^2}{r^2}+1\bigg),$$
$$\sigma_z = E\epsilon_z + 2\nu C = \text{const.}$$

Since ϵ_z is determined by the condition that $\int_a^b 2\pi r \sigma_z \, dr$ should be equal to the axial load L it follows that, except in the case of plane strain, ϵ_z depends on the state of stress in both plastic and elastic regions. The parameter C is eliminated by using the condition that the material immediately on the elastic side of the plastic boundary $r = c$ is on the point of yielding. Assuming that there is no work-hardening, and that σ_z remains the intermediate principal stress (the range over which this is true will be investigated *a posteriori*), Tresca's criterion furnishes the equation
$$\sigma_\theta - \sigma_r = Y \tag{33}$$
everywhere in the plastic region.

Hence $C = Yc^2/2b_0^2$, and
$$\left. \begin{aligned} \sigma_r &= -\frac{Yc^2}{2b_0^2}\left(\frac{b_0^2}{r^2}-1\right), \\ \sigma_\theta &= \frac{Yc^2}{2b_0^2}\left(\frac{b_0^2}{r^2}+1\right), \\ \sigma_z &= E\epsilon_z + \nu Y \frac{c^2}{b_0^2} \end{aligned} \right\} \quad (c \leqslant r \leqslant b_0). \tag{34}$$

The radial displacement is
$$u = -\nu \epsilon_z r + \frac{(1+\nu)Yc^2}{2Eb_0^2}\left\{(1-2\nu)r + \frac{b_0^2}{r}\right\}. \tag{35}$$

In the plastic region the equation of equilibrium combined with (33) leads to
$$\frac{\partial \sigma_r}{\partial r} = \frac{\sigma_\theta - \sigma_r}{r} = \frac{Y}{r}.$$
Therefore,
$$\left. \begin{aligned} \frac{\sigma_r}{Y} &= -\tfrac{1}{2} - \ln\left(\frac{c}{r}\right) + \frac{c^2}{2b_0^2}, \\ \frac{\sigma_\theta}{Y} &= \tfrac{1}{2} - \ln\left(\frac{c}{r}\right) + \frac{c^2}{2b_0^2} \end{aligned} \right\} \quad (a \leqslant r \leqslant c), \tag{36}$$
after using the condition for continuity of σ_r across $r = c$. The internal pressure is thus given by
$$\frac{p}{Y} = \ln\left(\frac{c}{a}\right) + \tfrac{1}{2}\left(1 - \frac{c^2}{b_0^2}\right). \tag{37}$$

This formula appears to have been first obtained by Turner[†] (1909). If the displacement of the internal surface is so small that we can neglect variations of the radius a in this formula, σ_r and σ_θ are determined by

[†] L. B. Turner, *Trans. Camb. Phil. Soc.* **21** (1909), 377; *Engineering*, **92** (1911), 115.

(36), parametrically through c, in terms of the pressure p without needing to calculate the deformation in the plastic region. In other words, σ_r and σ_θ are statically determined. Moreover, the relationship is *independent of the end conditions*. It is for these reasons that the use of Tresca's criterion greatly simplifies the problem; we shall see later that the same is not true when von Mises' criterion is used. The distribution

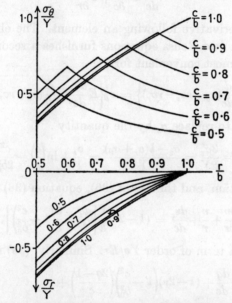

Fig. 14. Distribution of radial and circumferential stress components over transverse section of a partly plastic tube expanded by internal pressure (Tresca criterion; no strain-hardening).

of σ_r and σ_θ is shown in Fig. 14. On the other hand, σ_z can only be determined by the application of Reuss's equations, and therefore depends on the strain-history. This is the real difficulty in the problem, and one which has been avoided in most treatments by assumptions that are now known to be seriously in error.

The stress-strain relations of Reuss [(33) of Chapter II] are

$$d\epsilon_r = \frac{1}{E}(d\sigma_r - \nu\, d\sigma_\theta - \nu\, d\sigma_z) + \sigma'_r\, d\lambda,$$

$$d\epsilon_\theta = \frac{1}{E}(d\sigma_\theta - \nu\, d\sigma_z - \nu\, d\sigma_r) + \sigma'_\theta\, d\lambda,$$

$$d\epsilon_z = \frac{1}{E}(d\sigma_z - \nu\, d\sigma_r - \nu\, d\sigma_\theta) + \sigma'_z\, d\lambda.$$

By adding these we recover the compressibility relation

$$\frac{\partial v}{\partial r} + \frac{v}{r} + \frac{d\epsilon_z}{dc} = \frac{(1-2\nu)}{E}\frac{d}{dc}(\sigma_r+\sigma_\theta+\sigma_z), \qquad (38)$$

where v is the velocity defined in (10), and where, for shortness, we write

$$\frac{d}{dc} = \frac{\partial}{\partial c} + v\frac{\partial}{\partial r}$$

to denote the derivative following an element. The elimination of $d\lambda$ between the last two Reuss equations furnishes a second independent equation in the most convenient form:

$$\sigma_z'\left\{\frac{Ev}{r} - \frac{d}{dc}(\sigma_\theta-\nu\sigma_z-\nu\sigma_r)\right\} = \sigma_\theta'\left\{E\frac{d\epsilon_z}{dc} - \frac{d}{dc}(\sigma_z-\nu\sigma_r-\nu\sigma_\theta)\right\}. \qquad (39)$$

It is convenient to replace σ_z by the quantity

$$q = \frac{3\sigma_z'}{2Y} = \frac{\sigma_z - \frac{1}{2}(\sigma_r+\sigma_\theta)}{Y} = \frac{\sigma_z}{Y} + \ln\left(\frac{c}{r}\right) - \frac{c^2}{2b_0^2}. \qquad (40)$$

With this definition, and the use of (36), equation (38) becomes

$$\frac{\partial v}{\partial r} + \frac{v}{r} + \frac{d\epsilon_z}{dc} = (1-2\nu)\frac{Y}{E}\left[\frac{dq}{dc} - \frac{3}{c}\left(1-\frac{c^2}{b_0^2}\right)\right], \qquad (41)$$

with neglect of a term of order Yv/Er. Similarly, (39) reduces to

$$\{1-\tfrac{2}{3}(1-2\nu)q\}\frac{dq}{dc} + (1-2\nu)\left(1-\frac{c^2}{b_0^2}\right)\left(\frac{2q-1}{c}\right) +$$
$$+ \left\{\tfrac{4}{3}q + (1-2\nu)\frac{Y}{E}\right\}\frac{Ev}{Yr} + (\tfrac{2}{3}q-1)\frac{d}{dc}\left(\frac{E\epsilon_z}{Y}\right) = 0. \qquad (42)$$

A third relation between the unknowns ϵ_z, v, and q, is supplied by the end condition

$$\frac{L}{Y} = \int_a^{b_0} \frac{2\pi r\sigma_z\,dr}{Y} = \int_a^c 2\pi rq\,dr + \frac{\pi a^2 p}{Y} + \pi(b_0^2-c^2)\left\{\frac{E\epsilon_z}{Y} - \tfrac{1}{2}(1-2\nu)\frac{c^2}{b_0^2}\right\}, \qquad (43)$$

after using (37) and (40). The boundary conditions on $r = c$ are that v and σ_z should be continuous. Hence

$$\left.\begin{aligned}\left(\frac{Ev}{Y}\right)_{r=c} &= (1+\nu)\left\{1+(1-2\nu)\frac{c^2}{b_0^2}\right\} - \nu c\frac{d}{dc}\left(\frac{E\epsilon_z}{Y}\right); \\ q_{r=c} &= \frac{E\epsilon_z}{Y} - \tfrac{1}{2}(1-2\nu)\frac{c^2}{b_0^2}.\end{aligned}\right\} \qquad (44)$$

Equations (41), (42), (43), and (44) are the basis for a solution of the problem.

(iii) *Method of numerical integration of the general equations.* Except in special cases an analytic solution is out of the question, and the equations must be integrated numerically. The method follows naturally from the fact that the equations are hyperbolic (Appendix III), the characteristics in the (r, c) plane being the lines $c = $ constant, marking stages of the expansion, and the paths of the particles $dr - v\,dc = 0$ (Fig. 15). When the displacements are large, the positions of the second family of characteristics are not known to begin with, but have to be calculated as the integration proceeds. From (42) the differential relation along a characteristic of the family $dr - v\,dc = 0$ is

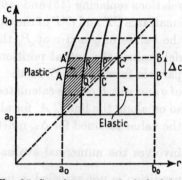

Fig. 15. Step-by-step method of solving equations for partly plastic tube, by integration along characteristics in the (r, c) plane.

$$\{1 - \tfrac{2}{3}(1 - 2\nu)q\}\,dq + (\tfrac{2}{3}q - 1)\,d\!\left(\frac{E\epsilon_z}{Y}\right) +$$
$$+ \left[\{\tfrac{4}{3}q + (1 - 2\nu)\frac{Y}{E}\}\frac{Ev}{Yr} + (1 - 2\nu)\left(1 - \frac{c^2}{b_0^2}\right)\!\left(\frac{2q-1}{c}\right)\right]dc = 0. \quad (45)$$

The differential relation along a characteristic of the family $dc = 0$ is obtained by eliminating dq/dc from (41) and (42):

$$\{1 - \tfrac{2}{3}(1 - 2\nu)q\}\,d\!\left(\frac{Ev}{Y}\right) +$$
$$+ \left[\{1 + \tfrac{2}{3}(1 - 2\nu)q\}\frac{Ev}{Yr} + 2v\,\frac{d}{dc}\!\left(\frac{E\epsilon_z}{Y}\right) + \frac{2(1+\nu)(1-2\nu)}{c}\!\left(1 - \frac{c^2}{b_0^2}\right)\right]dr = 0, \quad (46)$$

where it must be remembered that $d\epsilon_z/dc$ is independent of r. For numerical computation the differentials dq, $d\epsilon_z$, dv, dr, and dc, in (45) and (46), are replaced by first order differences Δq, etc., and their coefficients by mean values, between adjacent points of a characteristic network of arbitrarily fine mesh. The computation is carried out by the following step-by-step procedure. Suppose that the complete solution has been obtained up to the stage when the plastic boundary has reached a radius c (AB in Fig. 15), and that we wish to advance the solution by the next increment of expansion, during which the plastic boundary moves a further small distance Δc (to position $A'B'$). A value of the increment of axial strain, $\Delta\epsilon_z$, is guessed by extrapolation from the relation between ϵ_z and c so far determined. The boundary conditions

(44) on $r = c$ then give the values of q and v at C'. The difference relations replacing (45) and (46) are now solved for Δq and Δv across the small arcs CP and $C'P$ of the characteristics through C and C'. Knowing the values of q and v at P, the integration of $dr - v\,dc = 0$ along CP determines the actual position of P. This is the new position of the particle formerly at C on the plastic boundary. Similarly, the values of q and v at R can be calculated from the known values at P and Q, and so on along the line $C'A'$ for all elements in the plastic region. Finally, the value assumed for $\Delta \epsilon_z$ must be checked by substitution in (43) (this involves the numerical evaluation of the integral $\int_a^c rq\,dr$); if the end-condition is not satisfied, a new value must be guessed and the entire process repeated. There are several devices by which the procedure may be shortened, but even so it will be evident to the reader that the numerical solution generally demands considerable labour.

After the tube has become plastic throughout, c can naturally no longer be used as the time-scale. The external displacement is a convenient quantity to measure in practice, and may be used instead. As this stage of the expansion is of little practical interest (the pressure having passed its maximum) the modifications of (45) and (46) will not be discussed.†

3. Theory of the autofrettage process

(i) *General equations when the strains are small.* The numerical solution is appreciably shortened when the tube is of moderate thickness (b_0/a_0 less than 3 or 4, say) since the strains remain of elastic order of magnitude so long as the tube is only partly plastic. The displacements may then be neglected in the plastic region. This means that we can write a_0 for a in (37) and (43); for simplicity the subscript zero will be omitted henceforward. Furthermore, the curved set of characteristics $dr - v\,dc = 0$ can be regarded as straight ($r = $ constant) and so known beforehand. To the same order of approximation the operators d/dc and $\partial/\partial c$ are equivalent, and (10) may be replaced by

$$v = \frac{\partial u}{\partial c}. \tag{47}$$

By integration of (41), or directly, the compressibility equation is obtained in the simpler form

$$\frac{\partial u}{\partial r} + \frac{u}{r} + \epsilon_z = (1-2\nu)\frac{Y}{E}\left(q - 3\ln\frac{c}{r} + \frac{3c^2}{2b^2}\right). \tag{48}$$

† Details can be found in the papers referred to later.

The differential form of this along the lines $c = $ constant, replacing (46), is

$$d\left(\frac{Eu}{Y}\right)+\left[\frac{Eu}{Yr}+\frac{E\epsilon_z}{Y}+(1-2\nu)\left(3\ln\frac{c}{r}-\frac{3c^2}{2b^2}-q\right)\right]dr = 0. \qquad (49)$$

The differential relation (45), along $r = $ constant, simplifies to

$$[1-\tfrac{2}{3}(1-2\nu)q]\,dq+(\tfrac{2}{3}q-1)\,d\left(\frac{E\epsilon_z}{Y}\right)+\frac{4q}{3r}\,d\left(\frac{Eu}{Y}\right)+$$

$$+(1-2\nu)\left(\frac{2q-1}{c}\right)\left(1-\frac{c^2}{b^2}\right)dc = 0. \qquad (50)$$

The procedure for computation is similar to that already described for the general case.

The basic equations (36), (37), (43), (44), (49), and (50) have been written in non-dimensional form. On inspection it will be clear that the quantities σ_r/Y, σ_θ/Y, σ_z/Y, q, $E\epsilon_z/Y$, and Eu/Y, are functions only of $r/a, b/a, \nu$, and L/Y, but not of E. Thus, simple scaling of a single solution for a tube of given thickness and Poisson's ratio, subject to one of the three main end-conditions, supplies the solutions for similar tubes of material with the same Poisson's ratio but any yield stress Y and elastic modulus E. When the expansion is large, however, we can only transform from one solution to another if Y/E is the same, since the formulae for the stresses involve the varying radius a.

Furthermore, in plane strain the solution for a tube of one particular wall-ratio b_0/a_0 also furnishes the solutions for tubes of any lesser wall-ratio: this is a property curiously overlooked by most writers. The truth of the statement is self-evident physically since the stress and strain in any outer annulus of the tube are independent of the conditions in the remainder; the only link with the inner material is the radial pressure transmitted across the common interface, and it obviously makes no difference by what agency the pressure is considered to be applied. The mathematical expression of this property is that the equations are hyperbolic and that the boundary conditions depend only on the dimensions of the annulus and the pressure on its inner boundary. The solution for the open or closed end clearly does not have this property.

(ii) *Plane strain condition.* The first complete solution of these equations was obtained by Hill, Lee, and Tupper,[†] for tubes of wall-ratios $b/a \leqslant 2$ and Poisson's ratio 0·3, expanded under conditions of plane strain ($\epsilon_z = 0$). The end-condition (43) does not have to be satisfied at each stage of the expansion, and simply gives the tensile load L

[†] R. Hill, E. H. Lee, and S. J. Tupper, *Proc. Roy. Soc.* A, **191** (1947), 278.

needed to prevent longitudinal contraction. The trial-and-error process is therefore unnecessary, and the integration can be carried through direct. Fig. 16 illustrates the distribution of the axial stress σ_z at various

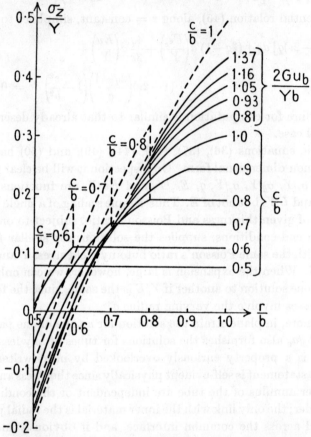

Fig. 16. Distribution of axial stress component in a partly plastic tube expanded by internal pressure under conditions of plane strain. The relevant part of the graph for a tube with internal radius a lies to the right of the abscissa a/b, where b is the external radius. The broken curves are obtained if the axial component of strain is disregarded in the plastic region.

stages of the flow, marked by the ratio c/b while the tube is partly plastic and by the external displacement u_b after it has become completely plastic. The stress distribution in a tube with a wall-ratio less than 2 is given, for various values of c/b, by the parts of the curves to the right of the abscissa a/b. While the tube is entirely elastic σ_z is constant over the tube, as we have seen. When the inner surface is on the point of yielding, it follows from (29) and (30) that σ_z has the value $\nu Y a^2/b^2$, corresponding to the horizontal line $c = a$. At a later stage σ_z is equal to $\nu Y c^2/b^2$ in the elastic region, according to (34); this corresponds to the horizontal segments for various values of c/a. σ_z is least on the inner surface and becomes compressive there after a comparatively small plastic expansion.

The broken curves in Fig. 16 represent the distribution calculated by Nadai[†] on the assumption that elastic compressibility could be neglected in the plastic region. According to the Reuss relation for the z direction this means that $\sigma_z = \frac{1}{2}(\sigma_r + \sigma_\theta)$, or $q = 0$. In view of the yield criterion (33), the deviatoric components of the elastic strain-increment are also zero. Since $\sigma_z = \nu(\sigma_r + \sigma_\theta)$ in the elastic region there is a discontinuity in σ_z across the plastic boundary of amount $(1-2\nu)c^2Y/2b^2$ (indicated by the broken vertical lines). This is $(1-2\nu)/2\nu$, or $66\frac{2}{3}$ per cent., by proportion. The discontinuity is only avoided by supposing that $\nu = \frac{1}{2}$ in the elastic region also. The overall accuracy of Nadai's solution improves as the expansion continues, since q approaches zero with increasing plastic strain[‡] (though not steadily in elements near the inner surface).

The autofrettage operator requires the relation between the pressure and the external displacement, in order to ensure that the tube, or gunbarrel, is everywhere of uniform and standard quality. If the expansion, observed at points along and around the barrel, is too great the forging is re-heat-treated. For the plane strain condition, and Tresca's yield criterion, the external expansion is given by (35) without further analysis:

$$\frac{Eu_b}{Yb} = (1-\nu^2)\frac{c^2}{b^2}. \qquad (51)$$

With (37) this determines u_b in terms of p. The value of u_b given by this formula is very little different from that for closed-end conditions (see (53) below).

(iii) *Approximation to von Mises' yield criterion.* The result that σ_z approaches the mean of the other two principal stresses with increasing plastic strain[§] is in agreement with the general conclusions of the analysis for simple compression in plane strain (Chap. IV, Sect. 5). The movement of the stress vector in the plane diagram is restricted to an angle of less than 14° near the direction representing pure shear

---

[†] A. Nadai, *Trans. Am. Soc. Mech. Eng.* **52** (1930), 193.

[‡] This verifies the initial assumption that σ_z is the intermediate principal stress.

[§] It is of historical interest to note that Saint-Venant, in the first treatment of the completely plastic tube, derived the equation $\sigma_z = \frac{1}{2}(\sigma_r + \sigma_\theta)$ from Lévy's stress-strain relations; see B. de Saint-Venant, *Comptes Rendus Acad. Sci. Paris*, **74** (1872), 1009 and 1083.

($\mu = 0$). This allows an approximation to von Mises' yield criterion by replacing the corresponding arc of the yield circle by an element of the tangent, and so writing

$$\sigma_\theta - \sigma_r = \frac{2Y}{\sqrt{3}}. \tag{52}$$

This is equivalent to Tresca's criterion with a modified yield stress, the factor m being given the value $2/\sqrt{3}$ (see p. 21). To express this more precisely we make use of the identity

$$\sum (\sigma_\theta - \sigma_r)^2 \equiv \tfrac{3}{2}(\sigma_\theta - \sigma_r)^2 + 2\left(\sigma_z - \frac{\sigma_\theta + \sigma_r}{2}\right)^2,$$

and rearrange von Mises' yield criterion in the form

$$\sigma_\theta - \sigma_r = \frac{2Y}{\sqrt{3}}(1-q^2)^{\frac{1}{2}}.$$

Now q is the difference between the values of σ_z in the exact solution (full curves in Fig. 16) and the values in Nadai's solution (broken curves). In the particular example we are considering, it will be seen that the numerically greatest value of q occurs on the plastic boundary at any stage in the expansion, and is equal to $(1-2\nu)c^2/2b^2$. The maximum cannot be very different in an accurate solution based on von Mises' criterion. Consequently, the local error introduced by the approximation (52) is never greater than about 2 per cent., while the overall error is much less.† We may conclude, therefore, that by replacing Y by $2Y/\sqrt{3}$ in the previous analysis we derive an excellent approximation to the solution for a metal yielding according to the law of von Mises. It is apparent that if the exact criterion were used the labour of a numerical solution would be considerably augmented, since the equations (36) no longer apply and all stress components depend on the strain-history.

(iv) *Closed-end condition.* According to the numerical solution of the previous section the load L that must be applied to prevent longitudinal contraction is less than $\pi a^2 p$, the corresponding closed-end force. The difference decreases as the expansion continues. This suggests that a closed tube will extend, but that the amount will be limited and the final stages of the expansion will be under approximate plane strain conditions. The result that the plane strain load tends to the closed-end force with increasing expansion may be shown in the following way.

† This has been verified by direct calculation of the solution for von Mises' criterion: see P. G. Hodge, Jr. and G. N. White, Jr., *Graduate Division of App. Math.*, Brown University, Tech. Rep. 29 (1949); *Journ. App. Mech.* **17** (1950), 180.

When the tube is completely plastic,

$$L = \int_a^b 2\pi r \sigma_z \, dr = Y \int_a^b 2\pi rq \, dr + \int_a^b \pi r(\sigma_r + \sigma_\theta) \, dr$$

$$= Y \int_a^b 2\pi rq \, dr + \int_a^b \pi r\left(2\sigma_r + r\frac{d\sigma_r}{dr}\right) dr$$

from the equilibrium equation. Since $q \to 0$ with increasing strain,

$$L \to \int_a^b \pi \frac{d}{dr}(r^2 \sigma_r) \, dr = \pi[r^2 \sigma_r]_a^b = \pi a^2 p,$$

which is just the closed-end condition.

A numerical solution for the closed-end condition has been obtained by Hill, Lee, and Tupper,† for $b/a = 2$ and Poisson's ratio $0\cdot3$. The distribution of σ_z is shown in Fig. 17. For comparison the quantity $\tfrac{1}{2}(\sigma_r + \sigma_\theta)$ is included at corresponding stages of the expansion (indicated by the broken curves). It is seen that q varies within even more restricted limits ($\pm \cdot 075$) than in the plane strain case. The assumption that σ_z is the intermediate principal stress is thereby justified; furthermore von Mises' criterion is approximated by (52) to within $0\cdot3$ per cent.

Various assumptions about the axial stress have been made in earlier treatments of the problem. For example, it has been supposed that σ_z has the uniform value $p/(b^2/a^2 - 1)$ everywhere in the tube.‡ A better assumption§ is that $q = 0$ (everywhere in the tube), or that σ_z is equal to $\tfrac{1}{2}(\sigma_r + \sigma_\theta)$; this is strictly true only when the material is incompressible, in which case the axial extension is zero. All elements are stressed in pure shear (with a superimposed hydrostatic stress). The end condition is satisfied (cf. the first paragraph), and the von Mises yield criterion reduces exactly to (52). The assumption gives σ_z correctly when all the tube is elastic (see (27)), but increasingly overestimates it in the elastic region up to a maximum of about 18 per cent. when $c = b$ (Fig. 17). σ_z is underestimated throughout nearly all the plastic region, the error being worst on the inner surface. On the other hand, a useful

---

† R. Hill, E. H. Lee, and S. J. Tupper, Ministry of Supply, Armament Research Dept., Theoretical Research Rep. 11/46.
‡ G. Cook, *Phil. Trans. Roy. Soc.* A, **230** (1931), 103.
§ G. Cook, *Proc. Inst. Mech. Eng.* **126** (1934), 407.

approximation to the external radial displacement is effected. The accurate value of u_b is given by

$$\frac{Eu_b}{Yb} = \frac{1}{Y}[\sigma_\theta - \nu(\sigma_r + \sigma_z)]_{r=b}$$

$$= \tfrac{1}{2}(2-\nu)\frac{c^2}{b^2} - \nu(q)_{r=b}, \quad \text{from (34) and (40).}$$

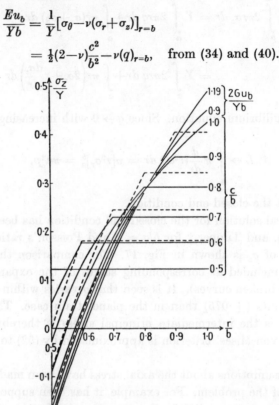

Fig. 17. Distribution of axial stress component in a partly plastic tube with a closed end and wall-ratio 2. The broken curves correspond to
$$\sigma_z = \tfrac{1}{2}(\sigma_r + \sigma_\theta).$$

$\nu(q)_{r=b}$ ranges from zero to -0.024 as the plastic region develops (see Fig. 17). Since the leading term varies correspondingly from 0·21 to 0·85, the formula

$$\frac{Eu_b}{Yb} = \tfrac{1}{2}(2-\nu)\frac{c^2}{b^2} \tag{53}$$

underestimates the true radial displacement (Fig. 18) by less than 3 per cent. The axial strain ϵ_z, calculated from (50) on the assumption that $q = 0$, or directly from the Reuss relations, will be found to vary across the plastic part of the tube. In the elastic part this fictitious axial strain is equal to $(1-2\nu)Yc^2/2Eb^2$, compared with the true value

$$\left\{\tfrac{1}{2}(1-2\nu)\frac{c^2}{b^2}+q\right\}\frac{Y}{E}.$$

The discrepancy (an overestimate) increases as the plastic region spreads, amounting to about 64 per cent. when $c = b$; the assumption therefore fails to give a useful estimate of ϵ_z. The true axial strain is shown in

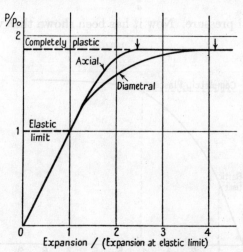

Fig. 18. Dependence of diametral and longitudinal expansions of a closed tube on the internal pressure (wall-ratio 2 and no strain-hardening).

Fig. 18 as a function of the autofrettage pressure and in Fig. 19 as a function of the external expansion. There is a discontinuity in gradient at the first moment of complete plasticity.

The elastic limit of the tube for applications of pressure subsequent to the initial overstrain is equal to the autofrettage pressure provided that there is no Bauschinger effect and that all elements of the tube recover elastically during the unloading. The condition for this is that p should be less than $2p_0$† (cf. the derivation of (17) for a spherical shell). It might be supposed that in real metals the presence of an appreciable Bauschinger effect after a small plastic strain would reduce the elastic limit of the tube below the autofrettage pressure. In fact, for reasons not well understood, autofrettaged tubes behave very nearly as though the Bauschinger effect were non-existent.‡

† L. B. Turner, *Trans. Camb. Phil. Soc.* **21** (1909), 377; *Engineering*, **92** (1911), 115.
‡ A. G. Warren, *Symposium on Internal Stresses in Metals and Alloys*, p. 209 (published by the Institute of Metals, 1947).

The accurate inclusion of work-hardening in the analysis would greatly complicate the solution, but a useful estimate of the pressure-expansion relation can be obtained by the following method, due to Nadai.† From the equation of equilibrium we derive the formula

$$p = \int_a^b (\sigma_\theta - \sigma_r) \frac{dr}{r}$$

for the internal pressure. Now it has been shown that von Mises' yield

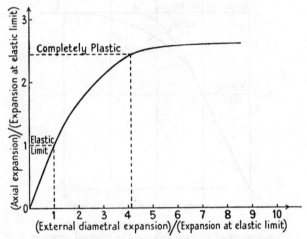

FIG. 19. Relation between the external diametral expansion and the longitudinal strain in a closed tube of wall-ratio 2 (no strain-hardening).

criterion can be replaced by the approximate relation $\sigma_\theta - \sigma_r \simeq 2Y/\sqrt{3}$ when the material does not harden. It may be expected that the relation will be equally valid when the material work-hardens, so that we can write

$$p \simeq \frac{2}{\sqrt{3}} \int_a^b Y \frac{dr}{r},$$

where Y is a function of the equivalent strain. Since the state of stress in any element is approximately a hydrostatic tension superposed on a pure shear, it is further supposed that Y is a function only of the maximum shear strain, to be determined from a torsion test on a thin-walled tube (where the shear stress τ is equal to $Y/\sqrt{3}$). Now the maximum (engineering) shear strain is $\gamma = \epsilon_\theta - \epsilon_r$, and this is proportional to $1/r^2$ in the elastic region. It is assumed, finally, that γ is proportional to $1/r^2$

† A. Nadai, *Plasticity*, chap. 29 (McGraw-Hill Book Co., 1931).

all through the tube at each stage of the expansion. Actually, $r^2\gamma$ increases as we pass from the plastic boundary to the bore, and according to the accurate solution the departure from constancy reaches 30 per cent. when the tube is completely plastic. However, it is found† that the error introduced into the calculated value of p by neglecting this variation is not more than one or two per cent., even when the stress-strain curve is well rounded. Hence, writing $\gamma = b^2\gamma_b/r^2$, where γ_b is the maximum shear strain on the external surface,

$$p = \int_{\gamma_b}^{\gamma_a} \tau(\gamma) \frac{d\gamma}{\gamma}, \qquad (54)$$

where $\gamma_a = b^2\gamma_b/a^2$ is the shear strain at the bore. From (35) and (53),

$$\gamma_b = (1+\nu)\frac{Yc^2}{Eb^2} = \frac{2(1+\nu)}{(2-\nu)}\frac{u_b}{b}, \quad \text{approximately.}\ddagger \qquad (55)$$

The pressure-expansion curve can then be calculated from (54) by numerical integration. A method for the converse calculation, that of deriving the stress-strain curve in shear from a known pressure-expansion relation, has been outlined by Shepherd.§

For a material with an upper yield-point Cook‖ has given the formula

$$p = 2s'\ln\frac{c}{a} + s\left(1 - \frac{c^2}{b^2}\right),$$

where s is the maximum shear stress in the material about to yield, and s' is the maximum shear stress throughout the plastic region; this may be proved by a simple modification of the equations leading to (37). There is a discontinuity across the plastic boundary in the circumferential stress. It is assumed that the plastic strains are restricted to the lower yield-point range of strain (this is usually of order 0·02, or some ten times the elastic strain at the upper yield-point). From experiments on closed tubes with wall-ratios b/a ranging from 1·17 to 4, Cook†† has shown that the value of s' deduced from the observed maximum pressure $2s'\ln b/a$ is in good agreement with that measured in a tension test. On the other hand, Morrison‡‡ has concluded that the lower yield stress is

---

† J. L. M. Morrison, *Proc. Inst. Mech. Eng.* **159** (1948), 81.
‡ Nadai assumed the material to be incompressible ($\nu = \frac{1}{2}$), so that $r^2\gamma$ was strictly constant.
§ W. M. Shepherd, ibid. 94.
‖ G. Cook, *Phil. Trans. Roy. Soc.* A, **230** (1931), 103.
†† G. Cook, *Proc. Inst. Mech. Eng.*, **126** (1934), 407; *Inst. Civil Eng.*, *Selected Engineering Paper No. 170* (1934); *Trans. Engineers and Shipbuilders in Scotland*, **81** (1937), 371.
‡‡ J. L. M. Morrison, op. cit.

governed by von Mises' criterion. It may be significant that the drop in stress was much greater in the mild steel used by Cook than in that used by Morrison. It is generally agreed, however, that the first yielding of mild steel is in accordance with Tresca's criterion, but only in the limited sense that under similar conditions the intermediate principal stress has no influence. The maximum shear stress is observed to be higher in torsion, flexure, and autofrettage, and Cook[†] is of the opinion that the local stress gradient influences yielding. This may also be the explanation of the scale effects found by Morrison[‡] in torsion.

(v) *Open-end condition.* At present an accurate numerical solution of the equations for a tube with open ends is not available. The solution is more difficult than for the closed tube in two respects.

In the first place, σ_z is not necessarily the intermediate principal stress. On general grounds it seems probable that for certain wall-ratios the axial stress may become locally less than the radial stress (i.e., more compressive). If this happens Tresca's criterion states that $\sigma_\theta - \sigma_z$, and not $\sigma_\theta - \sigma_r$, is constant. The fundamental equations are then entirely different from those we have been considering. However, it may be that the difference between σ_r and σ_z is so small that $\sigma_\theta - \sigma_r \simeq Y$ with sufficient accuracy. Also, the range of wall-ratios for which σ_r is the intermediate principal stress may not be usual in autofrettage practice.

The second difficulty appears to be more serious. The stress vector in the plane diagram must vary over a much wider range than for the closed tube. Consequently, von Mises' yield criterion may not be approximated by (52) with the required accuracy. In British design practice[§] this difficulty is met by assuming that $\sigma_\theta - \sigma_r = mY$, where m is an empirically determined quantity (cf. p. 21) ranging from 1 when b/a is small to $2/\sqrt{3}$ when b/a is very large.

(vi) *Solutions based on the Hencky stress-strain relations.* The analysis is less formidable when the stress-strain relations of Hencky ((36) of Chap. II) are adopted. By reason of the unique correspondence between stress and total strain which these relations assert, the difficulties involved in following the strain-history do not arise. The first complete solution in which compressibility is allowed for is due to Belayev and Sinitsky.[||]

[†] G. Cook, *Proc. Roy. Soc.* A, **137** (1932), 559; *Trans. Engineers and Shipbuilders in Scotland*, **81** (1937), 371.

[‡] J. L. M. Morrison, *Proc. Inst. Mech. Eng.* **142** (1940), 193.

[§] A. G. Warren, op. cit., p. 121.

[||] N. M. Belayev and A. K. Sinitsky, *Bull. Acad. Sci. U.R.S.S.* **2** (1938), 3; for the same theory extended to include strain-hardening, see ibid. **4** (1938), 21. See also W. W. Sokolovsky, *Prikladnaia Matematika i Mekhanika*, **7** (1943), 273; *Theory of Plasticity*, chap. 3 (Moscow, 1946).

A similar analysis has been independently formulated by Allen, and systematized for numerical calculation by Sopwith.†

Comparison of Sopwith's results with those of Figs. 16 and 17 shows that the Hencky relations lead to a distribution of axial stress which is not in error by more than about 5 per cent. of the maximum for both plane strain and closed-end conditions (the comparison for open ends is not yet possible). In the early stages of the expansion still closer agreement is found in the axial strain for the tube with closed ends, but the axial strain based on the Hencky relations eventually reaches a maximum value and then decreases. The reason for this rather striking difference is not difficult to understand. Both theories predict that σ_z tends to $\frac{1}{2}(\sigma_r+\sigma_\theta)$ with increasing plastic strain, and hence that the elastic component of the axial strain in each element approaches a constant value (since σ_r and σ_θ are constant after the tube becomes plastic, the displacements being small). But whereas the *increment* of axial plastic strain is zero according to the Reuss equations, it is the *total* axial plastic strain that is zero according to the Hencky equations. Consequently, in the former theory the axial strain asymptotically approaches a certain limiting value (in agreement with the experiments of Cook‡) while in the latter it must eventually reach a maximum and afterwards decrease.

It was proved in Section 6 of Chapter II that when the strain-path is such that the stress vector remains fixed in direction the Reuss and Hencky relations are identical. It seems, therefore, that the reason for the close agreement of the Reuss and Hencky theories in the expansion of a tube under plane strain or with closed ends is to be found in the circumstance that the stress vector is restricted to a small segment of the yield circle. A less good agreement would be expected for the open-end condition.

4. Expansion of a cylindrical cavity in an infinite medium

A solution in explicit terms can be found for the problem of the expansion of a cylindrical cavity from zero radius in an infinite medium.§ The axial strain is zero, and all dependent variables are functions of the single quantity $\theta = r/c$, since the configuration of stress and strain remains geometrically similar. We have also

$$c\frac{\partial}{\partial r} = \frac{d}{d\theta}, \qquad c\frac{\partial}{\partial c} = -\theta\frac{d}{d\theta}, \qquad c\frac{d}{dc} = (v-\theta)\frac{d}{d\theta}.$$

† D. N. de G. Allen and D. G. Sopwith, *Proc. 7th Int. Cong. App. Mech.*, London (1948). See also C. W. MacGregor, L. F. Coffin, Jr., and J. C. Fisher, *Journ. Franklin Inst.*, 245 (1948), 135.
‡ G. Cook, *Proc. Inst. Mech. Eng.* 126 (1934), 407.
§ R. Hill, *Journ. App. Mech.* 16 (1949), 295.

Equation (41) reduces to

$$v' + \frac{v}{\theta} = (1-2\nu)\frac{Y}{E}[(v-\theta)q' - 3], \qquad (56)$$

where the dash denotes differentiation with respect to θ. From (42)

$$[1 - \tfrac{2}{3}(1-2\nu)q](v-\theta)q' + (1-2\nu)(2q-1) + \left[\tfrac{4}{3}q + (1-2\nu)\frac{Y}{E}\right]\frac{Ev}{Y\theta} = 0. \qquad (57)$$

Now $\sigma_r + \sigma_\theta$ is zero in the elastic region (see (34)), and hence so are σ_z and q. We can expect that q will be very small near the surface of the cavity where the circumferential strain is infinitely large. It can be verified *a posteriori* that the numerically greatest value of q' occurs when $\theta = 1$, and that it is about 0·4. Since θ ranges from a/c to 1 in the plastic region, while v is of order Y/E or less, it is therefore a good approximation to neglect $(v-\theta)q'$ in comparison with 3 in (56). Integration of this equation with the boundary condition $v = (1+\nu)Y/E$ on $\theta = 1$ (see (44)) then gives

$$v = -\frac{3(1-2\nu)Y\theta}{2E} + \frac{(5-4\nu)Y}{2E\theta}; \qquad \frac{a}{c} \leqslant \theta \leqslant 1. \qquad (58)$$

Since $v = da/dc$ on the surface of the cavity and since c/a is constant, we derive immediately from (58) the expression

$$\frac{c}{a} = \left[\frac{2E}{(5-4\nu)Y}\right]^{\frac{1}{2}} \qquad (59)$$

to the usual order of approximation. Substituting (58) in (57), we obtain a differential equation for q:

$$q' - \frac{2(5-4\nu)q}{3\theta(\theta^2 - a^2/c^2)} + \frac{1-2\nu}{\theta} = 0, \qquad (60)$$

where, for simplicity, $\tfrac{2}{3}(1-2\nu)q$ has been neglected in comparison with unity in the coefficient of q' (the maximum value of q is usually less than 0·1). Integration, with the boundary condition $q = 0$ when $\theta = 1$, leads to

$$q = (1-2\nu)\left(1 - \frac{a^2}{c^2\theta^2}\right)^{2E/3Y} \int_\theta^1 \left(1 - \frac{a^2}{c^2\theta^2}\right)^{-(2E/3Y)} \frac{d\theta}{\theta} \qquad \left(\frac{a}{c} \leqslant \theta \leqslant 1\right). \qquad (61)$$

As mentioned on p. 115, the above distributions of v and q are also those round a cavity expanded from any *finite* radius. A numerical evaluation of the integral in (61) shows that q increases to a maximum near $r/c = 0·7$† as we move inward from the plastic boundary, and then

† The position of the maximum is insensitive to changes in E/Y; when $\nu = 0·3$ and $E/Y = 400$, the maximum value of q is 0·073, approximately.

decreases rapidly to zero at the surface. This is another example of the general principle that in plane deformation σ_z closely approaches the mean of the other two principal stresses after a plastic strain of a few times the yield-point strain. Since the elastic and plastic components of the strain are still comparable even when $r/c = 0.5$, it is understandable that q is only negligible compared with $\frac{1}{2}(\sigma_r+\sigma_\theta)/Y$ ($= \ln\theta$) in a relatively small part of the plastic region.

If the material yields according to von Mises' criterion, Y must be replaced by $2Y/\sqrt{3}$ in the above equations; the overall approximation is better than for a tube of moderate thickness. For most pre-strained metals c/a is between 10 and 15, while the internal pressure

$$p = \frac{Y}{\sqrt{3}}\left[1+\ln\frac{\sqrt{3}E}{(5-4\nu)Y}\right] \qquad (62)$$

is between about $3\cdot 3$ and $3\cdot 8Y$. An earlier analysis by Bishop, Hill, and Mott,† neglected volume changes in the plastic region, but, as for the spherical cavity, it is a better approximation to neglect volume changes universally.

When the material work-hardens, an analysis similar to that for the spherical cavity (Section 1 (v)) leads to a pressure

$$p = \frac{Y}{\sqrt{3}}\left[1+\ln\frac{E}{\sqrt{3}Y}\right] + \frac{2}{\sqrt{3}}\int_1^{c/a} H\left(\frac{1}{\sqrt{3}}\ln\frac{t^2}{t^2-1}\right)\frac{dt}{t}; \quad t = \frac{r}{a}; \qquad (63)$$

where $\sigma = Y+H(\epsilon)$ is the stress-strain curve in compression. If the rate of hardening is constant, so that $H = H'\epsilon$, then

$$p = \frac{Y}{\sqrt{3}}\left[1+\ln\frac{E}{\sqrt{3}Y}\right] + \frac{\pi^2 H'}{18}. \qquad (64)$$

† R. F. Bishop, R. Hill, and N. F. Mott, *Proc. Phys. Soc.* **57** (1945), 147.

VI

PLANE PLASTIC STRAIN AND THE THEORY OF THE SLIP-LINE FIELD

1. Assumption of a plastic-rigid material

IN the last two chapters we have been considering problems to which exact solutions are known, and where full allowance can be made for the elastic component of strain in the plastic region. In many of the problems of greatest practical interest, however, we are compelled by mathematical difficulties to disregard the elastic component of strain. For consistency we must also disregard the purely elastic strain in the non-plastic region. In effect, therefore, we work with a material that is rigid when stressed below the yield-point and in which Young's modulus has an infinitely great value. This hypothetical solid may be referred to as a *plastic-rigid* material, in contrast to the plastic-elastic material of Reuss.

It will have become clear by now that the distribution of stress in the plastic-rigid body is only likely to approximate that in a real metal under similar external conditions when the plastic material has freedom to flow in some direction. If the plastic material is severely constrained by adjacent elastic material (as in the expansion of a thick-walled tube, or in the bending of a beam), neglect of the elastic component of strain introduces serious errors into *certain* of the calculated stress components. On the other hand, even though an easy direction of flow is available, so that the elastic strain-increments soon become negligible throughout most of the plastic zone, there must still be a certain boundary layer, or transition region, bordering the elastic zone, in which the elastic and plastic strain-increments are comparable. The narrower this transition region, the better should be the overall approximation. Since the allowable error depends very much on the intended field of application of the solution, no more explicit rule can be laid down. In many technological forming processes (e.g. rolling, drawing, forging) experience shows that the assumption of a plastic-rigid material does not lead to any significant errors.

In the present chapter we shall be concerned with the behaviour of a plastic-rigid material under conditions of plane strain, and in particular with certain general properties of the stress and velocity distribution in the plastic region.

2. The plane strain equations referred to Cartesian coordinates

A state of plane strain is defined by the properties (i) that the flow is everywhere parallel to a given plane, say the (x, y) plane, and (ii) that the motion is independent of z. We are then dealing with a two-dimensional system (the page representing a typical (x, y) plane of flow), in which the non-vanishing velocity components are u_x and v_y, where the subscripts x, y, z are used to distinguish Cartesian components of velocity from the curvilinear components that will be employed later. Since τ_{xz} and τ_{yz} are zero by symmetry, σ_z is a principal stress.

If the hydrostatic component of the stress is compressive, a state of plane strain may be achieved by confining the specimen between (effectively) rigid blocks with faces parallel to the planes of flow. The two surfaces of contact must be well lubricated to allow the specimen to expand or contract freely. The dimensions of the specimen, and the forces or displacements applied to it, must be independent of the z coordinate. These conditions can generally only be exactly realized in a specially designed experiment. However, approximate states of plane strain are found in technological processes, for example when a wide sheet or block is reduced in thickness by passing it between cylindrical rolls or dies. If the percentage lateral spread is small in comparison with the percentage reduction in thickness, the sheet is deformed under approximately plane strain conditions, with the exception of narrow zones near the edges.

Since the volume of an element of plastic-rigid material does not alter, each incremental distortion in a state of plane strain consists of a pure shear. Hence, for the ideal isotropic material, the state of stress at each point is a pure shear stress τ, together with a hydrostatic pressure p (see p. 36, Chap. II). The stress σ_z normal to the planes of flow is therefore equal to $-p$, and the other principal stresses to $-p \pm \tau$. Expressed in Cartesian coordinates:

$$\sigma_z = \tfrac{1}{2}(\sigma_x + \sigma_y). \tag{1}$$

This relation, strictly true only for a plastic-rigid material undergoing deformation, is found to be very nearly satisfied for a plastic-elastic material after a comparatively small plastic distortion in plane strain, provided $\mu = 0$ when $\nu = 0$ (Lode's variables); examples of this, for the Reuss material, have been discussed in Chapter IV, Section 5, and in Chapter V, Section 3 (ii); it is rigorously true when Poisson's ratio is $\tfrac{1}{2}$. The theory of plane strain based on (1) is therefore often applicable to a part of the plastic region in a plastic-elastic material.

If there is no work-hardening, and if the yielding is not influenced by hydrostatic pressure, τ must be a constant which we shall denote by k, its precise value depending on the yield criterion. In Tresca's criterion, for example, k is equal to $Y/2$, while for von Mises' it is equal to $Y/\sqrt{3}$ (exactly as in a state of pure shear). All cases can be treated by the same theory if the yield criterion is written in the form

$$\tau^2 = \tfrac{1}{4}(\sigma_x-\sigma_y)^2+\tau_{xy}^2 = k^2, \tag{2}$$

and k is given the appropriate value in the final result. As usual, it is assumed that inertial stresses are negligible compared with k, and that the problem can be treated as quasi-static. In the absence of body forces, the equations of equilibrium reduce to

$$\frac{\partial \sigma_x}{\partial x}+\frac{\partial \tau_{xy}}{\partial y} = 0, \qquad \frac{\partial \tau_{xy}}{\partial x}+\frac{\partial \sigma_y}{\partial y} = 0. \tag{3}$$

The condition for zero volume change is

$$\frac{\partial u_x}{\partial x}+\frac{\partial v_y}{\partial y} = 0. \tag{4}$$

The slope θ of the principal axes of stress with respect to the x-axis is given (Appendix IV) by

$$\tan 2\theta = \frac{2\tau_{xy}}{\sigma_x-\sigma_y}, \tag{5}$$

while the slope θ' of the principal axes of strain-increment, or strain-rate, satisfies

$$\tan 2\theta' = \frac{\dfrac{\partial u_x}{\partial y}+\dfrac{\partial v_y}{\partial x}}{\dfrac{\partial u_x}{\partial x}-\dfrac{\partial v_y}{\partial y}}. \tag{6}$$

In an isotropic material the principal axes of stress and plastic strain-rate must coincide. Hence $\theta = \theta'$, and (5) and (6) give

$$\frac{2\tau_{xy}}{\sigma_x-\sigma_y} = \frac{\dfrac{\partial u_x}{\partial y}+\dfrac{\partial v_y}{\partial x}}{\dfrac{\partial u_x}{\partial x}-\dfrac{\partial v_y}{\partial y}}. \tag{7}$$

Since equations (4) and (7) are homogeneous in the velocities, the theory does not really involve the element of time and the calculated stresses are independent of the rate of strain; the progress of the deformation can be marked by any monotonically varying quantity, such as a load, an angle, or a characteristic length. The set of five equations (2), (3), (4), and (7), in the five unknowns σ_x, σ_y, τ_{xy}, u_x, v_y, is the basis for the calculation of the distribution of stress and strain in the plastic region in the plane problem, and was first formulated by Saint-Venant† in 1870. Saint-Venant used Tresca's criterion, and later derived equation (1) from Lévy's stress-strain relations.

It will be noticed that there are three equations involving only the three components of stress; when these have been found, the remaining two equations suffice to determine the velocity components. We can foresee, therefore, that two main types of plane strain problem will be encountered. In the first there will be a number of boundary conditions, involving only the stresses, sufficient to permit a determination of the plastic region and state of stress without considering the velocities. Such a problem is statically determined; the velocities can be calculated afterwards. In the second type there will usually be fewer boundary conditions in the stresses alone, and a compensatory greater number involving the velocities. The solutions for the stresses and velocities have then to be carried out together, and so this type of plane problem is not statically determined;‡ in certain statically undetermined problems *all* boundary conditions refer to stresses. The failure of many writers to apply this distinction in the solution of special problems has resulted in much unsatisfactory work; this is considered in detail later.

Finally, we need the equations for the stresses in the elastic region. The equations (3) of equilibrium are the same, but the yield criterion is replaced by the compatibility equation§

$$\left(\frac{\partial^2}{\partial x^2}+\frac{\partial^2}{\partial y^2}\right)(\sigma_x+\sigma_y) = 0. \tag{8}$$

This expresses the condition that the components of strain should be derivable from two continuous functions representing the x and y components of displacement. It is important to notice that the elastic

---

† B. de Saint-Venant, *Comptes Rendus Acad. Sci. Paris*, **70** (1870), 473; **74** (1872), 1009 and 1083.

‡ The term 'statically determined' was introduced by Hencky in 1923 to denote problems where there are as many equations in the stresses alone as unknown stresses, *irrespective of the boundary conditions*. The term has been used in this wider sense by many later writers, including the author. However, in view of the confusion that has existed in regard to the plane strain problem, it seems preferable, in an introduction to the subject, to redefine the term in the narrower sense, which is also more in accord with the literal meaning of the words. While in Hencky's sense, therefore, all problems of plane strain are statically determined, in the new sense only some are. This at once underlines the distinction between the two types of boundary-value problem, a distinction that the older terminology undoubtedly went far to obscure. The reader should take care, when consulting original papers, to ascertain which meaning is intended.

§ S. Timoshenko, *Theory of Elasticity*, p. 23 (McGraw-Hill Book Co., 1934). R. V. Southwell, *Introduction to the Theory of Elasticity*, chap. xii (Clarendon Press, 1936).

constants do not appear in the compatibility relation, which is therefore equally valid for the plastic-rigid material. We shall suppose that $\nu = \frac{1}{2}$, so that $\sigma_z = \frac{1}{2}(\sigma_x + \sigma_y)$ in the elastic region and hence throughout the whole body. Now for equilibrium the components of stress acting normally and tangentially over the plastic-elastic interface must be continuous. Furthermore, the material just on the elastic side of the interface must be on the point of yielding. The stress component parallel to the interface (i.e. acting normally on a plane element perpendicular to the interface) is then also continuous, and the plastic boundary coincides with one of the contours of maximum shear stress in the elastic region, namely the one along which the maximum shear stress has the value k.

Further consideration of the interrelation of the elastic and plastic states of stress will be left until later. The remainder of this chapter is concerned with the properties of the plastic stress and velocity equations, and the methods by which they are integrated.

3. The plane strain equations referred to the slip-lines

In order to devise methods of integration of the plastic equations we must first inquire whether they are hyperbolic.† Suppose that, starting from certain prescribed boundary conditions, values of the stresses and velocities have been obtained everywhere within the area bounded by a closed curve C, and that we wish to find when this solution cannot be extended outside C without additional information. In other words, we are seeking the condition that C should be a characteristic. Instead of applying the standard method, it is simpler to take advantage of the fact that the stress and velocity equations have the same form whatever the axes of reference. Let these be chosen so that the x-axis is directed along the outward normal at some point P on C, and the y-axis along the tangent (Fig. 20). We now fix our attention on the variation of the stress and velocity in the neighbourhood of P, and see whether the x derivatives of the stress and velocity components just outside C are uniquely determined by the basic equations (2), (3), (4), (7), and the known plastic state within C. If the x derivatives can be

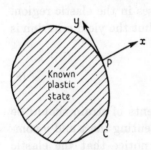

FIG. 20. Coordinate axes for investigation of characteristics in plane strain.

† The reader unfamiliar with this term will find it explained further in Appendix III. The present section is entirely self-contained, however.

so determined, the solution can be extended outwards through a further infinitesimal distance along the normal through P. On the other hand, if the equations fail to give any information about some, or all, of the derivatives, the solution cannot be extended in the neighbourhood of P unless further boundary conditions are specified outside C. If this is done, the derivatives will not necessarily be continuous.

For equilibrium σ_x and τ_{xy} must be continuous across C. Two values of σ_y can then be found to satisfy the yield criterion (2), one of these being, of course, the known value just inside C. We shall suppose here that σ_y is continuous.† Then the tangential derivatives $\partial\sigma_x/\partial y$, $\partial\sigma_y/\partial y$, and $\partial\tau_{xy}/\partial y$, are continuous across C. It follows from the equilibrium equations (3) that $\partial\sigma_x/\partial x$ and $\partial\tau_{xy}/\partial x$ are also continuous. To calculate $\partial\sigma_y/\partial x$ we differentiate the yield criterion to give

$$\tfrac{1}{4}(\sigma_x-\sigma_y)\left(\frac{\partial\sigma_x}{\partial x}-\frac{\partial\sigma_y}{\partial x}\right)+\tau_{xy}\frac{\partial\tau_{xy}}{\partial x}=0.$$

All quantities in this equation, except $\partial\sigma_y/\partial x$, are already known on both sides of C. If $\sigma_x \neq \sigma_y$ there is a unique solution for $\partial\sigma_y/\partial x$, which must therefore be continuous across C. However, if $\sigma_x = \sigma_y$, the equation gives no information about $\partial\sigma_y/\partial x$, and states that $\partial\tau_{xy}/\partial x = 0$ at P. By differentiating the yield criterion with respect to y it follows similarly that $\partial\tau_{xy}/\partial y = 0$ at P. From the equilibrium equations we have also that

$$\frac{\partial\sigma_x}{\partial x} = 0 = \frac{\partial\sigma_y}{\partial y} \quad \text{when} \quad \sigma_x = \sigma_y. \tag{9}$$

Now $\sigma_x = \sigma_y$ is the condition that the directions of maximum shear stress at P should be normal and tangential to C. Thus if the tangent at every point on C coincides with a maximum shear stress direction C is a characteristic for the stresses. The solution can only be continued across such a curve if further boundary conditions are prescribed; these may be such that $\partial\sigma_y/\partial x$ is discontinuous at each point. For a general set of Cartesian axes, not coinciding with the directions of maximum shear stress at the point under consideration, the normal derivatives of *all* the corresponding stress components are obviously discontinuous.

Since the density does not change, conservation of mass requires that the normal component of velocity must be continuous across any curve. On the other hand, *if* the boundary conditions could be suitably chosen, the tangential component could be discontinuous without introducing any *mathematical* inconsistency provided the plastic equations are satisfied on both sides of the curve. However, physical requirements

† Plastic states with discontinuities in the stress are examined in Sect. (8).

demand that we regard the plastic-rigid material as a real material in which the elastic moduli are allowed to increase without limit. A curve across which the tangential component of velocity is discontinuous must be regarded as the limit of a narrow transition region in which the rate of shear strain in the tangential direction is very large. Since directions of maximum shear stress and shear strain-rate coincide, the tangent at any point of the curve must be a maximum shear stress direction. However, we have still to prove that it is possible so to choose the boundary conditions that the tangential velocity component *can*, in fact, be discontinuous across such a curve. This is proved later (Sect. 6 (i)); for the moment we suppose that both components of velocity are always continuous. It follows that the derivatives $\partial u_x/\partial y$ and $\partial v_y/\partial y$ are continuous, and hence also $\partial u_x/\partial x$ by virtue of the incompressibility equation (4). Equation (7) serves to determine $\partial v_y/\partial x$ uniquely, unless $\sigma_x = \sigma_y$, in which case

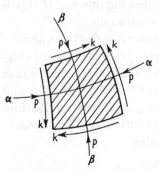

FIG. 21. Stresses on a small curvilinear element bounded by slip-lines; p = mean compressive stress, k = yield stress in shear.

$$\frac{\partial u_x}{\partial x} = 0 = \frac{\partial v_y}{\partial y} \qquad (10)$$

at P, and $\partial v_y/\partial x$ may be discontinuous.

Hence the characteristics of the stresses and the velocities coincide, and there are only two distinct characteristic directions at a point, viz. the directions of maximum shear stress or shear strain-rate (four would normally be expected from a set of four partial differential equations). In a state of plane strain the two orthogonal families of curves whose directions at every point coincide with those of the maximum shear strain-rate are known as *slip-lines*. To sum up: the equations of plane strain are hyperbolic, and the characteristics are the slip-lines.

This property compels us, if we wish to solve special problems, to treat the field of slip-lines as the fundamental unknown element to be determined. With this in view, we proceed to develop the differential relations holding along the slip-lines. Let the two families of slip-lines be labelled by the parameters α and β respectively; for shortness we shall speak of α-lines and β-lines. A small curvilinear element bounded by two pairs of neighbouring slip-lines is acted on by the system of stress components shown in Fig. 21. The normal stress on each face is equal to $-p$, where

p is the mean compressive stress $-\tfrac{1}{2}(\sigma_1+\sigma_2)$, σ_1 and σ_2 being the principal stresses $-p\pm k$ (from (1) it follows that p is also equal to $-\tfrac{1}{3}(\sigma_1+\sigma_2+\sigma_3)$, where $\sigma_3 \equiv \sigma_z$ is the third principal stress). The shear-stress components are equal to k. We distinguish the two families of slip-lines by the convention that, if the α- and β-lines are regarded as a pair of right-handed curvilinear axes of reference, the line of action of the algebraically greatest principal stress falls in the first and third quadrants. Evidently the state of stress at a point is completely specified if we know the orientation of the slip-lines and the value of p. Thus, if ϕ is the *anticlockwise* angular rotation of the α-line from the x-axis, the Cartesian components of stress are

$$\left.\begin{aligned}\sigma_x &= -p-k\sin 2\phi, \\ \sigma_y &= -p+k\sin 2\phi, \\ \tau_{xy} &= k\cos 2\phi.\end{aligned}\right\} \qquad (11)$$

These expressions, originally due to Lévy, may be immediately verified, either from Mohr's circle of stress (which is always of radius k in a plane plastic state), or from the equations transforming stress components from one set of axes to another (see Appendix IV).

Now regard a β-line as the curve C, and take the x- and y-axes coincident with the tangents to the α- and β-lines respectively. Substituting the expressions (11) into (9) we obtain

$$\left\{\frac{\partial}{\partial x}(p+k\sin 2\phi)\right\}_{\phi=0} = 0, \qquad \left\{\frac{\partial}{\partial y}(p-k\sin 2\phi)\right\}_{\phi=0} = 0.$$

Hence $\qquad \dfrac{\partial p}{\partial x}+2k\dfrac{\partial \phi}{\partial x} = 0, \qquad \dfrac{\partial p}{\partial y}-2k\dfrac{\partial \phi}{\partial y} = 0.$

This means that at the point P the tangential derivatives of the quantities $p+2k\phi$ and $p-2k\phi$ along the α- and β-lines, respectively, are zero. This obviously remains true if ϕ is measured relative to any fixed direction, not necessarily parallel to the tangent at P. Since P is any point it follows that

$$\left.\begin{aligned}p+2k\phi &= \text{constant on an } \alpha\text{-line,} \\ p-2k\phi &= \text{constant on a } \beta\text{-line.}\end{aligned}\right\} \qquad (12)$$

These relations are completely equivalent to the equilibrium equations. In general, of course, the values of the constants vary from one slip-line to another. The relations are usually attributed to Hencky[†] (1923),[‡]

[†] H. Hencky, *Zeits. ang. Math. Mech.* 3 (1923), 241.

[‡] Hencky derived the relations by referring the equilibrium equations to the slip-lines as curvilinear coordinates. J. Boussinesq, *Comptes Rendus Acad. Sci. Paris*, 74 (1872), 242 and 450, and *Ann. Ec. Norm. Sup.* 35 (1918), 70, and M. A. Sadowsky,

though similar ones, including these as a special case, were derived by Kötter† as early as 1903 for a plastic stress state in a soil; a soil is idealized as a plastic-rigid material whose yielding is influenced by hydrostatic pressure, and whose weight must be taken into account (see Chap. XI, Sect. 3).

To express the velocity equations in a similar way, we introduce velocity components u and v referred to the α- and β-lines (considered as a right-handed system of curvilinear coordinates). The equations analogous to (11) are

$$\left.\begin{array}{l} u_x = u\cos\phi - v\sin\phi, \\ v_y = u\sin\phi + v\cos\phi. \end{array}\right\} \quad (13)$$

Substituting these in (10):

$$\left\{\frac{\partial}{\partial x}(u\cos\phi - v\sin\phi)\right\}_{\phi=0} = 0, \qquad \left\{\frac{\partial}{\partial y}(u\sin\phi + v\cos\phi)\right\}_{\phi=0} = 0.$$

Proceeding in a way similar to the derivation of (12), we find that

$$\left.\begin{array}{l} du - v\,d\phi = 0 \text{ along an } \alpha\text{-line,} \\ dv + u\,d\phi = 0 \text{ along a } \beta\text{-line.} \end{array}\right\} \quad (14)$$

These equations are due to Geiringer‡ (1930). They are nothing more than a statement that the rate of extension along any slip-line is zero.

4. Geometry of the slip-line field

A field of slip-lines possesses several striking geometrical properties which are repeatedly employed in the solution of special problems.

(i) *Hencky's first theorem.* Consider a curvilinear quadrilateral $ABQP$ (Fig. 22) bounded by two α-lines, AP and BQ, and by two β-lines, AB and PQ. From (12):

$$p_Q - p_A = (p_Q - p_B) + (p_B - p_A) = 2k(2\phi_B - \phi_Q - \phi_A).$$

Also $\qquad p_Q - p_A = (p_Q - p_P) + (p_P - p_A) = 2k(\phi_Q + \phi_A - 2\phi_P).$

Therefore $\qquad\qquad \phi_Q - \phi_P = \phi_B - \phi_A.$ $\qquad(15)$

This is the expression of the necessary fact that the pressure difference between two points, calculated in the two possible ways, is the same. Equation (15) is known as Hencky's first theorem: it states that the angle

Trans. Am. Soc. Mech. Eng. **63** (1941), A–74, have used the trajectories of principal stress as curvilinear coordinates. The resulting equations are not useful for solving problems, since the slip-lines are the natural system of coordinates.

† F. Kötter, *Berlin Akad. Berichte* (1903), 229. For other accounts of the work of Hencky and Kötter, see A. Nadai, *Handbuch der Physik*, **6** (1928); H. Geiringer and W. Prager, *Ergebnisse d. exakten Naturwiss.* **13** (1934), 310; H. Geiringer, *Mém. Sci. Math.* **86** (1937).

‡ H. Geiringer, *Proc. 3rd Int. Cong. App. Mech. Stockholm*, **2** (1930), 185.

between two slip-lines of one family, where they are cut by a slip-line of the other family, is constant along their length. In other words, if we pass from one slip-line to another of the same family, along any intersecting slip-line, the angle turned through and the change in pressure are constant. Conversely, any two orthogonal families of curves with this

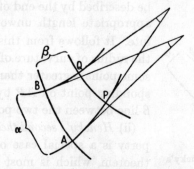

Fig. 22. Typical slip-line field, demonstrating Hencky's first theorem.

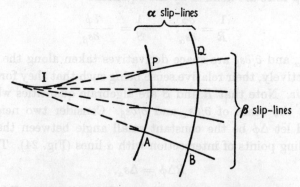

Fig. 23. Slip-lines of one family with a straight segment. Members of the other family intersecting the straight segments have a common evolute, to which the segments are tangential.

property constitute a possible slip-line field for a plastic mass in quasi-static equilibrium under certain boundary conditions; if the value of the hydrostatic pressure p is stated at one point it may be calculated everywhere in the field by means of either of the equations (12).

It follows immediately that if a segment AB of a β-line, say, is straight (Fig. 23), then so is the corresponding section PQ of any other β-line cut off by the α-lines through A and B. Notice that according to (12) and (14) p and v are both constant on each of the straight segments. If both families are straight in a certain region, the stress (but not necessarily the velocity) is uniform throughout the region (e.g. a state of simple shear). A further property of the field in Fig. 23 is that the straight segments all have the same length. This is most easily proved by noting that the α-lines cutting the segments have the same evolute I, since the β-lines are their common normals. An α-line cutting AB can therefore be described by the end of a taut string of an appropriate length unwound from the evolute. It follows from this construction that the radius of curvature of the slip-line BQ at some point is greater than that at the corresponding point on AP by the length of the β-line between the two points.

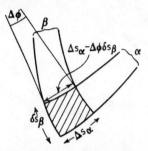

FIG. 24. Proof of Hencky's second theorem.

(ii) *Hencky's second theorem.* The last property is a special case of Hencky's second theorem, which is most directly proved as follows. The radii of curvature R and S of the α- and β-lines are defined by the equations

$$\frac{1}{R} = \frac{\partial \phi}{\partial s_\alpha}, \qquad \frac{1}{S} = -\frac{\partial \phi}{\partial s_\beta}, \tag{16}$$

where $\partial/\partial s_\alpha$ and $\partial/\partial s_\beta$ are space derivatives taken along the α- and β-lines respectively, their relative sense being such that they form a right-handed pair. Note that R and S are algebraic quantities whose signs depend on the sense of $\partial/\partial s_\alpha$ and $\partial/\partial s_\beta$. Consider two neighbouring β-lines and let $\Delta\phi$ be the constant small angle between them at the corresponding points of intersection with α-lines (Fig. 24). Then, from (16),
$$R\Delta\phi = \Delta s_\alpha,$$
where Δs_α is the length of the small arc of an α-line cut off by the two β-lines. Hence

$$\frac{\partial}{\partial s_\beta}(R\Delta\phi) = \frac{\partial}{\partial s_\beta}(\Delta s_\alpha) = -\Delta\phi$$

from the geometry of Fig. 24. Since $\Delta\phi$ is constant it may be taken outside the operator on the left-hand side, and we obtain

$$\left.\begin{aligned}\frac{\partial R}{\partial s_\beta} &= -1. \\ \text{Similarly,} \quad \frac{\partial S}{\partial s_\alpha} &= -1.\end{aligned}\right\} \tag{17}$$

These equations state that as we travel along a slip-line the radii of curvature of the slip-lines of the other family at the points of intersection change by the distance travelled.† The radius of curvature *diminishes* in magnitude as we travel towards the *concave* side of a slip-line, and all slip-lines of the same family are concave in the same direction. Hence, if the plastic state extends sufficiently far, the radii of curvature must eventually become zero. This means that neighbouring slip-lines run together, their envelope being a natural boundary of the analytic solution. A degenerate example of this, frequent in applications, is the field consisting of radii and concentric circular arcs, which obviously satisfies the condition (15). The envelope is here only a single point, which is also a singularity of the stress distribution.

An alternative form of (17), more useful for numerical computation, is derived by replacing the arc lengths Δs_α and Δs_β by $R\Delta\phi$ and $-S\Delta\phi$ respectively:

$$\left. \begin{array}{l} dS + R\,d\phi = 0 \text{ along an } \alpha\text{-line,} \\ dR - S\,d\phi = 0 \text{ along a } \beta\text{-line.} \end{array} \right\} \quad (18)$$

These relations are analogous to those previously obtained for the variation of p, u, and v, with the angle ϕ.

If the derivatives of the stresses are discontinuous across a slip-line, so also is the curvature of slip-lines of the other family. This may be proved in the following way. With the notation and coordinate axes of Section 3, let $\partial\sigma_y/\partial x$ be discontinuous across a β-line (whose tangent coincides with the y-axis). Now from (11)

$$\frac{\partial \sigma_y}{\partial x} = \left\{ \frac{\partial}{\partial x}(-p + k \sin 2\phi) \right\}_{\phi=0} = -\frac{\partial p}{\partial x} + 2k\frac{\partial \phi}{\partial x}.$$

But from (12) $$\frac{\partial p}{\partial x} + 2k \frac{\partial \phi}{\partial x} = 0.$$

Hence $$\frac{\partial \sigma_y}{\partial x} = -2 \frac{\partial p}{\partial x} = 4k \frac{\partial \phi}{\partial x}.$$

But $\partial\phi/\partial x \equiv \partial\phi/\partial s_\alpha = 1/R$ from (16). Thus, if the derivatives of the stresses are discontinuous across a β-line so is the curvature of the α-lines; similarly, if the stress derivatives are discontinuous across an α-line so is the curvature of the β-lines. Moreover, the jump in the radius of curvature is of constant amount all along the slip-line, since equation (17) holds in the fields on each side. Such a discontinuity in curvature is shown in Fig. 23 where the β-lines are straight on one side of AP, or BQ, and curved on the other. In general, the plastic region consists of a

† The equation $\partial(\Delta s_\alpha)/\partial s_\beta = -\Delta\phi$ may obviously be interpreted as stating that the distance between two *neighbouring* slip-lines varies linearly with their arc length; W. Prager, *Rev. Fac. Sci. Univ. Istanbul*, 4 (1939), 22.

number of subsidiary domains, across the boundaries of which the curvature of the slip-lines changes abruptly; this gives to the slip-line field a typical patchwork appearance.

5. The numerical calculation of slip-line fields

We now consider methods of constructing the slip-line field from known boundary conditions for the stresses, using the geometrical properties just established. Several distinct types of construction are

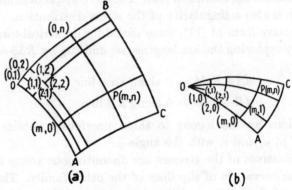

FIG. 25. First boundary-value problem; (a) given the positions of two intersecting slip-lines OA and OB, or (b) one slip-line OA and a point singularity O.

found to occur repeatedly in applications, usually in the same problem in different parts of the plastic region.

(i) *Two intersecting slip-lines given.* Suppose that the positions of the segments OA and OB of two intersecting slip-lines are known (Fig. 25 a); they may be regarded as the limit of a previously calculated slip-line field in some area to the left. It is required to calculate the slip-line field to the right of OA and OB, assuming that this region is also plastic. According to the theory of characteristics the field is uniquely determined within a curvilinear quadrilateral $OACB$ bounded by OA, OB, and the two intersecting characteristics (slip-lines) AC and BC through A and B. This is proved rigorously later; we are now concerned to formulate an approximate method of calculating the slip-line field. If one of the given slip-lines, say the β-line OB, is straight, all β-lines in $OACB$ are straight, as we have seen. In this case there is no difficulty in constructing the field: the β-lines are simply the normals to OA, and the α-lines are their orthogonal trajectories, each spaced a constant distance from OA. When both slip-lines are curved, suppose that OA is an α-line and let each of the segments OA and OB be subdivided into an arbitrary number of small arcs by the points $(1, 0), (2, 0),..., (m, 0),...,$ and $(0, 1), (0, 2),..., (0, n),...,$ respectively. The slip-lines through these base points form a network of which a general nodal point (m, n) is the meet of the slip-lines from the points $(m, 0)$ and $(0, n)$. By applying equation (15) we can immediately determine the angle ϕ for all points on the network: thus,

$$\phi_{m,n} = \phi_{m,0} + \phi_{0,n} - \phi_{0,0}.$$

It is often convenient to choose the base points a constant angular distance $\Delta\phi$ apart; the angle turned through in passing between any two adjacent nodal points is then constant and equal to $\Delta\phi$ everywhere in the field ($\Delta\phi$ denotes the *numerical* value of the change in angle). Such *equiangular nets* have a property that is very helpful in visualizing the distribution of the mean pressure p: one family of diagonal curves passing through opposite nodal points are contours of constant p, the other family being contours of constant ϕ.†

We now calculate the coordinates of the nodal points by an approximate step-by-step procedure. The position of the point $(1, 1)$ is determined by the conditions that it is the meet of two small arcs whose terminal slopes are known. The simplest, good, approximation consists in replacing each arc by a chord whose slope is the mean of the terminal slopes.‡ Thus, for the arc $(0, 1)$ to $(1, 1)$ we write

$$y_{1,1} - y_{0,1} = \{\tan \tfrac{1}{2}(\phi_{1,1} + \phi_{0,1})\}(x_{1,1} - x_{0,1}),$$

and for the arc $(1, 0)$ to $(1, 1)$

$$y_{1,1} - y_{1,0} = -\{\cot \tfrac{1}{2}(\phi_{1,1} + \phi_{1,0})\}(x_{1,1} - x_{1,0}),$$

where we have used the condition that the arcs should cut orthogonally at $(1, 1)$. $x_{1,1}$ and $y_{1,1}$ can be determined by solving these equations (or, less accurately, by geometrical construction). In an exactly similar way we can calculate the coordinates of $(2, 1)$ and $(1, 2)$. In general, having previously determined the coordinates of $(m-1, n)$ and $(m, n-1)$ the coordinates of (m, n) are calculated from the simultaneous equations

$$y_{m,n} - y_{m-1,n} = \{\tan \tfrac{1}{2}(\phi_{m,n} + \phi_{m-1,n})\}(x_{m,n} - x_{m-1,n}),$$

$$y_{m,n} - y_{m,n-1} = -\{\cot \tfrac{1}{2}(\phi_{m,n} + \phi_{m,n-1})\}(x_{m,n} - x_{m,n-1}).$$

The order in which the points are calculated is arbitrary, but in practice

† R. von Mises, *Zeits. ang. Math. Mech.* **5** (1925), 147.
‡ Much Russian work depends on a very crude approximation, namely that the arc is replaced by a chord whose slope is the *initial* value of ϕ; see, for example, W. W. Sokolovsky, *Theory of Plasticity*, p. 106 (Moscow, 1946). For the improvement effected by the present method, see, for example, P. S. Symonds, *Journ. App. Phys.* **20** (1949), 107.

it is found most convenient to calculate successively all the points on one slip-line before proceeding to the next. The hydrostatic pressure at the nodal points is found by the successive application of (12), or directly by the formula
$$p_{m,n} = p_{m,0} + p_{0,n} - p_{0,0}.$$
It is only necessary to know the value of p at one point, say the origin.

A method that is more accurate, and in some ways easier to apply, depends on solving the equations (18) for the curvatures R and S.† The base points are chosen so that the numerical change $\Delta\phi$ in ϕ between nodal points is constant. Equations (18) are replaced by the linear difference relations
$$S_{m,n} - S_{m-1,n} + \tfrac{1}{2}(R_{m,n} + R_{m-1,n})\lambda\Delta\phi = 0,$$
$$R_{m,n} - R_{m,n-1} - \tfrac{1}{2}(S_{m,n} + S_{m,n-1})\mu\Delta\phi = 0,$$
where λ and μ must be assigned the values $+1$ or -1 according to whether ϕ increases or decreases in travelling towards the point (m,n) along the respective slip-lines. Solving for $R_{m,n}$ and $S_{m,n}$, and retaining powers of $\Delta\phi$ only up to the second, there results
$$R_{m,n} = (1 - \tfrac{1}{4}\lambda\mu\Delta\phi^2)R_{m,n-1} + \tfrac{1}{2}\mu\Delta\phi(S_{m,n-1} + S_{m-1,n}) - \tfrac{1}{4}\lambda\mu\Delta\phi^2 R_{m-1,n},$$
$$S_{m,n} = (1 - \tfrac{1}{4}\lambda\mu\Delta\phi^2)S_{m-1,n} - \tfrac{1}{2}\lambda\Delta\phi(R_{m,n-1} + R_{m-1,n}) - \tfrac{1}{4}\lambda\mu\Delta\phi^2 S_{m,n-1}.$$
Since the coefficients on the right-hand sides are the same for all nodal points, the step-by-step calculation of R and S is very rapid compared with the calculation of x and y in the first method. The values of $S_{m,0}$ on OA and of $R_{0,n}$ on OB are obtained by numerical integration of the respective equations (18) along OA and OB, starting from the initial values $S_{0,0}$ and $R_{0,0}$ of the radii of curvature of OB and OA at O. Note that R and S may be discontinuous across OB and OA, respectively. Having evaluated R and S at the nodal points the coordinates x and y are calculated from the formulae

$$\left.\begin{aligned} x_{m,n} - x_{0,n} &= \int_{\phi_{0,n}}^{\phi_{m,n}} \cos\phi \, ds_\alpha = \int_{\phi_{0,n}}^{\phi_{m,n}} R\cos\phi \, d\phi, \\ y_{m,n} - y_{0,n} &= \int_{\phi_{0,n}}^{\phi_{m,n}} \sin\phi \, ds_\alpha = \int_{\phi_{0,n}}^{\phi_{m,n}} R\sin\phi \, d\phi. \end{aligned}\right\} \quad (19)$$

It is sufficient simply to use mean values of the integrands between the

---

† R. Hill, E. H. Lee, and S. J. Tupper, Ministry of Supply, Armament Research Department, Theoretical Research Rep. 28/45; also briefly reported by R. Hill and E. H. Lee, *Proc. 6th Int. Cong. App. Mech.*, Paris (1946). A similar, but less accurate, method consists in using only the initial values of R and S; see, for example, W. Prager, *Journ. Aero. Sci.* **15** (1948), 253.

nodal points if $\Delta\phi$ is small enough, say 5° or ·08727 radian. Alternatively, the integration may be carried out along a β-line in terms of S.

A third method consists in the introduction of quantities $\bar{x}$ and $\bar{y}$ defined† by the equations

$$\left.\begin{array}{l} \bar{x} = x\cos\phi + y\sin\phi, \\ \bar{y} = -x\sin\phi + y\cos\phi. \end{array}\right\} \qquad (20)$$

The geometrical interpretation of $\bar{x}$ and $\bar{y}$ is shown in Fig. 26: they are the coordinates of the point P under consideration referred to axes

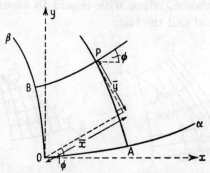

FIG. 26. Definition of the quantities $\bar{x}, \bar{y}$, used in one method of constructing a slip-line field.

passing through the origin O and parallel to the slip-line directions at P. The variation of $\bar{y}$ along an α-line satisfies the equation

$$d\bar{y} = (-\sin\phi\, dx + \cos\phi\, dy) - (x\cos\phi + y\sin\phi)\, d\phi$$
$$= -\bar{x}\, d\phi,$$

since $dy = \tan\phi\, dx$ along an α-line. Similarly, the variation of $\bar{x}$ along a β-line is given by

$$d\bar{x} = (\cos\phi\, dx + \sin\phi\, dy) + (-x\sin\phi + y\cos\phi)\, d\phi$$
$$= \bar{y}\, d\phi,$$

since $dy = -\cot\phi\, dx$ along a β-line. Hence

$$\left.\begin{array}{l} d\bar{y} + \bar{x}\, d\phi = 0 \text{ along an } \alpha\text{-line,} \\ d\bar{x} - \bar{y}\, d\phi = 0 \text{ along a } \beta\text{-line.} \end{array}\right\} \qquad (21)$$

No use has been made of the special character of the slip-line field in this derivation; the equations (21) are true for any two orthogonal families of curves. We now introduce the geometrical property (15), which determines ϕ at all points of the network. It is again advantageous to take a

† See S. Christianovitch, *Matematicheski Sbornik*, **1** (1936), 511, who attributes the introduction of these variables to S. G. Mikhlin.

network with a constant numerical increment $\Delta\phi$. By analogy with (18) we can immediately write down the equations

$$\bar{x}_{m,n} = (1-\tfrac{1}{4}\lambda\mu\Delta\phi^2)\bar{x}_{m,n-1}+\tfrac{1}{2}\mu\Delta\phi(\bar{y}_{m,n-1}+\bar{y}_{m-1,n})-\tfrac{1}{4}\lambda\mu\Delta\phi^2\bar{x}_{m-1,n},$$
$$\bar{y}_{m,n} = (1-\tfrac{1}{4}\lambda\mu\Delta\phi^2)\bar{y}_{m-1,n}-\tfrac{1}{2}\lambda\Delta\phi(\bar{x}_{m,n-1}+\bar{x}_{m-1,n})-\tfrac{1}{4}\lambda\mu\Delta\phi^2\bar{y}_{m,n-1}.$$

x and y are calculated from $\bar{x}$ and $\bar{y}$ by the equations

$$x = \bar{x}\cos\phi - \bar{y}\sin\phi,$$
$$y = \bar{x}\sin\phi + \bar{y}\cos\phi.$$

There is little to choose, either with regard to accuracy or to speed, between this method and the last.

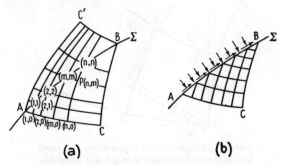

Fig. 27. Second boundary-value problem: given (a) all three stress components along a segment AB of a curve Σ, or (b) the external stress acting on a plastic surface.

A very important special case is obtained if the radius of curvature of one of the given slip-lines, say OB, is allowed to become indefinitely small, while the change in ϕ between O and B is held constant. O is then a singularity through which pass all α-lines within a certain angular span (Fig. 25b). The angle $\phi_{m,n}$ corresponding to a general point $P(m,n)$ is given by the same formula as before, where $\phi_{0,n}$ now represents the angle between the α-lines OA and OP at O. The method of calculating the coordinates of P is unchanged. The field defined by OA and the singularity at O can be formally continued round O through any desired angle; in practice the limit is set by other boundary conditions. Notice that the radius of curvature S of all β-lines is zero at O, so that, from (17) or (18), all α-lines have the same curvature at O.

(ii) *The stresses given along a certain curve (Cauchy problem)*. Suppose that all three stress components are given along a segment AB of a curve Σ (Fig. 27a) and that they satisfy the yield criterion. Assuming that the material in a sufficiently large area around AB is plastic, it is desired to calculate the slip-line field. From the theory of characteristics it is known that the conditions given on AB are sufficient to define uniquely the field within a curvilinear quadrilateral $ACBC'$ bounded by the two pairs of intersecting slip-lines through A and B, namely AC, BC, and AC', BC'. This is proved rigorously later; we now formulate an unambiguous step-by-step method of calculating the field.

Two possibilities arise: Σ may lie in the first and third of the quadrants formed by any pair of slip-lines intersecting on Σ, or it may lie in the second and fourth quadrants. For definiteness, we shall consider the first possibility, so that AC and $C'B$ are α-lines. Choose AC and AC' as curvilinear coordinate axes and divide AB by points $(1,1)$, $(2,2)$,..., (m,m),..., (n,n),..., into arbitrarily small segments; these define a net of slip-lines. The pressure $p_{n,m}$ and angle $\phi_{n,m}$ at a general nodal point $P(n,m)$ are then related to the given values at the points of intersection of the slip-lines through P with AB by the equations

$$p_{n,m} - p_{m,m} = 2k(\phi_{m,m} - \phi_{n,m}),$$
$$p_{n,m} - p_{n,n} = 2k(\phi_{n,m} - \phi_{n,n}).$$

Hence
$$\left.\begin{array}{l} p_{n,m} = \tfrac{1}{2}(p_{m,m}+p_{n,n})+k(\phi_{m,m}-\phi_{n,n}), \\ \phi_{n,m} = \dfrac{1}{4k}(p_{m,m}-p_{n,n})+\tfrac{1}{2}(\phi_{m,m}+\phi_{n,n}). \end{array}\right\} \quad (22)$$

Since the three stress components are given along Σ, the values of p and ϕ are uniquely determined there. From the last two equations p and ϕ can then be calculated at any nodal point.

We now calculate in turn the coordinates of the nodal points nearest to Σ, namely $(1,0)$, $(2,1)$, $(3,2)$,..., and $(0,1)$, $(1,2)$, $(2,3)$,... . Any one of three methods may be adopted, analogous to those described in the previous section. In the first method the elementary arcs of the slip-lines connecting $(1,0)$, say, to the adjacent points $(0,0)$ and $(1,1)$ on Σ are replaced by straight lines whose slopes are mean values assigned from the terminal values of ϕ. Having found the positions of nodal points adjacent to Σ we proceed to the calculation of the next two rows of points: $(2,0)$, $(3,1)$, $(4,2)$,..., and $(0,2)$, $(1,3)$, $(2,4)$,... . The boundary-value problem involved in this, and in successive stages, has already been considered in (i). The second method requires the determination of R and S, using the finite difference equivalents of (18) described in (i). The starting values of R and S on Σ must be obtained from the given variation of p and ϕ. If $\partial/\partial s$ denotes the tangential derivative along Σ in the sense from A to B,

$$\frac{\partial}{\partial s} = \cos\theta\frac{\partial}{\partial s_\alpha} + \sin\theta\frac{\partial}{\partial s_\beta},$$

where θ, at any point on Σ, is the anti-clockwise rotation from the α-direction to the tangent to Σ (Fig. 27a). Hence

$$\frac{\partial \phi}{\partial s} = \frac{\cos \theta}{R} - \frac{\sin \theta}{S}, \quad \text{from (16),}$$

and

$$\frac{\partial}{\partial s}\left(\frac{p}{2k}\right) = -\frac{\cos \theta}{R} - \frac{\sin \theta}{S}, \quad \text{from (12) and (16).}$$

The slip-line curvatures at points on Σ are therefore given in terms of the known variation of p and ϕ along Σ by the relations

$$\left.\begin{array}{l}\dfrac{1}{R} = -\dfrac{1}{2\cos\theta}\dfrac{\partial}{\partial s}\left(\dfrac{p}{2k} - \phi\right), \\[6pt] \dfrac{1}{S} = -\dfrac{1}{2\sin\theta}\dfrac{\partial}{\partial s}\left(\dfrac{p}{2k} + \phi\right).\end{array}\right\} \quad (23)$$

The sequence of calculations of R and S at nodal points is the same as in the first method. The coordinates are finally obtained from equations analogous to (19). The third method depends on the introduction of $\bar{x}$ and $\bar{y}$ as in (20).

It is obvious that if the given stress components on Σ are such that Σ is a slip-line, the quadrilateral degenerates into AB taken twice, so that the slip-line field cannot be determined in any neighbourhood of Σ without other information. This is a restatement of the fact that a slip-line is a characteristic.

The present boundary-value problem is most frequently met with when Σ is part of the external surface of a specimen on which given external forces are acting (Fig. 27b). Two stress components only are then known *a priori*; namely the normal and shear stresses acting on the surface. Two values of the third component can be found to satisfy the yield criterion (for example, a free surface may be in a plastic state of tangential compression or tension); the conditions of the problem will decide which alternative is correct. Only that part of the slip-line field $ACBC'$ lying inside the surface is relevant, namely the curvilinear triangle ACB (supposing C to be the interior vertex).

In general it is not possible to choose the nodal points subdividing Σ such that the net is equiangular. Evidently, this can only be done if Σ is itself a contour of constant p or of constant ϕ. A common example of the former occurs when Σ is a free surface, that is, a surface not under the action of external forces. The latter occurs rarely; a possible situation is that of a straight boundary acted on by a constant frictional stress and any distribution of normal stress.

A problem which is the converse of the present one is the following: given a slip-line OA, construct a free surface passing through O (and cutting OA at $45°$), assuming the material between Σ and OA to be plastic. The solution is unique; an equiangular net, with nodal points falling on the surface, can be used in the calculation since the surface is a contour of constant p.

(iii) *One slip-line given together with a curve along which ϕ is known (mixed problem).* Let OA (Fig. 28) be a given segment of a slip-line, say an α-line. OA can be thought of as the boundary of a previously

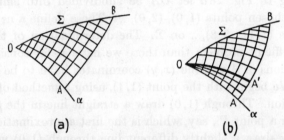

FIG. 28. Third boundary-value problem: given a slip-line OA and a curve Σ along which the inclinations of the slip-lines are known. There is a singularity at O in (b) but not in (a).

determined slip-line field in some area to the left. A curve Σ, along which ϕ is known (but *not* p), is also given; the interior angle between the two curves at their point of intersection O is *acute*. On the assumption that the region between OA and Σ is plastic it is required to calculate the slip-line field there. This is uniquely determined in the triangular region bounded by OA, Σ, and the β-line AB through A (provided, of course, that the given conditions are such that slip-lines of one family do not run together to form a natural boundary).

There are two possibilities to be considered. If the given value of ϕ on Σ at O is equal to the value of ϕ on OA at O, we have the configuration of Fig. 28 a. On the other hand, if the two values of ϕ are not equal, there is a singularity at O. If, now, the α-direction on Σ at O lies in the interior angle formed by the tangents at O to Σ and OA, the configuration of Fig. 28 b is obtained (the same applies, *mutatis mutandis*, if OA is a β-line). The field OAA' is defined by OA and the singularity at O; its construction has already been described in (i). OA' is the α-line whose direction at O coincides with the given α-direction on Σ; the angle between OA and OA' is therefore equal to the difference of the two ϕ-values at O. The problem is thus reduced to that of determining the field defined by OA' and the given condition along Σ; this is identical with the problem of Fig. 28 a. If, however, the α-direction on Σ at O lies outside the interior angle formed by the tangents at O, there is no slip-line field satisfying the given conditions; the region between OA and Σ cannot therefore be wholly plastic. We shall see later that a slip-line field can be found if OA is allowed to be an *envelope* of α-lines, but in that case the given conditions are not sufficient to determine the field uniquely; moreover, p becomes infinite at O.

Returning to Fig. 28 a let OA be subdivided into small arcs by arbitrarily chosen points $(1,0)$, $(2,0)$,.... These define a network with nodal points $(1,1)$, $(2,2)$,... on Σ. The determination of these points is a more difficult problem than those we have so far considered since the calculations of ϕ and the (x,y) coordinates have to be carried out together. We begin with the point $(1,1)$, using a method of successive approximation. Through $(1,0)$ draw a straight line in the β-direction, cutting Σ in a point P', say, which is the first approximation to $(1,1)$. We can now draw a slightly different line through $(1,0)$ with a slope corresponding to the mean of the values of ϕ at P' and $(1,0)$. This gives a new point of intersection P'' with Σ, the second approximation to $(1,1)$. The process is repeated until the difference between successive approximations becomes less than the nominal accuracy of the calculations. The points $(2,1)$, $(3,1)$,... can then be calculated successively in the usual way, the whole procedure beginning again with the point $(2,2)$.

It is clearly only possible to choose the base points on OA so that the net is equiangular when ϕ is constant along Σ; in this case there is no need to use the method of successive approximation since ϕ and the coordinates can be found separately. An example of this situation occurs if the specimen is in contact with a straight rigid surface along which the frictional shear stress is constant. In particular, if the surface is smooth, the slip-lines meet it at 45° and the problem is identical with that of finding the field defined by OA and its mirror image in Σ.

A similar, but rather more general, boundary-value problem involves a slip-line and a curve along which the stresses are subject to some condition. Such a condition is equivalent to a relation between p and ϕ. The calculation of p has therefore to be carried out simultaneously with that of ϕ and the (x,y) coordinates. The same principles are involved, but the computation is heavier. An example of this boundary-value problem occurs when the material is in contact with a rigid surface along which the coefficient of friction is constant.

6. The numerical calculation of the velocity distribution

The components of velocity are found from Geiringer's equations (14), or from their finite difference equivalents, once the slip-line field is known. If the problem is statically determined, the slip-line field is defined uniquely by the stress boundary conditions; the velocity boundary conditions are then just sufficient for calculating the velocity distribution. This type of plane problem presents no great difficulty. If, however, the problem is not statically determined there are insufficient stress boundary conditions to define the slip-line field uniquely. Uniqueness is obtained by choosing the one field that also satisfies the velocity boundary conditions. There are more of these than would be needed merely to determine the velocities *if* the slip-line field were known. In other words, the velocity boundary conditions impose restrictions on the slip-line field itself. In general, the approach to such problems must be a process of trial and error. A plastic region and associated slip-line field, satisfying all stress boundary conditions, are assumed. The corresponding velocity distribution is then computed, using only those velocity boundary conditions that are necessary for this purpose (there is generally some freedom of choice here). The solution can now be tested to see whether the remaining velocity boundary conditions are satisfied. If not, the slip-line field must be modified and the procedure repeated until the agreement is sufficiently close. This is obviously a very laborious process, and so far no problems requiring this trial and error method appear to have been solved. Solutions have been confined to statically undetermined problems where the correct slip-line field can be decided in advance by exact analysis.

At present, however, we are concerned with methods for calculating the velocity components relative to a given slip-line field, known or assumed. Since the slip-lines are characteristics for the velocity equations, the boundary-value problems are analogous to those already described for the stresses.

(i) *Normal components of velocity given along intersecting slip-lines.* Let OA and OB (Fig. 25a) be segments of an α- and a β-line, respectively. Let the velocity component v be given along OA, and the component u along OB. By the direct application of (14) we can immediately derive u along OA and v along OB. If one family of slip-lines is straight, say the β-lines, v is constant on each and is therefore a function of ϕ only. Hence the change in u between two β-lines, namely $\int v\, d\phi$, is also constant. When both families are curved we replace (14) by its finite difference equivalents referred to the previously calculated network of slip-lines:

$$u_{m,n}-u_{m-1,n} = \tfrac{1}{2}(v_{m,n}+v_{m-1,n})(\phi_{m,n}-\phi_{m-1,n}),$$
$$v_{m,n}-v_{m,n-1} = -\tfrac{1}{2}(u_{m,n}+u_{m,n-1})(\phi_{m,n}-\phi_{m,n-1}).$$

These equations determine u and v at the nodal point (m, n) in terms of the values of u and v at the points $(m-1, n)$ and $(m, n-1)$. If the network is equiangular the equations may be solved to give

$$u_{m,n} = (1-\tfrac{1}{4}\lambda\mu\Delta\phi^2)u_{m-1,n}+\tfrac{1}{2}\lambda\Delta\phi(v_{m,n-1}+v_{m-1,n})-\tfrac{1}{4}\lambda\mu\Delta\phi^2 u_{m,n-1},$$
$$v_{m,n} = (1-\tfrac{1}{4}\lambda\mu\Delta\phi^2)v_{m,n-1}-\tfrac{1}{2}\mu\Delta\phi(u_{m,n-1}+u_{m-1,n})-\tfrac{1}{4}\lambda\mu\Delta\phi^2 v_{m-1,n},$$

by analogy with the previous equations for R, S and $\bar{x}$, $\bar{y}$.

Since the normal components of velocity can be arbitrarily prescribed, without inconsistency, along two intersecting slip-lines, it follows that it is possible to have a discontinuity in the tangential component of velocity across a slip-line. Such a discontinuity must be regarded as the limit of an infinitely great shear-strain rate. Thus the component v can be discontinuous across a β-line ($\partial v/\partial s_\alpha \to \infty$), and the component u across an α-line ($\partial u/\partial s_\beta \to \infty$). It follows from (14) that *the jump in u or v is constant* along the respective slip-lines. These discontinuities are frequently encountered in applications, and usually spring from a sharp corner pressed into the material. They are a feature peculiar to a non-hardening plastic-rigid body, and correspond to what in a real material would be a more-or-less narrow transition region where the shear-strain rate is very large; it is evident that such transition regions tend to be less sharp in a metal with a high rate of work-hardening. A small element of material crossing a velocity discontinuity abruptly changes its direction of motion, and undergoes a sudden finite shear parallel to the discontinuity.

(ii) *Both components of velocity given along a certain curve.* Suppose that u and v are given along the arc AB of a curve Σ which is not a slip-line and which is not cut twice by the same slip-line (Fig. 27 a). Then the velocity solution is uniquely defined in the curvilinear quadrilateral $ACBC'$ formed by the slip-lines through the end-points of the segment. u and v are calculated at nodal points by the equations in (i).

This type of boundary-value problem always arises at the plastic-elastic boundary. According to our original assumptions the material in the elastic region moves as a rigid body. If the velocity is to be continuous, the material just on the plastic side of the boundary must have the same velocity as the rigid material. Regarding AB in Fig. 27 a as the plastic boundary, the upper side being the elastic region, we know u and v along AB. Hence the velocity distribution is uniquely defined within ABC. The plastic material in ABC therefore moves as a rigid whole attached to the elastic material, since such a motion obviously satisfies the velocity equations and the boundary conditions. It follows that, in all plane strain problems, we can distinguish two parts of the plastic region whose common interface consists of one or more slip-lines: a part in which the material is actually undergoing deformation, and a part which is rigid and is prevented from deforming by the constraint of the bordering elastic material. It frequently happens that the tangential velocity component is discontinuous across the slip-lines separating the rigid and deforming parts of the plastic region. The rigid part corresponds to what in a real material would be a region where the plastic strains are of the same order of magnitude as the elastic strains.

(iii) *The normal component of velocity given along a slip-line, together with a boundary condition along an intersecting curve.* Let v be given along an α-line OA (Fig. 28 a) and let Σ be some curve intersecting OA at O and along which there is a given boundary condition $f(u, v) = 0$ to be satisfied. For example, Σ might be a fixed rigid surface so that the velocity component normal to Σ must be zero. The problem has a unique solution provided there is no intervening slip-line field corresponding to a singularity at O. If this were so (Fig. 28 b) the velocity solution could not be constructed in the field OAA' unless the value of u were given along the β-line AA' (problem (i)).

Referring to Fig. 28 a, we have the following equations for the unknown components $u_{1,1}$ and $v_{1,1}$ at the point $(1, 1)$ on Σ:

$$v_{1,1} - v_{1,0} = -\tfrac{1}{2}(u_{1,1} + u_{1,0})(\phi_{1,1} - \phi_{1,0}),$$
$$f(u_{1,1}, v_{1,1}) = 0.$$

If f is linear in u and v, as it will generally be, these equations can be solved directly. Having found u and v at $(1, 1)$ we can use them to calculate u and v at $(2, 1)$ (problem (i)), and so on successively for the points $(3, 1), (4, 1), \ldots$. The procedure begins again with the point $(2, 2)$.

7. Analytic integration of the plane strain equations

So far we have only examined approximate methods of integrating the plane strain equations. We now derive exact expressions for the coordinates, slip-line curvatures, and velocity components at any point of a field in terms of the boundary values. The expressions contain integrals which can only be evaluated in finite form if the boundary conditions are especially simple. In the majority of special problems the integrals would have to be evaluated numerically and this would generally be more troublesome than the approximate methods already described. Consequently the exact analysis is of limited practical value. It is mainly useful for testing the relative accuracy of approximate methods.

We introduce the curvilinear coordinates (α, β) of a point P referred to an α-line OA and a β-line OB as curvilinear axes (Fig. 26). The value of ϕ at P is taken, as usual, to be the angle between the α-direction at O and the α-direction at P, measured anti-clockwise. The parameters (α, β) at P are then defined as follows:

α = value of ϕ at the point where the β-line through P cuts the α base-line OA;

β = value of ϕ at the point where the α-line through P cuts the β base-line OB.

Thus α is constant along β-lines, and β is constant along α-lines; in particular, $\alpha = 0$ is the β base-line and $\beta = 0$ is the α base-line, just as with ordinary Cartesian coordinates. The correspondence between a pair of values (α, β) and a point (x, y) is unique except where one family of slip-lines is straight; for example, if a β-line is straight the same values of α and β correspond to all points of the line.

From Hencky's first theorem (15) it follows that

and that
$$\left. \begin{array}{c} \phi = \alpha + \beta \\ \dfrac{p}{2k} = \beta - \alpha + \dfrac{p_0}{2k}, \end{array} \right\} \tag{24}$$

where p_0 is the value of p at the origin O. Conversely,

$$2\alpha = \frac{p_0}{2k} - \left(\frac{p}{2k} - \phi\right), \qquad 2\beta = -\frac{p_0}{2k} + \left(\frac{p}{2k} + \phi\right), \tag{25}$$

Since $d\phi = d\alpha$ on an α-line, and $d\phi = d\beta$ on a β-line, the equations (14) can be rewritten in the form

$$\frac{\partial u}{\partial \alpha} - v = 0, \qquad \frac{\partial v}{\partial \beta} + u = 0, \tag{14'}$$

except when one family of slip-lines is straight. Similarly, equations (18) and (21) become

$$\frac{\partial S}{\partial \alpha} + R = 0, \qquad \frac{\partial R}{\partial \beta} - S = 0; \tag{18'}$$

VI. 7] INTEGRATION OF PLANE STRAIN EQUATIONS 153

and $\qquad \dfrac{\partial \bar{y}}{\partial \alpha} + \bar{x} = 0, \qquad \dfrac{\partial \bar{x}}{\partial \beta} - \bar{y} = 0.$ (21')

From (14'): $\qquad \dfrac{\partial^2 u}{\partial \alpha \partial \beta} = \dfrac{\partial v}{\partial \beta} = -u, \qquad \dfrac{\partial^2 v}{\partial \alpha \partial \beta} = -\dfrac{\partial u}{\partial \alpha} = -v.$

From (18'): $\qquad \dfrac{\partial^2 R}{\partial \alpha \partial \beta} = \dfrac{\partial S}{\partial \alpha} = -R, \qquad \dfrac{\partial^2 S}{\partial \alpha \partial \beta} = -\dfrac{\partial R}{\partial \beta} = -S.$

From (21'): $\qquad \dfrac{\partial^2 \bar{x}}{\partial \alpha \partial \beta} = \dfrac{\partial \bar{y}}{\partial \alpha} = -\bar{x}, \qquad \dfrac{\partial^2 \bar{y}}{\partial \alpha \partial \beta} = -\dfrac{\partial \bar{x}}{\partial \beta} = -\bar{y}.$

Thus all the quantities $u, v, R, S, \bar{x}, \bar{y}$, separately satisfy the same differential equation:
$$\dfrac{\partial^2 f}{\partial \alpha \partial \beta} + f = 0. \qquad (26)$$

This is the well-known 'equation of telegraphy', whose connexion with the plane strain problem was first pointed out in 1923 by Carathéodory and Schmidt.† It is *not* satisfied in a field where one family of slip-lines is straight. Solutions corresponding to such a field are therefore not given by this equation and are sometimes called 'lost' solutions.

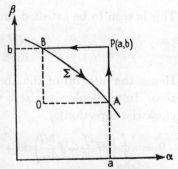

FIG. 29. Representation of slip-lines and other curves in the (α, β) plane for Riemann's method of integration.

The equation is integrated by the method of Riemann. Consider the problem of Cauchy (Sections 5 (ii) and 6 (ii)) where the values of f and one of its derivatives, say $\partial f/\partial \alpha$, are given along a curve Σ which is not a slip-line. The other derivative $\partial f/\partial \beta$ is immediately calculable in terms of $\partial f/\partial \alpha$ and the space derivative along Σ:

$$\partial f/\partial s = (\cos \theta/R)\, \partial f/\partial \alpha - (\sin \theta/S)\, \partial f/\partial \beta,$$

where θ is the angle between an α-line and the tangent to Σ. As originally formulated, the boundary conditions were that the pairs of quantities (u, v), (R, S), or $(\bar{x}, \bar{y})$, should be given along Σ. This is clearly equivalent to specifying a member of a pair and its derivative. For example, if u and v are given, we know $\partial u/\partial \alpha$ and $\partial v/\partial \beta$ from (14'); conversely if u and $\partial u/\partial \alpha$ (or $\partial u/\partial \beta$) are given we can calculate v.

It is helpful to regard the problem in terms of elements in the (α, β) plane (Fig. 29), instead of in the real (x, y) plane. In the (α, β) plane the slip-lines are represented by straight lines parallel to the axes, while Σ

† C. Carathéodory and E. Schmidt, *Zeits. ang. Math. Mech.* 3 (1923), 468.

transforms into a certain curve. Let $P(a, b)$ be the point at which it is required to calculate the value of f in terms of the given conditions along AB, where AP and BP represent the two slip-lines through P, namely $\alpha = a$ and $\beta = b$. Riemann's method depends on the possibility of finding a particular analytic solution of (26) with special properties that will be laid down later. This solution is known as the Green's function for this particular differential equation and boundary-value problem; it is denoted by $G(\alpha, \beta)$. We remark first that the expression

$$\left(G\frac{\partial f}{\partial \alpha} - f\frac{\partial G}{\partial \alpha}\right) d\alpha + \left(f\frac{\partial G}{\partial \beta} - G\frac{\partial f}{\partial \beta}\right) d\beta$$

is a perfect differential. The necessary and sufficient condition for this is that

$$\frac{\partial}{\partial \beta}\left(G\frac{\partial f}{\partial \alpha} - f\frac{\partial G}{\partial \alpha}\right) = \frac{\partial}{\partial \alpha}\left(f\frac{\partial G}{\partial \beta} - G\frac{\partial f}{\partial \beta}\right).$$

This is seen to be satisfied since f and G are solutions of (26), so that

$$G\frac{\partial^2 f}{\partial \alpha \partial \beta} = -Gf = f\frac{\partial^2 G}{\partial \alpha \partial \beta}.$$

Hence the integral of the above expression round any closed curve is zero. In particular it is zero when taken round APB (say, in an anticlockwise direction):

$$0 = \oint_{APB} \left\{\left(G\frac{\partial f}{\partial \alpha} - f\frac{\partial G}{\partial \alpha}\right) d\alpha + \left(f\frac{\partial G}{\partial \beta} - G\frac{\partial f}{\partial \beta}\right) d\beta\right\}$$

$$= \int_{AP} \left(f\frac{\partial G}{\partial \beta} - G\frac{\partial f}{\partial \beta}\right) d\beta + \int_{PB} \left(G\frac{\partial f}{\partial \alpha} - f\frac{\partial G}{\partial \alpha}\right) d\alpha +$$

$$+ \int_{AB} \left\{\left(G\frac{\partial f}{\partial \alpha} - f\frac{\partial G}{\partial \alpha}\right) d\alpha + \left(f\frac{\partial G}{\partial \beta} - G\frac{\partial f}{\partial \beta}\right) d\beta\right\}.$$

Suppose that the function $G(\alpha, \beta)$ is such that $G = 1$ on AP and BP; then $\partial G/\partial \beta = 0$ on AP and $\partial G/\partial \alpha = 0$ on BP. The first two integrals can now be evaluated explicitly and we have

$$0 = (f_A - f_P) + (f_B - f_P) + \int_{AB} \left\{\left(G\frac{\partial f}{\partial \alpha} - f\frac{\partial G}{\partial \alpha}\right) d\alpha + \left(f\frac{\partial G}{\partial \beta} - G\frac{\partial f}{\partial \beta}\right) d\beta\right\},$$

or $\quad f_P = \tfrac{1}{2}(f_A + f_B) + \tfrac{1}{2} \int_{AB} \left\{\left(G\frac{\partial f}{\partial \alpha} - f\frac{\partial G}{\partial \alpha}\right) d\alpha + \left(f\frac{\partial G}{\partial \beta} - G\frac{\partial f}{\partial \beta}\right) d\beta\right\}. \quad (27)$

This is the required expression for f, or for any one of u, v, R, S, $\bar{x}$, $\bar{y}$.

It remains to find the particular function G. It may be shown that this is
$$G(a,b,\alpha,\beta) \equiv J_0[2\sqrt{\{(a-\alpha)(b-\beta)\}}], \qquad (28)$$
where $J_0(z)$ is the Bessel function of zero order defined by the equations
$$J_0'' + \frac{J_0'}{z} + J_0 = 0, \qquad J_0(0) = 1, \qquad J_0'(0) = 0, \qquad (29)$$
where dashes denote differentiation of $J_0(z)$ with respect to z. The expansion of J_0 in ascending powers (convergent for all z) is
$$J_0(z) = \sum_0^\infty (-)^n \frac{(\tfrac{1}{2}z)^{2n}}{(n!)^2} = 1 - \frac{z^2}{4} + \frac{z^4}{64} - \dots.$$

The dependence of the function G on the particular point P under consideration is emphasized by the notation $G(a,b,\alpha,\beta)$. We can verify that the function (28) satisfies the differential equation (26) by direct differentiation:
$$\frac{\partial G}{\partial \alpha} = J_0' \frac{\partial z}{\partial \alpha} = -J_0' \sqrt{\left(\frac{b-\beta}{a-\alpha}\right)},$$
$$\frac{\partial^2 G}{\partial \alpha \partial \beta} = -J_0'' \frac{\partial z}{\partial \beta}\sqrt{\left(\frac{b-\beta}{a-\alpha}\right)} + \frac{J_0'}{2\sqrt{\{(a-\alpha)(b-\beta)\}}} = J_0'' + \frac{J_0'}{z},$$
$$\frac{\partial^2 G}{\partial \alpha \partial \beta} + G = J_0'' + \frac{J_0'}{z} + J_0 = 0.$$

Furthermore, when $\alpha = a$, or $\beta = b$, $z = 0$ and $G \equiv J_0(0) = 1$; these are the boundary conditions that G was required to satisfy along AP and BP. It should be noticed that the curvatures of the slip-lines may be such that $(a-\alpha)(b-\beta)$ is negative. Although the argument of J_0 is then complex, J_0 itself is still real, as can be seen from the power series. For purposes of numerical computation we have to use tables of the function
$$I_0(z) \equiv J_0(iz).$$

The corresponding formula for the boundary-value problem of Sections 5 (i) and 6 (i) can be obtained by taking the curve Σ to be the two slip-lines OA and OB along which the values of f are given. We have
$$f_P = \tfrac{1}{2}(f_A + f_B) + \tfrac{1}{2}\int_0^a \left(G\frac{\partial f}{\partial \alpha} - f\frac{\partial G}{\partial \alpha}\right)d\alpha + \tfrac{1}{2}\int_0^b \left(G\frac{\partial f}{\partial \beta} - f\frac{\partial G}{\partial \beta}\right)d\beta,$$

or $f(a,b) = J_0\{2\sqrt{(ab)}\}f(0,0) + \int_0^a J_0[2\sqrt{\{(a-\alpha)b\}}]\frac{\partial f}{\partial \alpha}d\alpha +$
$$+ \int_0^b J_0[2\sqrt{\{a(b-\beta)\}}]\frac{\partial f}{\partial \beta}d\beta \qquad (30)$$

by integrating by parts.

The slip-line field defined by two slip-lines which are circular arcs of equal radii is of considerable importance in applications. If we choose the (x,y) axes of reference as shown in Fig. 30, α is negative and β positive over the whole field. Suppose we wish to calculate the (x,y) coordinates of a network of slip-lines by the (R, S) method of Section 5 (i). R is equal to $-r$ on OA, and S is equal to $-r$ on OB. Hence, from (18'), $S = -r(1-\alpha)$ on OA, and $R = -r(1+\beta)$ on OB. Inserting these

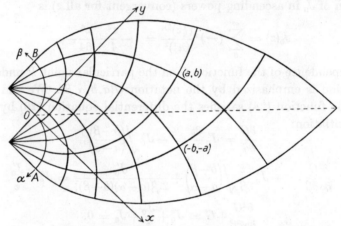

Fig. 30. Slip-line field defined by equal circular arcs OA and OB. The net shown is equiangular, with a 15° interval. For accurate values of the nodal points see Table I at the end of the book.

boundary values in (30), we obtain the following formula for the value of R at a general nodal point (a, b):

$$-\frac{1}{r}R(a,b) = J_0\{2\sqrt{(ab)}\} + \int_0^b J_0[2\sqrt{\{a(b-\beta)\}}]\,d\beta.$$

Put $z = 2\sqrt{\{a(b-\beta)\}}$, so that

$$\int_0^b J_0[2\sqrt{\{a(b-\beta)\}}]\,d\beta = \frac{1}{2a}\int_0^{2\sqrt{ab}} zJ_0(z)\,dz = \frac{1}{2a}\int_0^{2\sqrt{ab}} d(zJ_1),$$

where $J_1(z)$ is the Bessel function of the first order ($J_1 = -J_0'$). Hence

$$-\frac{1}{r}R(a,b) = J_0\{2\sqrt{(ab)}\} + \sqrt{\frac{b}{a}}J_1\{2\sqrt{(ab)}\},$$

or

$$\left|\frac{R}{r}\right| = I_0\{2\sqrt{(|ab|)}\} + \sqrt{\left|\frac{b}{a}\right|}I_1\{2\sqrt{(|ab|)}\}. \tag{31}$$

This form must be used in calculations since ab is negative; the values of I_0 and I_1 can be obtained from tables. From considerations of symmetry the value of S at (a, b) is equal to the value of R at $(-b, -a)$, the reflection of (a, b) in the axis of symmetry. Formula (31) can be used to check the accuracy of the approximate method for calculating R and S described in Section 5 (i). It is found that if an equiangular network is used, with a 5° interval ($\Delta\phi = \cdot 08727$ radians), the error is less than 0·1 per cent. over the field defined by circular arcs of 45° span. The method is therefore capable of a high order of accuracy, even with a relatively coarse network. Equation (19) has been used to calculate values of (x, y) at points of a 5° network over a field defined by 90° circular arcs. The results are given in Table 1 at the end of the book.†

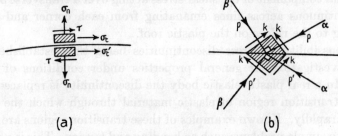

Fig. 31. Plastic stress discontinuity across a surface.

This field occurs in many of the problems whose solutions are described in the following chapters.

For an account of other properties of the equations of Hencky and Geiringer and the derivation of certain special solutions (mainly of academic interest) the reader is referred to the monographs by Geiringer and Prager‡ (1934), and by Geiringer§ (1937).

8. Discontinuities in the stress

It has been shown that there may be discontinuities across slip-lines in the velocity, the gradient of the velocity, and the gradient of the stress. We have not yet examined the possibility that the stress itself may be discontinuous across some curve. Considerations of equilibrium demand only that the stress component σ_n normal to the curve (Fig. 31 a) and the shear stress τ parallel to the curve should be the same on both sides; the components σ_t and σ'_t acting parallel to the curve may be different. The reader is already familiar with one example, namely in the bending of a plastic-rigid sheet in plane strain (Chap. IV, Sect. 6). The

† Calculations of S. J. Tupper and the author.
‡ H. Geiringer and W. Prager, *Ergebnisse d. exakten Naturwiss.* 13 (1934), 310.
§ H. Geiringer, *Mém. Sci. Math.* 86 (1937).

longitudinal stress is tensile on one side of the central (neutral) axis and compressive on the other. As the bending torque is increased the central strip of elastic material shrinks and finally vanishes. When this happens the longitudinal stress on the tension side is everywhere of amount $2k$ while on the compression side it is $-2k$; there is therefore a discontinuity of amount $4k$ along the neutral axis. (We may note, parenthetically, that the sheet cannot begin to bend until it is entirely plastic, owing to the rigid core.) A second example of a stress discontinuity (though not in plane strain) has been previously noticed in the torsion of a plastic-rigid bar whose contour has sharp corners (Chap. IV, Sect. 8 (vi)). The tangential component of the shear stress acting over a transverse section is discontinuous across lines emanating from each corner and corresponding to the ridges on the plastic roof.

The possibility of stress discontinuities being thus established, we must investigate their general properties under conditions of plane strain. In a real plastic-elastic body the discontinuity is replaced by a narrow transition region of elastic material through which the stress changes rapidly. Known examples of these transition regions are so far limited to simple problems such as bending and torsion. This is not surprising since they are probably mostly produced under boundary conditions for which an exact solution for the stress distribution would be prohibitively difficult. Within recent years, however, problems of increasing complexity have been solved with the aid of the plastic-rigid body approximation, and the possibility of encountering stress discontinuities in future work must be envisaged. The properties of stress discontinuities were apparently first systematically investigated by Prager.†

The amount $\sigma_t - \sigma_t'$ of the discontinuity in tangential stress is restricted by the condition that the material on both sides of the line is plastic. If σ_n and τ are regarded as given, possible values of σ_t and σ_t' satisfy the yield criterion
$$(\sigma - \sigma_n)^2 + 4\tau^2 = 4k^2.$$
Thus
$$\sigma = \sigma_n \pm 2\sqrt{(k^2 - \tau^2)}.$$
The tangential stress may therefore have one of two values (corresponding in Mohr's representation of stress to the two circles of radius k through the point (σ_n, τ)). Hence, if there is a discontinuity, it is of amount $4\sqrt{(k^2 - \tau^2)}$. The jump in the mean compressive stress p is evidently half this, namely $2\sqrt{(k^2 - \tau^2)}$.

† W. Prager, *R. Courant Anniversary Volume*, p. 289 (Interscience Publishers, New York, 1948).

It is more interesting and useful to regard the discontinuity in terms of the slip-lines. Consider the equilibrium of a quadrilateral element with sides perpendicular to the four slip-line directions through the point under consideration (Fig. 31 b). By resolving in the tangential direction it follows that the element must be symmetrical about the tangent, or, in other words, that the α (or β) directions on either side are reflections in the curve. If θ is the acute angle between an α-direction and the tangent to the curve, it follows by resolving in the normal direction that
$$|p-p'| = 2k \sin 2\theta. \tag{32}$$
In Fig. 31 b the algebraically greater value of p is on the upper side of the curve ($p > p'$); if it were on the lower side the α-lines would point to the left instead of to the right (this configuration is obtained simply by turning the page through 180°).

The curvatures of the slip-lines change abruptly across a stress discontinuity. From equation (23) the curvatures $1/R'$ and $1/S'$ on the lower side of the curve in Fig. 31 b are given by
$$\frac{1}{R'} = -\frac{1}{2\cos\theta}\frac{\partial}{\partial s}\left(\frac{p'}{2k}-\phi'\right), \qquad \frac{1}{S'} = -\frac{1}{2\sin\theta}\frac{\partial}{\partial s}\left(\frac{p'}{2k}+\phi'\right).$$
On the upper side the anti-clockwise rotation of the α-direction from the tangent is $-\theta$, and so
$$\frac{1}{R} = -\frac{1}{2\cos\theta}\frac{\partial}{\partial s}\left(\frac{p}{2k}-\phi\right), \qquad \frac{1}{S} = \frac{1}{2\sin\theta}\frac{\partial}{\partial s}\left(\frac{p}{2k}+\phi\right).$$
Hence
$$\frac{1}{R'}-\frac{1}{R} = \frac{1}{2\cos\theta}\frac{\partial}{\partial s}\left[\left(\frac{p-p'}{2k}\right)-(\phi-\phi')\right],$$
$$\frac{1}{S'}+\frac{1}{S} = \frac{1}{2\sin\theta}\frac{\partial}{\partial s}\left[\left(\frac{p-p'}{2k}\right)+(\phi-\phi')\right].$$
But $p-p' = 2k\sin 2\theta$ and $\phi-\phi' = 2\theta$, and so
$$\left.\begin{aligned}\frac{1}{R'}-\frac{1}{R} &= -2\sin\theta\tan\theta\frac{d\theta}{ds}, \\ \frac{1}{S'}+\frac{1}{S} &= 2\cos\theta\cot\theta\frac{d\theta}{ds}.\end{aligned}\right\} \tag{33}$$
These equations relate the jumps in the curvatures to the variation of the angle θ along the discontinuity.

The question now arises as to whether there can be a discontinuity in tangential velocity across a stress discontinuity. That this cannot be so, if the plastic-rigid material is regarded as a plastic-elastic material in which Young's modulus is allowed to increase without limit, may be demonstrated by the following argument.† Suppose, if possible, that a narrow transition region could exist in a plastic-elastic body such that the stress and velocity change rapidly across it. For equilibrium it is necessary that the shear-stress component and the normal stress component must be very nearly constant through the region (and in the limit exactly so). It is therefore impossible that all the transition region should be plastic, since the stress component parallel to the line of the transition region varies rapidly, by hypothesis. Thus, most of the change in stress must take place through a central elastic strip. On the other hand, the change in velocity must clearly take place through a plastic strip; this requires the slip-lines there to be directed effectively along and perpendicular to the line of the transition region (since the shear strain-rate becomes infinite in the limit). This leads to an inconsistency since the stress in the plastic strip must be very nearly constant and equal to the stress just outside the transition region—to which, by hypothesis, the slip-lines are inclined at some finite angle. From this standpoint, then, slip-line fields involving simultaneous stress and velocity discontinuities‡ cannot be allowed. Such fictitious fields are, however, occasionally useful as approximations when the correct solution is not known (see pp. 172 and 220).

† Letter by the author to Prof. W. Prager, Brown University (23 Oct. 1948).
‡ A. Winzer and G. F. Carrier, *Journ. App. Mech.* 15 (1948), 261.

VII
TWO-DIMENSIONAL PROBLEMS OF STEADY MOTION

1. Formulation of the problem

IN this and the two following chapters the ideas and techniques of Chapter VI are applied to the solution of a variety of problems of plane strain. We begin with problems of steady motion, in which the stress and velocity do not vary at any fixed point (referred to some coordinate system which may be at rest or moving uniformly). This condition is fulfilled in continuous processes of shaping metals, for example rolling, drawing, and extrusion. The approximation by the plastic-rigid body is likely to be good since the strains are usually large; the error due to the neglect of work-hardening can be largely removed by a simple correction factor, as we shall see. While the two-dimensional theory is strictly valid only for wide blocks or sheets, as in strip-rolling, it gives a useful qualitative picture of the deformation in such processes as the drawing of wire or the extrusion of rod; it will be shown, too, that the theoretical results, suitably modified, frequently provide a close estimate of the energy consumption.

Problems of steady motion differ from those we have so far discussed in that the distributions of stress and velocity are presumably independent of the manner in which a steady state is reached. A proof is lacking, but, provided the external conditions are invariable, it appears unlikely that the final steady state could be affected by variations at the outset (for example, in the shape of the end of a length of strip introduced into the roll gap). However, when the conditions under which a steady state is established cannot be *completely* specified, it is conceivable that the previous history may have an influence. For example, the shape of the dead metal separating off at the beginning of direct extrusion must partly depend on the physical properties of the end of the billet; since the dead metal directs the flow, very much as a rough die, it must control the final steady state. Without an explicit uniqueness theorem we cannot be certain that the conditions under which we seek a steady state necessarily ensure a unique solution. However, reference to experiment will indicate when we can reasonably expect our conditions to be definitive.†

† This argument, a common one in applied mathematics, tacitly assumes that the basic equations *completely* parallel the actual physical behaviour.

Assuming, now, that the steady state is unique, and that we do not need to trace the previous development of the plastic region, the stress and velocity at a fixed point are functions only of position; the element of time, or progress of the deformation, is absent. The plastic region is fixed in space (or can be made so by imposing a uniform velocity on the whole system); rigid material enters on one side, is deformed while passing through, and leaves on the other in a uniform stream which becomes rigid again as it unloads. A problem of steady motion is, by its very nature, statically undetermined since the maintenance of a uniform flow of material, to and from the plastic region, imposes restrictions on the shape and position of the plastic boundary.

We now formulate the problem precisely. The plastic region consists, as was shown in Section 6 (ii) of Chapter VI, of a part which is rigidly held by the non-plastic material, and a part where deformation is occurring. The slip-line boundary separating the two parts must, in the first instance, be constructed to satisfy all conditions in stress and velocity that *directly* concern the zone of plastically deforming material. The restrictions on the stress distribution in this zone may include, for example, frictionless surfaces or prescribed external loads; the usual restrictions on the velocity are that it must have a zero component normal to fixed boundaries, and that the component normal to the plastic-rigid boundary must be compatible with the rigid-body motion of material outside the zone. In general (cf. Chap. VI, Sect. 6) we must construct the slip-line field by a trial-and-error process, first choosing a field that satisfies the stress conditions, and then examining whether it is associated with a velocity solution that is consistent with the velocity conditions[†] and which implies that the rate of work done on every plastic element is positive.

It may happen that these boundary values alone do not uniquely determine the slip-line field or that the assumption as to the general position of the plastically deforming zone is incorrect (for example, in sheet-drawing we do not know *a priori* whether a standing wave of plastic material is formed ahead of the die). These questions are decided by the consideration that the rigid material (which may be plastic,

† Most writers on steady-motion problems have unfortunately neglected to do this; see, for instance, W. W. Sokolovsky, *Theory of Plasticity* (Moscow, 1946); K. H. Shevchenko, *Prikladnaia Matematika i Mekhanika*, 5 (1941), 439, and 6 (1942), 381; *Izvestia Akad. Nauk, SSSR*, 3 (1946), 329; F. K. Th. van Iterson, *Plasticity in Engineering* (Blackie, 1947); E. Siebel, *Journ. Iron and Steel Inst.* 155 (1947), 526; G. F. Carrier, *Quart. App. Math.* 6 (1948), 186; H. I. Ansoff, *Graduate Division of App. Math.*, Brown University, Tech. Reps. 14 and 23 (1948). In general their proposed slip-line fields can be shown to be incorrect; examples are discussed later.

FORMULATION OF THE PROBLEM

non-plastic, or unloaded) must be able to sustain the forces on its perimeter. Thus, the validity of the slip-line field is finally proved by demonstrating the existence of a steady-state distribution of stress in the rigid material such that the yield limit is nowhere exceeded.

It appears to be impossible to formulate an explicit step-by-step method for the solution of problems of steady motion; the element of trial and error can hardly be eliminated. Each problem has its peculiarities and these must first be appraised qualitatively. Physical intuition having indicated the probable general location of the plastic region, the position of a section of some slip-line is assumed; it should be a slip-line from which, with the stress boundary conditions, a plastic field can be built up unambiguously. The field is extended, unless halted by an inconsistency, until it seems large enough to permit the overall deformation required by the steady state; in rolling, for example, the plastic region must clearly extend right through the sheet, since the material leaving the rolls moves faster than it did before entering. A slip-line field having been guessed, we construct the velocity solution, beginning where sufficient boundary values are available. Having obtained the distribution of velocity we examine whether it is compatible with the boundary values not so far used; since the problem is statically undetermined there are always more boundary conditions than are needed to define the velocity solution for an *assumed* field. If all conditions cannot be satisfied the original choice of slip-line must be modified. When all conditions directly affecting the plastic zone are satisfied we have next to verify that the rate of plastic work is everywhere positive, and have finally to examine the associated stress distribution in the rigid material. If everything is consistent we have found a possible steady state.

With a little practice and experimenting the reader will soon gain experience in choosing a favourable starting slip-line, while known solutions will suggest possible patterns of slip-lines. Nevertheless, each new problem usually presents fresh features, and there are still many problems of practical importance whose solutions are not known, even qualitatively.

2. Sheet-drawing†

(i) *Drawing through a smooth die*. It is supposed that the die has straight rigid walls, with an included angle of 2α, and that there is no

† This section follows closely the paper by R. Hill and S. J. Tupper, *Journ. Iron and Steel Inst.* 159 (1948), 353.

friction, so that slip-lines meet the wall at 45°. We are seeking the steady-state configuration in which the thickness of the sheet, initially H, is reduced to h by pulling the sheet through the die under a force directed along the central axis. There is evidently a steady state in which the sheet passes symmetrically through the die, and we shall suppose that the surface remains undistorted until it reaches the die (Fig. 32). For the starting slip-line let us choose the one through A meeting the wall at 45°, and assume its shape over a section AC, where C is its intersection

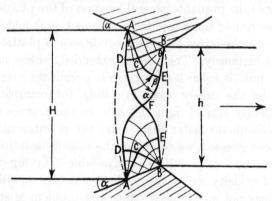

Fig. 32. Slip-line field and plastic region for drawing through a smooth wedge-shaped die giving a moderate reduction in thickness.

with the slip-line through B meeting the wall at 45°. The slip-line field is thereby uniquely determined within ABC (third boundary-value problem; Chapter VI, Section 5 (iii)). Now A and B must clearly be singularities for the stress distribution; this fact enables the field to be continued round A and B, as far as is necessary, to form the regions ACD and BCE (first boundary-value problem, special case; Chap. VI, Sect. 5 (i)). The slip-lines CD and CE then define a region $CDEF$ (first boundary-value problem). Since the zone of plastically deforming material must extend through the sheet, the point F lies on the centre line, so that it is common to the two plastic zones spreading symmetrically from opposite sides of the die. Furthermore, since there is no reason (at this stage) for supposing that the stress is discontinuous, the slip-lines at F must be inclined at 45° to the centre line. It is easy to see that these restrictions on F fix the angles CAD and CBE, hitherto left arbitrary. For the moment, the fractional reduction in thickness, $r = (H-h)/H$, is taken sufficiently small to make this construction possible.

A trial slip-line field having been assumed, we proceed to calculate the velocity distribution, the remainder of the sheet being taken as rigid. If U is the speed of drawing, the speed of approach to the die is Uh/H since there is no volume change and the flow is steady. The normal component of velocity is therefore known on the slip-lines† ADF and BEF, and the solution may be begun in region $CDEF$ (first boundary-value problem; Chap. VI, Sect. 6 (i)).‡ This solution gives the normal components on CD and CE, which, with the components on AD and BE, define the solutions in ACD and BCE. From these we obtain the normal components on AC and BC, defining the velocity in ABC. In general the calculated velocity in ABC would not be found to be tangential to the wall along AB. For this boundary condition to be satisfied the starting slip-line AC must be correctly chosen, and in this way the velocity boundary conditions impose restrictions on the slip-line field.

In this instance we are able to show by direct analysis, without any calculations, that all conditions are satisfied when AC is straight. All slip-lines in ABC are then straight, and the slip-lines in ACD and BCE are straight lines and concentric circular arcs (Chap. VI, Sect. 4 (i)). Now by the equations of Geiringer ((14) of Chap. VI) the velocity component along any straight slip-line is a constant, and the change in the normal component in passing between two straight slip-lines is also constant. Since, however, the material to the left of ADF is rigid, its component of velocity normal to AD is everywhere the same; the normal component is therefore constant on each slip-line through A, and in particular on AC; this is true whatever the distribution of velocity along CD. Similarly, the normal component is constant on BC. It follows that the velocity is uniform throughout ABC. However, since Geiringer's equations imply zero change of volume, and since the inflow across ADF has been made equal to the outflow across BEF by basing the solution on an entry speed Uh/H, there is no *net* flow of material across AB. Thus, if AC is straight, the velocity in ABC would, when calculated, be found to be tangential to the wall. It is remarkable that this proof has been carried through without needing to calculate, in detail, either the slip-line field or the distribution of velocity in $CDEF$. It will be seen that there is a tangential discontinuity of velocity of amount $rU/\sqrt{2}$ across ADF and BEF; the possibility of such a discontinuity was

† This will be used as a shorthand expression for 'the component, normal to the slip-line, of the velocity at a point on the slip-line'.

‡ We do not, as yet, need to distinguish the α and β families; the distribution of velocity calculated from the Geiringer equations is independent of the convention, since the equations merely express zero rate of extension along the slip-lines.

established in Chapter VI, Section 6 (i). The sense in which elements are sheared when crossing the discontinuity requires that ADF be an α-line, in order that the work of distortion should be positive. A proof that the rate of work is positive everywhere in the field has not been given, but it is reasonable to assume that this would be found to be so.

Since the slip-lines CD and CE are circular arcs of equal radius, the field $CDEF$ is that of Fig. 30. By applying Hencky's first theorem it is apparent that the angles $CAD = \psi$, and $CBE = \theta$, must satisfy the relation
$$\theta - \psi = \alpha \tag{1}$$
if the slip-lines at F are to make 45° with the centre-line. Since the coordinates of an equiangular network with a 5° mesh have been calculated (Table 1, p. 350), it is convenient to take F to be a nodal point of this network. According to (1) this is so only when the semi-angle of the die is an integer multiple of 5°. For each of these die-angles we can then calculate the reductions corresponding to a series of values of θ and ψ, at intervals of 5° from zero upwards. The values of θ and ψ for any other reduction can be obtained by interpolation. The maximum reduction for which the present field is valid, for geometrical reasons alone, is
$$r = 1 - \frac{h}{H} = \frac{2\sin\alpha}{1 + 2\sin\alpha}, \tag{2}$$
corresponding to $\psi = 0$, $\theta = \alpha$ (the points E and F coincide).

We have still to determine the stress distribution in the plastic zone, and for that we require the value of the mean compressive stress p_0 at one point of the field, say F. This is provided by the condition that the sheet is drawn without back-pull, that is, no external force is applied to the part of the sheet approaching the die. We have, therefore, to equate to zero the longitudinal component of the total force acting over ADF; this is found by integrating along ADF the resolved components of the pressure p and the shear stress k, using Hencky's equations ((12) of Chap. VI) to determine the value of p at any point in terms of p_0. Having found p_0 in this way, we can calculate the stress at any point in the field. Since both families of slip-lines in ABC are straight, the pressure q on the die wall is distributed uniformly; its dependence on die-angle and reduction in thickness is shown in Fig. 33. The force T (per unit width) needed to draw the sheet is equal to the longitudinal component $q(H-h)$ of the thrust on the die, and the mean tensile stress t in the drawn sheet to $qr/(1-r)$ (Fig. 34). The right-hand extremities of the curves correspond to the reduction (2), which is peculiar in that an explicit formula can be obtained for the drawing stress. By considering

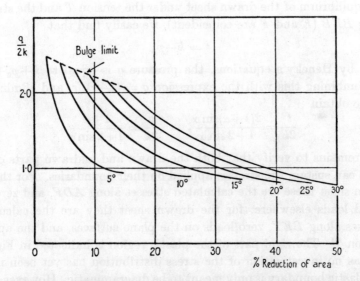

Fig. 33. Relation between the die-pressure and the reduction in thickness for various semi-angles in drawing through a smooth wedge-shaped die (no back-pull and no work-hardening).

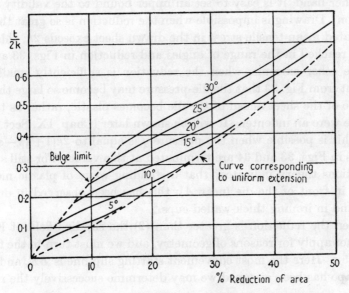

Fig. 34. Relation between the mean drawing stress and the reduction in thickness for various semi-angles in drawing through a smooth wedge-shaped die (no back-pull and no work-hardening). The right-hand broken curve corresponds to uniform deformation, and the left-hand curve to incipient bulging.

the equilibrium of the drawn sheet under the tension T and the stresses along BEB (E and F are coincident), we easily find that
$$t = k - p_0.$$
Also, by Hencky's equations, the pressure q is equal to $k+p_0+2k\alpha$. On combining this with the expression $q = (1-r)t/r$, and eliminating p_0, we obtain
$$\frac{t}{2k} = \frac{2(1+\alpha)\sin\alpha}{1+2\sin\alpha}, \qquad \frac{q}{2k} = \frac{1+\alpha}{1+2\sin\alpha}. \tag{3}$$

It remains to verify that both the drawn and undrawn parts of the sheet can sustain the stresses applied on their boundaries. For the undrawn sheet these are the calculated stresses along ADF, and zero external loads elsewhere; for the drawn sheet they are the calculated stresses along BEF, zero loads on the plane surfaces, and the applied tension T. The rigid part of the plastic region is indicated in Fig. 32, but, as no investigation of the stress distribution has yet been made, the plastic boundary is only meant to be diagrammatic. However, from calculations of the *average* shear stress over oblique sections, it seems that, in general, the yield stress is unlikely to be locally exceeded. On the other hand, it is easy to set an upper bound to the validity of the solution. Drawing is impossible when the reduction is so great that the calculated mean tensile stress in the drawn sheet exceeds $2k$; this limit is not reached in the range of angles and reduction in Figs. 33 and 34. At the other extreme, when the reduction is sufficiently small, it is evident from Fig. 33 that the die-pressure may become so large that the surface of the sheet ahead of the die becomes plastic, rather as though the die were an indenter. It will be shown later (Chap. IX, Sect. 5 (ii)) that this is possible when the pressure q is equal to $2k(1+\tfrac{1}{2}\pi-\alpha)$; the curves in Figs. 33 and 34 are terminated at this value. For still smaller reductions theory indicates† that a standing wave of plastic material forms in front of the die (near A); this has been observed in drawing bars and in ironing thick-walled cups.‡

When the reduction is greater than (2) the slip-line field of Fig. 32 does not apply for reasons of geometry, and we must turn to the field of Fig. 35 a. Here the most convenient starting slip-line is BE, and when its shape has been assumed we may determine successively the regions

† R. Hill, *Dissertation*, p. 158 (Cambridge 1948), issued by Ministry of Supply, Armament Research Establishment, as Survey 1/48.

‡ Measurements of the wave form, and an investigation of its dependence on die-angle and reduction, have been made by G. C. Briggs and H. W. Swift, *Motor Industry Research Association*, Rep. 1947/R/4.

BCE, ABC, $CEDF$, $ACFG$, and AGH (first and third boundary-value problems). It may be demonstrated that the velocity conditions are satisfied when BE is straight (and therefore inclined at 45° to the centre line); the slip-lines of the same family in BEC are then also straight, as are both families in ABC and the AC family in $ACFG$. The velocity solution must be begun in AGH, the only region where two boundary conditions are available, and continued in the order $ACFG$, ABC, $CEDF$, and BCE. The solution is possible only when the calculated velocity on BE is compatible with the rigid-body movement of

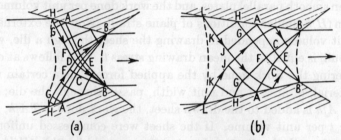

Fig. 35. (a) Slip-line field for somewhat larger reductions. (b) Incorrect field for still larger reductions, with an accompanying discontinuity in velocity along the arrowed slip-lines.

the drawn sheet; this is so if the velocity component normal to BE is the same at all points. By the argument used previously it may be shown that the normal component of velocity is constant on each of the straight slip-lines in $ACFG$, and hence that the velocity in ABC is uniform (third boundary-value problem). Thus, the normal component of velocity is constant on BC and therefore on all slip-lines of the same family in BEC, in particular BE; this completes the proof. There is a tangential discontinuity in velocity, initiated at H, and extending along the slip-lines $HGFDB$; there is no discontinuity across BE. A qualitative application of Hencky's theorem shows that the pressure on the die is constant over BA and rises steadily over AH. Little practical interest attaches to a detailed knowledge of the stresses, since the drawing force is given sufficiently accurately by the simple theory of Sachs (see below).

For still greater reductions (Fig. 35 b) it is natural to expect that the slip-line field would be the continuation of that in Fig. 35 a.† In general,

---

† This was independently proposed by W. W. Sokolovsky, *Theory of Plasticity*, p. 184 (Moscow, 1946), who, however, was unaware of the necessity to justify the slip-line field by considering the velocity solution. Sokolovsky did not suggest a field for smaller reductions.

however, this cannot be true since the velocity discontinuity transmitted from the die entrance along the arrowed slip-lines finishes (undiminished) somewhere along the exit slip-line. This is obviously incompatible with the rigid-body motion of the drawn sheet. So long as the discontinuity terminates on the exit corner of the die, as it does for certain reductions and die-angles, the solution is valid, but otherwise the solution fails; the correct field is not yet known.

(ii) *Deformation and efficiency.* The most efficient, or ideal, means of reducing the thickness of a sheet would be by compressing it uniformly between smooth parallel plates, and the work done per unit volume would be $2k\ln(H/h)$ under conditions of plane strain. Now the external work per unit volume expended in drawing the sheet through a die, without back-pull, is equal to the mean drawing stress t. This follows at once by considering the work done by the applied force T as a certain volume of material, of area A and unit width, passes through the die; since a length A/h is added to the drawn sheet, the work done is TA/h, that is, T/h or t per unit volume. If the sheet were compressed uniformly in passing through the die, the drawing stress would be $2k\ln(H/h)$; this is represented by the broken curve in Fig. 34. However, uniform compression is, of course, prevented by the constraint of the rigid material; the difference between the broken and solid curves corresponds to the non-useful work expended in distortions which do not contribute to the final reduction of thickness. When the reduction is small and the die-angle large, the efficiency, defined as the ratio of the ideal work to the actual work, falls as low as 50 per cent. On the other hand, when the reduction is large the curves for each die-angle tend to approach the broken curve, and the efficiency is correspondingly high. However, this does not necessarily imply that the actual deformation in drawing closely approximates uniform compression.

Consider, for example, the distortion of a square grid scribed on a longitudinal section of the sheet (Fig. 36); this has been calculated for $\alpha = 15°$ and a reduction of 34·1 per cent., corresponding to the special configuration (2). The method is to calculate first the trajectories for particles on the longitudinal lines, and then to find the successive arc-lengths covered in equal times. It will be seen that the distortion consists essentially of successive shears over two oblique directions in the element, the first on crossing the entry slip-line ACE, and the second, in the opposite sense, while crossing the exit slip-line BE; the strain sustained while passing through the region BCE is comparatively slight. Even though the strain-path is so different from uniform compression the work of distortion (Fig. 34) is practically ideal. There appears to be no experimental investigation of the distortion in sheet drawing; the cusp in the transverse lines after drawing,† due to the velocity discontinuity, would probably not be observed in a real metal, where work-hardening tends to diffuse zones of intense shear. Apart from this feature, apparently peculiar to plane strain, there is a qualitative

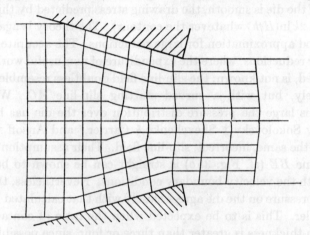

FIG. 36. Distortion of a square network in drawing through a smooth wedge-shaped die of semi-angle 15°, giving a reduction of 34·1 per cent.

resemblance to the distortion found by Taylor and Quinney‡ in grids scribed on split wires.

The earliest theory of the stress distribution in drawing, due to Sachs§ (1927), was based on the assumptions that all the material between the planes of entry and exit is plastic, that the principal axes of stress are everywhere parallel and perpendicular to the axis of the sheet, and that the stress is uniform over any transverse section. These lead to the formula $t = 2k\ln(H/h)$,‖ which we have seen to be only a good approximation when the reduction is sufficiently large. Körber and Eichinger†† have attempted to allow for the expenditure of non-useful work by

---

† It may be shown, by considering the distortion of a small element on the centre line, that the total included angle 2χ of the cusp is such that $\cot\chi = 4\sin^2\alpha$.

‡ G. I. Taylor and H. Quinney, *Journ. Inst. Metals*, **49** (1932), 187.

§ G. Sachs, *Zeits. ang. Math. Mech.* **7** (1927), 235. Sachs' theory was for wire drawing; this is the plane strain analogue.

‖ The same formula is obtained if the material in the die is regarded as part of a ring which is being pulled inwards by a uniform internal tension, the centre of the ring being the vertex of the angle made by the walls of the die. The lines of principal stress are then radii and concentric circular arcs, with this vertex as centre.

†† F. Körber and A. Eichinger, *Mitt. Kais. Wilh. Inst. Eisenf.* **22** (1940), 27.

adding a term in α to represent the work of shearing an element as it enters and leaves the die; their formula implies that the curve for each die-angle is spaced parallel to the ideal curve, which is still a poor approximation.

Sachs and Klingler[†] have slightly extended the elementary theory to drawing through rough dies of any shape, particularly a circular contour (if the die is smooth, the drawing stress predicted by this theory is equal to $2k\ln(H/h)$ whatever the contour). The theory is again likely to be a good approximation for large reductions. The accurate solution for smaller reductions, where the expenditure of non-useful work cannot be neglected, is not known; the slip-line field doubtless resembles Fig. 32 qualitatively, but with a curved starting slip-line AC. When the reduction is large the pressure distribution over the die has been calculated by Sokolovsky,[‡] Shevchenko,[§] Carrier,[||] and Ansoff,[††] all of whom use the same, incorrect, slip-line field. Their assumption that the exit slip-line BE (cf. Fig. 35 b) is straight can be shown to be incompatible with the velocity boundary conditions. Nevertheless, the distribution of pressure on the die agrees closely with that calculated by Sachs and Klingler. This is to be expected when the ratio of contact-arc to mean strip-thickness is greater than three or four, since possible plastic states of stress are severely restricted by the stress conditions on the die alone and cannot differ much from each other. There is no *a priori* reason to suppose that the drawing stress calculated from an arbitrarily assumed field is less inaccurate than that of Klingler and Sachs. There is little point in making extensive calculations of the small error in the elementary theory unless it be done with reasonable accuracy.

(iii) *Allowance for work-hardening.* It is not difficult to allow for the contribution of strain-hardening to the drawing force by a method that is at once sufficiently accurate and easy to apply.[‡‡] We have already proved that when there is no friction the mean plastic work per unit volume is equal to the drawing stress t. Now for a non-hardening material, obeying the Lévy–Mises relations, the work per unit volume is equal to $Y \int \overline{d\epsilon}$, where $\int \overline{d\epsilon}$ is the equivalent strain (equation (25) of

---

[†] G. Sachs and L. J. Klingler, *Journ. App. Mech.* **14** (1947), A–88.
[‡] W. W. Sokolovsky, *Theory of Plasticity*, p. 192 (Moscow, 1946).
[§] K. H. Shevchenko, op. cit., p. 162.   [||] G. F. Carrier, op. cit., p. 162.
[††] H. I. Ansoff, op. cit., p. 162. Ansoff also treated the problem by approximating the smooth contour by a polygon; he proposed a solution in terms of stress discontinuities (chap. vi, sect. 8), but since the velocity is also discontinuous the solution is invalid although it is a fairly good approximation.
[‡‡] A complicated correction has been introduced into Sachs' theory by E. A. Davis and S. J. Dokos, *Journ. App. Mech.* **66** (1944), A–193.

Chap. II). We can therefore define a mean equivalent strain for sheet drawing equal to t/Y, or $t/\sqrt{3}k$. Let us assume that, to a first approximation, the same mean equivalent strain $t/\sqrt{3}k$ is imparted by the die whatever the strain-hardening characteristics of the material. Then the drawing stress, or the mean work per unit volume, is equal to the area under the equivalent stress-strain curve (the true-stress versus logarithmic-strain curve in compression) up to the strain $t/\sqrt{3}k$. This strain is

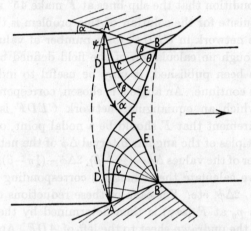

Fig. 37. Slip-line field and plastic region for drawing through a rough wedge-shaped die giving a moderate reduction in thickness.

equal to the appropriate ordinate in Fig. 34 multiplied by the factor $2/\sqrt{3}$; in particular, the equivalent strain corresponding to uniform *plane* compression is $(2/\sqrt{3})\ln(H/h)$ (this should be contrasted with the equivalent strain in *free* compression which, as we have seen, is identical with the logarithmic strain $\ln(H/h)$). The yield stress of the drawn sheet in uniaxial tension is equal to the ordinate on the stress-strain curve at the mean equivalent strain.

(iv) *Drawing through a rough die.* Let μ be the coefficient of friction between the material and the die, and let q' be the pressure on the die at any point. According to equation (11) of Chapter VI the acute angle β between the wall and the β-line is given by

$$\cos 2\beta = \mu q'/k. \tag{4}$$

Fig. 37 shows the slip-line field, which is analogous to that for the smooth die (Fig. 32). It may be proved by the argument used previously that the velocity boundary conditions are satisfied when the slip-line AC is straight. The pressure q' on the die is therefore uniform, and hence, by equating the drawing force to the longitudinal component of the thrust on the die, the mean drawing stress is found to be

$$t' = (1+\mu\cot\alpha)rq'/(1-r). \tag{5}$$

From Hencky's first theorem we have also

$$\theta-\psi = \tfrac{1}{4}\pi+\alpha-\beta, \tag{6}$$

which is the condition that the slip-lines at F make 45° with the centre line. A prerequisite for the solution of this problem is the coordinates of the slip-line network in $CEDF$, for a number of values of the ratio AC/BC. Although no calculation of the field defined by unequal circular arcs has been published, it may be useful to indicate how the solution might continue. An angle β is chosen, corresponding to a ratio AC/BC for which an equiangular network $CEDF$ is known. The practical requirement that F should be a nodal point, or that θ and ψ should be multiples of the angular interval $\Delta\phi$ of the network, restricts α to one or other of the values $\Delta\phi-(\tfrac{1}{4}\pi-\beta)$, $2\Delta\phi-(\tfrac{1}{4}\pi-\beta)$, etc. Choosing a value of α we calculate the reductions corresponding to values of ψ equal to 0, $\Delta\phi$, $2\Delta\phi$, etc. For each of these reductions the mean compressive stress p_0 at F (say) is then determined by the condition for equilibrium of the undrawn sheet to the left of ADF. An application of Hencky's equations gives q' directly. Finally μ is obtained from (4), and t' from (5). The result is a series of values of t' for a specific α and a series of paired values of r and μ. By starting with another value of β we can obtain a second series of paired values; in general, however, α would be different. This can evidently be avoided only if the particular set of ratios AC/BC are such that the interval between the corresponding values of β is $\Delta\phi$. Supposing this to be so, we can assemble, by repetition, values of t' for the given die-angle and paired values of μ and r covering an area in the (μ, r) plane. Interpolation will then yield the drawing stress for arbitrary values of the reduction and coefficient of friction.

We now turn to an approximate method of estimating the effect of die friction on the drawing stress. This may be based on the theoretical indication† that the pressure on the die is but little affected by friction provided μ is less than about 0·1, which is usually so in good drawing practice. We assume, therefore, that

$$q' \sim q, \tag{7}$$

and hence, from (5),
$$t' \sim (1+\mu\cot\alpha)t, \tag{8}$$

† See R. Hill and S. J. Tupper, op. cit., p. 163.

where q and t are the die-pressure and drawing stress for the die when smooth (corrected for work-hardening, if necessary). Sachs and other writers have taken account of friction in a more complicated way, but their theories retain the defect already noticed for smooth dies.

For a given reduction the frictional contribution to the drawing load decreases as the die-angle is increased; on the other hand the contribution from the non-useful work increases. For a given reduction and coefficient of friction there will, therefore, be an optimum die-angle for which the drawing load is least. This can be directly determined by using (8) in conjunction with Fig. 34. For a non-hardening material and a coefficient of friction of 0·05, for example, it is found that the relation between the optimum half-angle $\bar{\alpha}$ and the fractional reduction r can be closely represented by the relation $\bar{\alpha}° = 40r$ $(r < 0·5)$; when $\mu = 0·10$ the optimum is about 10 per cent. greater. The optimum angle is less for an annealed metal than it is for a pre-strained metal; precise values for any given stress-strain curve can be found by applying the correction (iii) for hardening to t, before substitution in (8). Since the load rises sharply for die-angles less than the optimum, and only gradually for those greater, it is better when grinding dies to err on the side of a too-large angle.

(v) *The influence of back-pull.* Suppose that a tensile force F (per unit width) is applied to the sheet approaching the die. If there is no die friction all boundary conditions in the zone of plastically-deforming material are evidently satisfied if a hydrostatic tensile stress $f = F/H$ is added at every point, the slip-line field being unaltered. This is equally true for a work-hardening material, provided the law of hardening is independent of the hydrostatic component of stress. The rigid part of the plastic zone is, of course, altered, and it is assumed that the rigid material is able to sustain the new system of boundary stresses. The pressure on the die is reduced by f (a change tending to minimize die wear), while the drawing stress is augmented by f, and the drawing force by fh or Fh/H. It is conventional to define a factor b, expressing the dependence of the drawing force T and die load Q on the back-pull F according to the relations

$$T = T_0+(1-b)F, \qquad Q = T_0-bF, \qquad (9)$$

where T_0 is the drawing force (and die load) when no back-pull is applied. Hence, when the friction is zero, $b = 1-h/H = r$. The amount of back-pull that can be applied is limited by the onset of necking in the drawn sheet when $T = 2kh$.

If, however, the die is rough the problem is very much more difficult. A change in the pressure on the die alters the angle of intersection of the slip-lines with the wall (equation (4)), and with it the whole slip-line field. The effect of back tension on the stress distribution is therefore not merely additive. Failing an accurate solution we may obtain an approximation[†] for the back-pull factor by introducing the assumption (7), and so writing

$$Q' = (1+\mu \cot \alpha)(H-h)q' \sim (1+\mu \cot \alpha)(H-h)q = (1+\mu \cot \alpha)Q.$$

Hence, if T' and T'_0 are the drawing forces for a rough die with, and without, back-pull,

$$T' = Q'+F \sim (1+\mu \cot \alpha)Q+F$$
$$= (1+\mu \cot \alpha)(T_0-rF)+F, \text{ from (9)},$$
$$\sim T'_0+(1-b)F,$$

where
$$b = (1+\mu \cot \alpha)r. \qquad (10)$$

This may be compared with a formula due to Lunt and MacLellan:[‡]

$$b = 1-(1-r)^{1+\mu \cot \alpha}. \qquad (11)$$

This follows either from Sachs' theory or from the assumption that the plastic material is drawn towards the virtual apex of the die (see footnote on p. 171); the same expression is obtained if work-hardening is included. The formulae (10) and (11) agree (to the first order) when μ is small, while (10) predicts the greater b when μ is large; however, both rest on assumptions which are likely to be less accurate when μ is large.

The efficiency of back-pull drawing depends on whether the work done against the back-pull by the sheet can be usefully recovered. Let us suppose that this is possible, in order to set an upper limit to the efficiency. The external energy expended per unit volume of material is then $t'-f$, which, from (9), is equal to

$$t'_0+\left(\frac{1-b}{1-r}\right)f-f = t'_0-\left(\frac{b-r}{1-r}\right)f.$$

Adopting the formula (10) for b, it follows that there is a saving of work equal to $f\mu \cot \alpha r/(1-r)$ per unit volume. The saving is proportional to the back-pull, and is greater the rougher the die.

(vi) *Application of the plane strain theory to wire-drawing.* At present there exists no analysis of the stress distribution in wire-drawing (a

---

[†] R. Hill, *Journ. Iron and Steel Inst.* **161** (1949), 41.

[‡] R. W. Lunt and G. D. S. MacLellan, *Journ. Inst. Metals*, **72** (1946), 65. Their analysis is for wire-drawing; this is the plane strain analogue. See also G. D. S. MacLellan, *Journ. Iron and Steel Inst.*, **158** (1948), 347.

problem of axial symmetry) of an accuracy comparable with that for sheet-drawing. There is available only the theory of Sachs† which neglects, as we have seen, the non-useful work of distortion. Sachs' formula for the drawing stress is

$$t = Y\left(1 + \frac{\tan\alpha}{\mu}\right)\left[1 - \left(\frac{a}{A}\right)^{\mu \cot\alpha}\right],$$

where A and a are the initial and final cross-sectional areas, respectively. By allowing μ to tend to zero, we recover the expression $t = Y\ln(A/a)$, corresponding to uniform extension. Note that

$$\left(\frac{a}{A}\right)^{\mu\cot\alpha} = \exp\left(\mu\cot\alpha\ln\frac{a}{A}\right) \sim 1 - \mu\cot\alpha\ln\frac{A}{a}.$$

The circumstance that Sachs's theory is a good approximation for small die-angles and sufficiently large reductions, both for wire- and sheet-drawing, suggests the assumption that t/Y (for a wire) is the same function of the reduction in *area* $(1-a/A)$ as $t/2k$ (for a sheet) is of the reduction in *thickness* $(1-h/H)$. Thus, when the die is smooth and the material non-hardening, the drawing stress for a reduction in area r and an angle α is equal to Y times the ordinate in Fig. 34 corresponding to the same angle and a reduction in thickness r. This empirical rule is obviously consistent for reductions where Sachs' theory is accurate, and should at least reproduce the qualitative dependence of the non-useful work on die-angle at lower reductions. Evidence for this is provided by the data of Linicus and Sachs‡ for brass wires. They estimated the friction by using rotated dies, and also obtained stress-strain curves for the drawn wires; they were thus able to make rough corrections for friction and work-hardening, finally deriving the relation between drawing stress and reduction in area for a non-hardening material drawn through a smooth die. The agreement with Fig. 34 is remarkably close, for example in the spacing of the curves for different die-angles, in the characteristic inflexion, in the asymptotic approach to Sachs's curve, and even in magnitude.

Friction may be included by the factor (8); this leads to a slightly greater load than Sachs' formula in the range where the latter is most accurate. Strain-hardening can be allowed for by the method (iii) (the equivalent strain is equal to the appropriate ordinate of Fig. 34). The

---

† For an account of Sachs' theory, and empirical modifications of it, see R. W. Lunt and G. D. S. MacLellan, op. cit., p. 176. A wide review of wire-drawing literature has been given by G. D. S. MacLellan, op. cit., p. 176.

‡ W. Linicus and G. Sachs, *Mitt. deutschen Materialprüfungsanstalten*, 16 (1931), 38.

existence of an optimum die-angle giving minimum load has been observed in many experiments, for example by Linicus and Sachs and by Weiss,† who fitted his data for copper wires by the relation $\tilde{\alpha}° = 30r$ (coefficient of friction not known). The derivation of the formula for the back-pull factor can be taken over unchanged. Although the linear dependence of the load on the back-pull is well confirmed, the lack of an experimental determination of μ prevents a test of the formula for b.‡

A direct determination of μ, under actual conditions, demands a knowledge of the mean pressure on the die. This is more difficult to obtain in wire-drawing than in sheet-drawing, where the two halves of the die are separate and the resultant load can be measured directly. The difficulty can be overcome, in principle, by the use of a die split in two along a plane through the axis; the force needed to hold the halves together during drawing is to be measured. Let Q_s be the splitting force, that is, the transverse component of the reaction between the wire and one-half of the die, and let $Q(= T'-F)$ be the longitudinal component of the reaction on the whole die. Then

$$Q = q'(A-a)(1+\mu\cot\alpha), \qquad (12)$$

where q' is the mean pressure on the die. A resolution in the transverse direction of forces on one half of the die gives

$$Q_s = \frac{q'}{\pi}(A-a)(\cot\alpha-\mu). \qquad (13)$$

Combining (12) and (13), we obtain a formula for μ in terms of measurable quantities:

$$\mu = \frac{1-\pi(Q_s/Q)\tan\alpha}{\tan\alpha+\pi(Q_s/Q)}. \qquad (14)$$

3. Ironing of a thin-walled cup§

The particular ironing process to be analysed here is one where the wall thickness of a cup is reduced while the internal diameter is kept constant. This is achieved by forcing the cup, mounted on a tightly fitting punch, through a die (Fig. 38a). The circumferential, or hoop, strain imparted to an annular element on the inside of the wall is exactly zero, while it increases steadily for annular elements of increasing initial radius. When the wall thickness is a sufficiently small fraction of the

---

† L. Weiss, *Zeits. Metallkunde*, **19** (1927), 61. See also E. L. Francis and F. C. Thompson, *Journ. Inst. Metals*, **46** (1931), 313.

‡ See the critical review of the literature on back-pull by J. G. Wistreich, *Journ. Iron and Steel Inst.* **157** (1947), 417; the effects on mechanical properties are also discussed.

§ R. Hill, ibid. **161** (1949), 41.

cup diameter, however, the hoop strain is everywhere negligible compared with the strain component resulting from the reduction in thickness. Thus, if W and w are the wall thicknesses before and after ironing, and d is the punch diameter, the maximum hoop strain is roughly $2(W-w)/d$, while the reduction in thickness involves a strain not less than $\ln(W/w)$. For the practical range of reductions, say up to 50 per cent., the hoop strain is always about a fraction $2w/d$ of the strain resulting from the change of thickness. If d is not less than, say, $30W$ it should be a good approximation to treat the problem as one of plane strain,

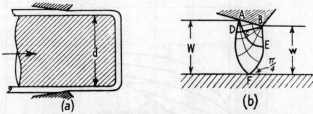

FIG. 38. (a) Ironing a thin-walled tube or cup. (b) The slip-line field when the deformation is treated as approximately plane strain.

and also to neglect the contribution of the hoop stress in the equations of equilibrium.

We shall be concerned only with the steady state, which is established if there is no friction between the wall and the punch, and if the length of the cup is sufficiently great. It is evident that the problem is identical with that of sheet-drawing if we regard the surface of the punch as the centre line of the sheet (Fig. 38 b). The punch load, transmitted to the wall of the cup through its base, is

$$L = \pi w(d+w)t',$$

where t' is the mean tension in the ironed wall. From (8) this is

$$L = \pi w(d+w)(1+\mu \cot \alpha)t,$$

where t is to be read from Fig. 34. It is customary in ironing practice to regard L as a function of the *absolute* reduction for a given initial wall thickness. An appropriate non-dimensional measure of the load would then be $L/2\pi dWk$, and we may write, with sufficient accuracy,

$$\frac{L}{2\pi dWk} = (1+\mu \cot \alpha)(1-r)\frac{t}{2k}. \tag{15}$$

$L/2\pi dWk$ is shown in Fig. 39, when $\mu = 0$, as a function of $r = 1-w/W$ (this is a suitable non-dimensional measure of the absolute reduction since W is being held constant). All curves would ultimately pass through a maximum and return to zero. The broken curve $(1-r)\ln\{1/(1-r)\}$, representing the load for uniform compression, is included for comparison; it reaches a maximum value of $1/e$ (~ 0.368) for a fractional reduction $1-1/e$ (~ 0.632). Also shown is the locus $t = 2k$ which represents the limit set by yielding of the ironed wall and intersects the lower curve at its maximum. The effect of the factor $(1+\mu\cot\alpha)$ is to

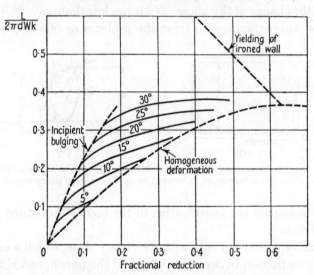

Fig. 39. Ironing load versus reduction in thickness for various semi-angles (smooth die and punch; no work-hardening).

make the curves cross, indicating an optimum die-angle for which the load is least (see (iv) above). The general effect of work-hardening is to increase the slopes of the curves without altering their relative positions.

If the punch is not perfectly smooth, frictional stresses will be induced where the wall and the punch move with different speeds, that is, to the left of F where the punch moves faster than the wall (the ironed wall is obviously carried along with the punch). The presence of this friction is recognizable by the decrease of the load, since progressively less of the wall remains un-ironed. The problem is not then strictly one of steady motion, since the slip-line field must be continually altering. However, we can easily form a rough estimate of the influence of the friction. If the slip-line field intersects the punch in one point only, the frictional stresses are equivalent to a negative back-pull. The frictional force F on the punch therefore decreases the tension in the ironed wall by the amount $(1-b)F$, where b is the back-pull factor defined in equation (9).

Consequently, the load on the punch is increased, not by the full value of the friction, but only by a fraction bF. In practice the frictional force F would have to be regarded as an unknown quantity to be determined experimentally. When the reduction is so large that the slip-line field meets the surface of the punch over a finite length, the frictional stresses can no longer be regarded as equivalent to a negative back-pull. An analysis based on a simple extension of Sachs' theory of drawing, which should be fairly accurate in this range of reductions, has been given by Sachs, Lubahn, and Tracy.†

4. Sheet extrusion‡

(i) *Extrusion through a wedge-shaped die.* In direct extrusion the metal billet is held in a container and forced through a die by a ram (Fig. 40).

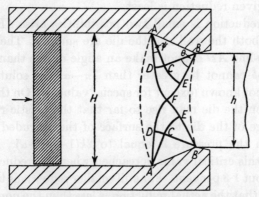

Fig. 40. Slip-line field for direct extrusion through a wedge-shaped die.

We are concerned now with the extrusion of a sheet through a wedge-shaped die of total included angle 2α, the thickness being reduced from H to h. The process is clearly very similar to that of sheet-drawing, the main difference being that in the one process the sheet is pushed, and in the other pulled, through the die. Thus, all boundary conditions directly affecting the plastic zone are satisfied by the slip-line field used in the problem of sheet-drawing (cf. Figs. 32 and 37). The rigid part of the plastic region and the stresses in the non-plastic material are, however, different, owing to the new distribution of surface forces. When the die

---

† G. Sachs, J. D. Lubahn, and D. P. Tracy, *Journ. App. Mech.* 11 (1944), A–199.
‡ R. Hill, *Journ. Iron and Steel Inst.* 158 (1948), 177. Incorrect slip-line fields have been proposed by F. K. Th. van Iterson, *Plasticity in Engineering* (Blackie, 1947), and by E. Siebel, *Journ. Iron and Steel Inst.* 155 (1947), 526; neither writer examined the associated velocity solution.

is smooth, the state of stress in the deforming part of the plastic region differs from that in sheet-drawing only by the addition of a hydrostatic pressure, of amount equal to the drawing stress t (thereby, the resultant external force on the extruded sheet is reduced to zero). Hence, if there is no friction between the container wall and the sheet, the pressure of extrusion (defined as load divided by area of ram) is equal to the mean drawing stress for the same die. Thus, the extrusion pressure for a smooth die may be read directly from Fig. 34, while the pressure on the die is equal to H/h times the pressure on the die in drawing (Fig. 33). The contributions of die friction and strain-hardening may be estimated by the methods described in Sections 2 (iii) and 2 (iv). The contribution of wall friction to the load is simply additive in this range of reductions. As with sheet-drawing there is an optimum die-angle for which the load to achieve a given reduction is least.

When the reduction is small two possibilities arise. For simplicity suppose that both the wall and the die are smooth. Then, on the one hand, the slip-line AD cannot make an angle of less than 45° with the wall, and so ψ cannot be greater than $\tfrac{1}{2}\pi-\alpha$; the solution for lower reductions is not known (except for special values†). On the other hand, the pressure on the die may rise so far that the plastic region spreads round the edge of the die to the surface of the extruded sheet; this is possible when the pressure is equal to $2k(1+\tfrac{1}{2}\pi+\alpha)$. For example, when $\alpha = 5°$ this critical value is reached when the reduction has been lowered to about 1·8 per cent. For still smaller reductions the theoretical implication is that the actual reduction is less than the nominal, due not merely to elastic recovery but to plastic flow round the die corner. It is furthermore implied that recovery may be complete, and that most of the billet then passes through undeformed, plastic distortion occurring only near the surface. However, the theory may cease to be valid for very small reductions, since the effects of elastic compressibility may not be negligible.

(ii) *Inverted extrusion through a square die.* The extrusion is said to be inverted when the ram is held fixed and the die is forced into the billet (alternatively, the container and the billet can be moved together towards a stationary die). A frictional resistance between the billet and the container is thereby prevented. When the die is square (or wedge-shaped with a large included angle) it is observed that some material in the corner between the wall and the die is held back and is not extruded with the rest of the billet. The boundary of this material,

† R. Hill, *Journ. Iron and Steel Inst.* **156** (1947), 513.

called dead metal, is marked by a narrow zone of intense shear (a definite fracture cannot usually be seen, because any gap or crack which opens is at once closed by the severe hydrostatic pressure). The dead metal separates off from the main billet during the early stages of extrusion and thereafter constrains the flow as if it were a perfectly rough die (no doubt its shape is modified immediately after fracture). Since the surface of fracture must evidently depend, among other variables, on the

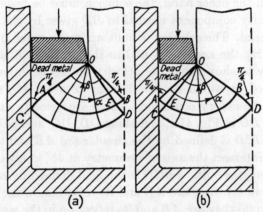

Fig. 41. Slip-line field for inverted extrusion through a square die giving a reduction of (a) more than 50 per cent., and (b) less than 50 per cent.

friction over the die, the temperature and speed of extrusion, and the physical properties of the end of the billet, it is impossible to predict its position by theory. Nevertheless, for a useful estimate of the extrusion pressure, we may assume a boundary for the dead metal which allows a simple solution and at the same time is reasonably close to what is actually observed. Fortunately, it has been shown that the extrusion pressure is but little affected by the choice of boundary, within fairly wide limits.†

Possible slip-line fields are shown in Fig. 41 for reductions greater, and less, than 50 per cent. respectively. The boundary of the dead metal must be a slip-line, since the maximum frictional shear-stress is induced there. This slip-line meets the container at 45° since, by hypothesis, there is no frictional stress on the wall. We now verify that the velocity boundary conditions are satisfied when OB is straight, so that the slip-lines in OAB are radii and circular arcs. It is supposed for ease of description that the die is stationary. In Fig. 41 a the velocity solution must be begun in $ACDE$, since this is the only region where two boundary

† R. Hill, op. cit., p. 181.

conditions are known; these are that the normal component of velocity should be zero on AC, while on CD it should be equal to the normal component of the speed of the billet. The solution continues in the order BED, AOB. The velocity component normal to AO is uniform (in fact zero) and hence, by the now-familiar theorem, the normal component is constant on each radius through O, in particular OB. The solution is therefore compatible with the rigid-body motion of the extruded sheet. In Fig. 41 b, on the other hand, the solution must be begun in $BECD$, using the velocity component normal to BD given by the speed of the extruded material. The solution is built up in the order $BECD$, ACE, BEO, AEO. By the argument used in the problem of sheet-drawing (p. 165) it may be shown that the velocity on AO is uniform and that, when calculated, it would be found to be zero on AO, so satisfying the remaining boundary condition. There are velocity discontinuities along CD and DO in Fig. 41 a, and along OD, DC, and CO in Fig. 41 b.

The field $ABCD$ is defined by the circular arc AB and the condition that slip-lines intersect the axis of symmetry at 45° in (a), or the wall at 45° in (b). Thus, for both, $ABCD$ is the field defined by equal circular arcs (Fig. 30); in (a) these are AB and its reflection in the axis of symmetry, while in (b) they are AB and its reflection in the wall. In (a) OB is inclined at 45° to the axis of symmetry, and the angle AOB is chosen so that the slip-line OAC intersects the wall at 45°; in (b) OA makes 45° with the wall, and the angle AOB is chosen so that OBD intersects the axis of symmetry in 45°. When the reduction is sufficiently small (approximately 8 per cent.) the plastic region extends round O to the free surface of the extruded billet. At the other extreme, geometrical requirements prevent the application of the present solution beyond the reduction for which OA is tangential to the die.

The value of the mean compressive stress is determined at one point (on OB, say) by the condition that the resultant longitudinal force on the extruded sheet should be zero; Hencky's equations then give the stress at any point in the field. It is interesting to notice that the mean pressure on the die for a fractional reduction r in (a) is equal to the mean pressure on the die for a reduction $(1-r)$ in (b); this is a simple consequence of the property that the respective fields $ABCD$ are reflections of each other. The extrusion pressure P is shown in Fig. 42 as a function of the reduction in thickness $(1-h/H)$; when the reduction is 50 per cent., $P = k(1+\tfrac{1}{2}\pi)$. k here represents the yield stress in shear at the mean temperature and speed of extrusion; billets are generally extruded hot, at temperatures where the work-hardening is small. The broken curve in the Figure is the pressure $2k\ln(H/h)$ corresponding to uniform compression. The relative expenditure of non-useful work is seen to be considerable; it is roughly the same as that observed by Eisbein and Sachs† for cylindrical billets of brass.

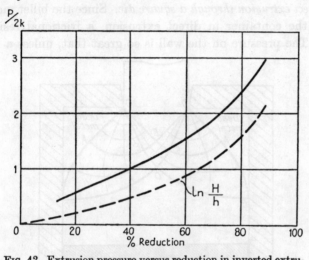

Fig. 42. Extrusion pressure versus reduction in inverted extrusion through a square die. The broken curve corresponds to uniform deformation.

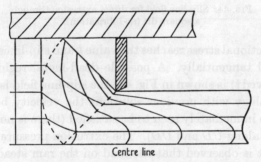

Fig. 43. Distortion of a square grid in inverted extrusion through a square die giving 50 per cent. reduction.

The distortion of a square grid has been calculated for a reduction of 50 per cent.‡ (Fig. 43). Apart from the cusp on the axis of symmetry, there is a close resemblance with the distortion observed in small-scale experiments on cylindrical billets,§ particularly the severe dragging-back

† W. Eisbein and G. Sachs, *Mitt. deutschen Materialprüfungsanstalten*, **16** (1931), 67.
‡ For details, see the paper by Hill, op. cit., p. 181.
§ C. E. Pearson, *The Extrusion of Metals* (Chapman and Hall, 1944).

of the surface layers. The cusp is less pronounced when the reduction is small, the acute angle between the axis and the deformed transverse lines of the grid being $\cot^{-1}(H/h-1)^2$ when the reduction is less than 50 per cent. There is no cusp for the field in Fig. 44.

(iii) *Direct extrusion through a square die.* Since the billet moves relatively to the container in direct extrusion, a frictional resistance is induced. The pressure on the wall is so great that, unless a lubricant

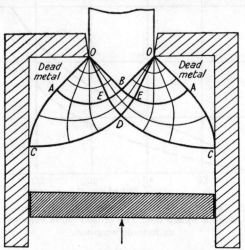

Fig. 44. Slip-line field for direct extrusion through a square die (no lubrication).

is used, the frictional stress reaches the value k and slip-lines of one family meet the wall tangentially. A possible dead-metal region (similar to what is observed†) is shown in Fig. 44; the slip-line field is self-explanatory. By analogy with previous solutions the velocity boundary conditions can be immediately seen to be satisfied (there is now no velocity discontinuity along CD and OD). The extrusion pressure has not been calculated. It is observed that the load on the ram steadily decreases during extrusion, attaining its minimum value shortly before the completion of extrusion. A rapid rise then follows, owing to the difficulty of ejecting the remaining disk. It is found that the minimum load is roughly equal to the steady-state load in inverted extrusion.

5. Piercing

In this process a billet is held in a container and hollowed out by a punch. When the problem is regarded as one of plane strain the only

† C. E. Pearson, op. cit., p. 185.

difference between the steady states of inverted extrusion and piercing with a smooth container is that the positions of the central axis and the wall are interchanged, provided the dead metal (if present) is identical in shape. We shall be concerned only with piercing by a flat punch. It is assumed that a false head of dead metal, shaped like a 90° wedge, becomes attached to the punch and is retained there during the steady-state period; this presumably requires a certain minimum degree of friction on the punch. The slip-line field for a reduction of less than 50

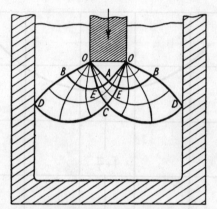

Fig. 45. Slip-line field for piercing with a rough flat die (reduction less than 50 per cent.)

per cent. is shown in Fig. 45; this should be compared with Fig. 41 b. When the container is not perfectly lubricated, the angles at which the slip-lines meet the wall must be suitably modified.

When there is no wall friction, the pressure on the punch is clearly equal to (extrusion pressure)/(reduction in thickness), the extrusion pressure being that corresponding to the same reduction in inverted extrusion. The relation between the punch pressure P and the reduction in thickness, i.e. (punch width)/(width of container), has been derived from Fig. 42 and is shown in Fig. 46. The pressure has a minimum value of $2k(1+\tfrac{1}{2}\pi)$ for a 50 per cent. reduction, and the curve is symmetrical about this point (in the range of validity of the solution). Such a flat minimum was obtained by Siebel and Fangmeier† in experiments on the piercing of cylindrical lead billets with cut-back flat punches. They found that the pressure rose rapidly towards the steady-state value, and that the rise was steepest for large reductions. The steady state persisted until the punch had penetrated to within half its width from

† E. Siebel and E. Fangmeier, *Mitt. Kais. Wilh. Inst. Eisenf.* **13** (1931), 28.

the base of the container. In the present problem the punch pressure needed to begin piercing would be $2k(1+\tfrac{1}{2}\pi)$ for reductions of less than 50 per cent. (see Chap. IX, Sect. 5 (i)), while for reductions greater than 50 per cent. piercing would not begin until the steady-state pressure was applied. This is in broad agreement with the observations of Siebel and Fangmeier, if allowance is made for the rounding of the pressure-penetration curve by elastic strains and work-hardening. When the

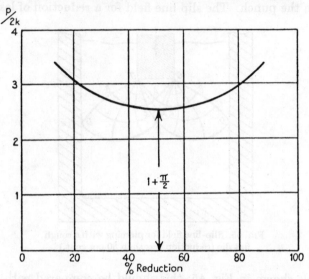

Fig. 46. Relation between die-pressure and reduction in thickness in piercing.

reduction is very small the elastic compressibility becomes important and the steady-state pressure should approximate that needed for deep punching into an infinite medium. If we accept the theoretical result of Bishop, Hill, and Mott (p. 106), the maximum pressure should be about $4 \cdot 5 \times 2k$ (the exact value depends on the elastic constants). It is not clear what the limiting pressure should be at the other extreme as the reduction is indefinitely increased.

6. Strip-rolling

(i) *Scope of the theory*. The following account of theories of rolling is restricted to the cold rolling of strip. This is the term applied to thin sheets with a thickness not more than about 0·2 inches, and a width/thickness ratio of at least ten. The strip is reduced in thickness by passing it between two parallel cylindrical rolls whose radius is of the order of one hundred times the thickness of the strip. Since the width of the strip is large compared with the length of the arc of contact between strip and rolls, the constraint of the non-plastic material inhibits lateral spread (it is normally less than 1 or 2 per cent.). The deformation of the strip is therefore essentially plane strain, with the exception of narrow zones near the edges. The calculation of stresses in the rolling of rectangular bars, where the spread cannot be neglected, has hardly progressed beyond the empirical stage.

The accurate solution of the problem of strip-rolling is not yet known, and even the qualitative appearance of the slip-line field has still to be demonstrated. Most present theories assume, in effect, that the material is uniformly compressed while passing between the rolls. Orowan[†] has formulated an approximate method (see Sect. 8 below) of allowing for the expenditure of work in non-uniform distortion, which indicates[‡] that the correction is very small when the arc of contact is greater than the mean thickness of the strip (cf. Sachs' theory of sheet-drawing). Fortunately, apart from skin passes, the reductions given in industrial practice are such that the arc of contact is rarely less than three times as great as the strip thickness. In this range the error in the elementary theory is likely to be smaller than the limits within which many of the variables can be specified in practice, for example the physical properties of the strip, the frictional conditions, and the elastic distortion of the rolls, to name only a few. It is worth while, therefore, to describe the elementary theory in some detail.

(ii) *Elastic distortion of the rolls.* In any process of shaping metals the apparatus is elastically distorted, and this must be allowed for in an accurate calculation of the load and the energy expenditure. It is particularly necessary to make this allowance in strip-rolling where the rolls are flattened so much that the arc of contact is often increased by some 10 or 20 per cent., and even doubled when the draft is very small. The only practicable method of estimating roll-flattening appears to be that due to Hitchcock,[§] who replaced the actual pressure distribution over the roll by an elliptical one giving the same total load. It is known from the theory of elastic contact between bodies, due to Hertz, that the roll surface under this pressure distribution is deformed into a cylindrical

---

[†] E. Orowan, *Proc. Inst. Mech. Eng.* **150** (1943), 140.
[‡] D. R. Bland and H. Ford, ibid. **159** (1948), 144.
[§] J. H. Hitchcock, *Am. Soc. Mech. Eng. Research Publication* (1930), Appendix I. A more detailed investigation of the distortion of the arc of contact has shown that Hitchcock's formula cannot be bettered without considerable computation: D. R. Bland, *Proc. Inst. Mech. Eng.* (to be published).

arc of larger radius† (the change in shape due to the frictional forces is probably very small and is neglected). The shape of the flattened roll is thus expressible by one parameter only, namely the radius R' of the deformed arc of contact. This is a great mathematical, and indeed a practical, convenience. The accuracy of Hitchcock's method is not known, but it is difficult to see how his assumption could be improved in any simple or useful way.

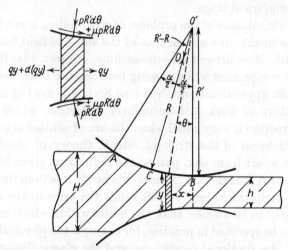

Fig. 47. Elastic distortion of a cylindrical roll with (inset) the stresses acting on a slice of the strip. O is the spindle axis and O' is the centre of curvature of the deformed arc of contact (after Hitchcock).

If R is the radius of the undeformed roll, the change in curvature due to a force P per unit width acting over an arc-length L is

$$\frac{1}{R}-\frac{1}{R'}=\frac{P}{cL^2}, \qquad (16)$$

where c is a constant depending on the elastic properties of the rolls and is equal to $\pi E/16(1-\nu^2)$. If P is measured in tons per inch and lengths in inches, c has a value of about $2 \cdot 9 \times 10^3$ tons per square inch for steel rolls.

Referring to Fig. 47, which represents a transverse section through one roll, let O be the centre of the roll, or the point where the line of the spindle axis intersects the section, and let O' be the centre of the circular arc of contact AB. Since the assumed pressure distribution is symmetrical about the mid-point C of the arc of contact, it follows that O' lies on CO

† See S. Timoshenko, *Theory of Elasticity*, p. 349 (McGraw-Hill Book Co., 1934).

produced. If we neglect that part of the arc of contact due to the elastic recovery of the rolled strip, the roll surface must be horizontal at the point of exit B. Thus $O'B$ is vertical. Let the angular arc of contact be denoted by α (angle $AO'B$); in strip-rolling α is generally less than 0·15 radians. B is evidently a distance $\frac{1}{2}(R'-R)\alpha$, approximately,† ahead of the line through the two roll centres; this is a fraction $\frac{1}{2}(1-R/R')$ of the arc of contact. The length of the arc of contact, and the distance apart of the centres of the rolls, is determined by the reduction given to the strip. If δ is the draft, or the difference $H-h$ between the initial and final thicknesses of the strip, it is easy to show by geometry that the length L of the arc of contact AB is approximately equal to $\sqrt{(R'\delta)}$. Substitution in (16) gives

$$R' = R\left(1+\frac{P}{c\delta}\right), \tag{17}$$

which is Hitchcock's formula for the radius of the deformed arc of contact. The angular measure α of the deformed arc, with respect to its centre, is L/R' or $\sqrt{(\delta/R')}$ (note, also, that $L = \delta/\alpha$).

(iii) *General considerations.* Since the speed of the strip is increased in passing through the rolls there should be a point N on the arc of contact where the local elements of the roll and strip move with the same speed; this is called the neutral point. To the entry side of N, where the roll moves faster than the strip, frictional stresses are induced which draw the material into the roll gap; to the exit side of N, the frictional stresses oppose the delivery of the strip, which here moves faster than the roll. In a steady state the strip adjusts its speed relatively to the speed with which the rolls are driven, in such a way that the external stresses acting on the strip are in equilibrium. This determines the position of the neutral point.

Let θ be the angular position of a generic point on the arc of contact referred to $O'B$, and let ϕ be the angular position of the neutral point N. If $p(\theta)$ is the distribution of normal pressure‡ on the roll, a small element of the surface, of length $R'\,d\theta$, is acted on by a normal force $pR'\,d\theta$ (per unit width) and a tangential frictional force $\mu p R'\,d\theta$ (μ is less than 0·1 in cold rolling with a lubricant, and so the frictional stress μp is almost always less than the shear yield stress k).§ The equation of

---

† In this and subsequent formulae $\sin \alpha$ is replaced by α, and $\cos \alpha$ by unity, with an error not greater than 1 per cent.

‡ p must not be confused with the mean compressive stress, for which the same symbol has previously been used in Hencky's equations.

§ The modification of the theory when μ is so large that the frictional stress attains the value k over part of the arc of contact (as in hot-rolling) has been given

equilibrium for the horizontal forces acting on the strip is

$$T = 2\int_0^\alpha p\sin\theta R'\,d\theta - 2\left(\int_\phi^\alpha \mu p\cos\theta R'\,d\theta - \int_0^\phi \mu p\cos\theta R'\,d\theta\right),$$

where $T = T_f - T_b$ is the difference of the front and back tensions (per unit width) applied to the strip. Replacing $\sin\theta$ by θ, and $\cos\theta$ by 1, to a sufficient approximation (since α is generally less than 0·15 radians) we obtain

$$\frac{T}{2R'} = \int_0^\alpha p\theta\,d\theta - \mu\left(\int_\phi^\alpha p\,d\theta - \int_0^\phi p.d\theta\right). \tag{18}$$

The resultant vertical force P per unit width acting on the roll is given by

$$P = \int_0^\alpha p\cos\theta R'\,d\theta + \left(\int_\phi^\alpha \mu p\sin\theta R'\,d\theta - \int_0^\phi \mu p\sin\theta R'\,d\theta\right).$$

The terms in brackets is negligible to the present order of approximation, and so

$$\frac{P}{R'} = \int_0^\alpha p\,d\theta. \tag{19}$$

The torque G per unit width acting on each roll is found by taking moments about the spindle axis O for forces on the roll. The lever arm of the force $pR'\,d\theta$ is approximately $\pm(R'-R)(\tfrac{1}{2}\alpha-\theta)$, depending on which side of C the element lies. The lever arm of the force $\mu p R'\,d\theta$ is R, to the same order of approximation. Hence

$$\frac{G}{RR'} = \mu\left(\int_\phi^\alpha p\,d\theta - \int_0^\phi p\,d\theta\right) + \left(\frac{R'}{R}-1\right)\int_0^\alpha p(\theta-\tfrac{1}{2}\alpha)\,d\theta. \tag{20}$$

The contribution of the moment of the normal pressure is generally omitted in existing theories of rolling, either as being zero when roll-flattening is disregarded, or otherwise negligible. This is a reasonable approximation if the roll-flattening is small, but the correction is desirable in the last passes of a series where α is small and R'/R is large. Another expression for the torque, which is often useful, can be derived from (18) and (20) by eliminating the term in μ common to both equations:

$$G = R'^2\int_0^\alpha p\theta\,d\theta - \tfrac{1}{2}RT - \tfrac{1}{2}(R'-R)\alpha P. \tag{21}$$

by E. Orowan, *Proc. Inst. Mech. Eng.* **150** (1943), 140. A. Nadai, *Journ. App. Mech.* **6** (1939), A–55, has considered frictional conditions such that the frictional stress is proportional to the relative speed of slip.

If the moment of the normal pressure is neglected, the formula for the torque is

$$\bar{G} = RR'\int_0^\alpha p\theta\, d\theta - \tfrac{1}{2}RT.$$

Hence $\quad G = \bar{G}\left[1 - \left(\dfrac{R'}{R}-1\right)\left(\dfrac{\alpha RP}{2\bar{G}}-1-\dfrac{RT}{2\bar{G}}\right)\right].$ (21′)

It will be shown later that G is less than $\bar{G}$ (see equation (33)).

The distribution of the pressure over the roll must, of course, be found by considering the plastic state of stress in the strip. When $p(\theta)$ is known, the roll-force P is determined from equation (19), and the torque G from equation (20) or (21). The rate of external work performed on the rolls and strip is

$$2G\omega + T_f U - T_b U\frac{h}{H},$$

where ω is the angular velocity of the rolls and U is the exit speed of the strip. It is conventional to define a quantity f, known as the forward slip, expressing the relative difference in the speeds of strip and roll at the point of exit:

$$f = (U - R\omega)/R\omega = \frac{U}{R\omega} - 1. \qquad (22)$$

Except when front tension is applied, f is normally less than 0·05 in strip-rolling. In unit time a volume Uh (per unit width) of strip is rolled, and so the work per unit volume may be expressed as

$$W = \frac{2G}{(1+f)Rh} + t_f - t_b, \qquad (23)$$

where $t_f = T_f/h$ is the mean front tension stress and $t_b = T_b/H$ is the mean back tension stress. Not all this energy is used in plastic deformation, since a part is dissipated by the friction over the arc of contact, in proportion at each point to the difference in the speeds of the roll and strip.

A relation between the forward slip f and the neutral angle ϕ may be obtained without a detailed analysis if we neglect the variation of velocity across any vertical section. If h_n is the strip thickness at the neutral point the mean speed over the corresponding vertical section is Uh/h_n. This, by the above assumption, is equal to $R\omega$, and so

$$h_n/h \sim U/R\omega = 1+f,$$

from (22). But $h_n - h = R'\phi^2$, with neglect of higher order terms in ϕ,

and hence
$$\phi \sim \sqrt{\left(\frac{fh}{R'}\right)}, \qquad \frac{\phi}{\alpha} \sim \sqrt{\left\{\frac{(1-r)f}{r}\right\}}. \tag{24}$$

This formula is not valid for very large reductions.

(iv) *Elementary theory of the stress distribution.* Consider the stresses on a thin slice of the strip contained between two vertical sections distant x and $x+dx$ from the exit plane $O'B$ (Fig. 47). Let $y(x)$ be the thickness of the strip at any point. Let qy be the resultant *tensile* force acting over a vertical section; q is then the mean, taken over the section, of the normal component of stress in the horizontal direction. The two forces $pR'\,d\theta$, acting on the ends of the slice, have a horizontal resultant of amount $2pR'\sin\theta\,d\theta$, or $p\,dy$. The two forces $\mu pR'\,d\theta$ have a resultant of amount $2\mu pR'\cos\theta\,d\theta$, or $2\mu p\,dx$. The equation of equilibrium of the slice is therefore
$$d(qy)+p\,dy\pm 2\mu p\,dx = 0, \tag{25}$$
where the upper sign refers to the exit side of the neutral point N, and the lower sign to the entry side.

So far the analysis has been exact; we now introduce the assumption that characterizes the elementary theory:
$$p+q = 2k, \tag{26}$$
where k, as usual, denotes the yield stress in shear. This equation may be derived from the yield criterion by making three separate assumptions, namely (i) that the stress is uniformly distributed over a vertical section, (ii) that p and q are the magnitudes of the principal stresses at any point on the arc of contact, and (iii) that the plastic region extends everywhere between the planes of entry and exit. In many presentations the second assumption is replaced by the statements that the principal stress axes are everywhere horizontal and vertical, and that the magnitude of the vertical principal stress is then approximately equal to p. However, it seems preferable to make only the one postulate (26). If the material work-hardens, k is to be regarded as a function of x. Assuming that the deformation is approximately uniform compression, the value of $2k$ at a section x is the ordinate of the compressive stress-strain curve, obtained under conditions of plane strain, corresponding to the abscissa $\ln(H/y)$. Alternatively, $2k$ is equal to $2/\sqrt{3}$ times the ordinate of the uniaxial stress-strain curve at the abscissa $(2/\sqrt{3})\ln(H/y)$ (this is the equivalent strain; see p. 30). When k varies, equation (25) must be integrated by a small-arc process. A simple method of allowing for work-hardening is described later.

The theory of strip-rolling expressed in equations (25) and (26) is due to von Karman† (1925). Assuming, now, that k is constant, the elimination of q gives

$$\tfrac{1}{2}y\frac{dp}{dx}\mp\mu p = k\frac{dy}{dx}. \qquad (27)\ddagger$$

The boundary conditions are that $q = t_b$ and $p = 2k-t_b$ when $x = L$; $q = t_f$ and $p = 2k-t_f$ when $x = 0$. Since the equation is first order these boundary conditions uniquely define the two solutions starting from the points of exit and entry, and corresponding respectively to the upper and lower sign. The solution must be continued until a value of x is reached where the two values of p are identical; this is the neutral point (the equation of equilibrium of the strip as a whole (18) is automatically satisfied). Since the arc of contact is circular,

$$y = h+\frac{x^2}{R'}, \qquad \frac{dy}{dx} = \frac{2x}{R'},$$

to the usual approximation for small angles of contact. Substitution in (27) gives

$$\tfrac{1}{2}\left(h+\frac{x^2}{R'}\right)\frac{dp}{dx} = 2k\frac{x}{R'}\pm\mu p. \qquad (28)$$

Introducing the non-dimensional quantities

$$\xi = \frac{x}{\sqrt{(R'h)}} = \sqrt{\left(\frac{R'}{h}\right)}\theta = \sqrt{\left(\frac{r}{1-r}\right)}\frac{\theta}{\alpha}, \qquad \eta = \frac{p}{2k},$$

we obtain

$$\tfrac{1}{2}(1+\xi^2)\frac{d\eta}{d\xi} = \xi\pm a\eta, \qquad a = \mu\sqrt{\left(\frac{R'}{h}\right)} = \frac{\mu}{\alpha}\sqrt{\left(\frac{r}{1-r}\right)}, \qquad (29)$$

where r is the fractional reduction δ/H, and a is a parameter whose value is normally between 0·5 and 2·0 in strip-rolling (R'/h is normally in the range 50–500). The boundary conditions are

$$\eta = 1-\frac{t_f}{2k}, \quad \xi = 0; \qquad \eta = 1-\frac{t_b}{2k}, \quad \xi = \sqrt{\left(\frac{r}{1-r}\right)}.$$

In equation (29) the upper sign is to be taken on the exit side of the neutral point, and the lower sign on the entry side. Following the

† Th. von Karman, *Zeits. ang. Math. Mech.* 5 (1925), 139.

‡ This equation is applicable (with the upper sign) to sheet-drawing or extrusion through a die of any contour $y(x)$. When the die is wedge-shaped ($dy/dx = 2\tan\alpha$) the equation leads to the plane strain analogue of Sachs' formula (p. 177) for the drawing stress. A. T. Tselikov, *Metallurg* 6 (1936), 61 (Russian), has based a theory of rolling on an approximation to the arc of contact by its chord; for an account in English, see L. R. Underwood, *Sheet Metal Industries*, (February, 1946), 288.

standard method for solving linear first-order equations we find that

$$\eta = e^{-2a\psi}\left[\left(1-\frac{t_b}{2k}\right)e^{2a\psi_r} - 2\int_\psi^{\psi_r} e^{2a\psi}\tan\psi\,d\psi\right]$$

(entry side, $\psi_r \geqslant \psi \geqslant \psi_n$),

$$\eta = e^{2a\psi}\left[\left(1-\frac{t_f}{2k}\right) + 2\int_0^\psi e^{-2a\psi}\tan\psi\,d\psi\right]$$

(exit side, $\psi_n \geqslant \psi \geqslant 0$),

(30)

where

$$\psi = \tan^{-1}\xi = \tan^{-1}\sqrt{\left(\frac{R'}{h}\right)}\theta, \qquad \psi_r = \tan^{-1}\sqrt{\left(\frac{r}{1-r}\right)} = \sin^{-1}\sqrt{r},$$

and ψ_n, the value of ψ corresponding to the neutral point, satisfies the equation

$$e^{-2a\psi_n}\left[\left(1-\frac{t_b}{2k}\right)e^{2a\psi_r} - 2\int_{\psi_n}^{\psi_r} e^{2a\psi}\tan\psi\,d\psi\right]$$
$$= e^{2a\psi_n}\left[\left(1-\frac{t_f}{2k}\right) + 2\int_0^{\psi_n} e^{-2a\psi}\tan\psi\,d\psi\right].$$

As μ decreases, the neutral point moves towards the point of exit. The least value of μ for which rolling is possible under the given tensions corresponds to $\psi_n = 0$, or to the solution $a = \bar{a}$ of the equation

$$\left(1-\frac{t_b}{2k}\right)e^{2\bar{a}\psi_r} - 2\int_0^{\psi_r} e^{2\bar{a}\psi}\tan\psi\,d\psi = \left(1-\frac{t_f}{2k}\right). \qquad (31)$$

From (19) the roll-force is

$$P = R'\int_0^\alpha p\,d\theta = 2k\sqrt{(R'h)}\int_0^{\sqrt{\{r/(1-r)\}}} \eta\,d\xi.$$

Since the radius R' of the deformed arc of contact depends on P through Hitchcock's formula (17), it can only be found by successive approximation. Thus, a trial value of R' is first used in a calculation of P; the corresponding value of R' given by Hitchcock's formula will generally be different from the trial value, which must therefore be modified. When compatible values of P and R' have been determined, the torque is found from (21):

$$G = R'^2\int_0^\alpha p\theta\,d\theta - \tfrac{1}{2}RT - \tfrac{1}{2}(R'-R)\alpha P$$
$$= 2kR'h\int_0^{\sqrt{\{r/(1-r)\}}} \eta\xi\,d\xi - \tfrac{1}{2}RT - \tfrac{1}{2}(R'-R)\alpha P.$$

The integrals in (30) and the value of ψ_n can only be evaluated by approximate methods. Trinks† has computed a set of solutions when the applied tensions are zero, and has presented the roll-force and peak pressure (but not the torque) graphically in terms of the parameters r and μ/α (the nominal accuracy is uncertain). Nadai‡ has calculated the pressure distributions for various values of r, μ/α, and the front and back tensions, but makes the approximation $\tan\psi \sim \psi$ in order to complete the integration explicitly. A thorough examination of the overall accuracy of Nadai's method does not appear to have been made, but in a few instances it has been found that the roll-force is in error by 5–10 per cent.§ A closer approximation, with an error less than 2 per cent. if no tensions are applied, has been suggested by Bland and Ford.‖ Their formula for the pressure distribution is

$$\left. \begin{aligned} \eta &= \left(1-\frac{t_b}{2k}\right)(1-r)e^{2a(\psi_r-\psi)}\sec^2\psi \quad \text{(entry side)}, \\ \eta &= \left(1-\frac{t_f}{2k}\right)e^{2a\psi}\sec^2\psi \quad \text{(exit side)}. \end{aligned} \right\} \tag{32}$$

When $t_f = 0 = t_b$, an inspection of (30) reveals that their approximation is equivalent to writing

$$2\int_0^\psi e^{-2a\psi}\tan\psi\,d\psi = \int_0^\psi e^{-2a\psi}\cos^2\psi\,d(\tan^2\psi) \sim \int_0^\psi d(\tan^2\psi),$$

and

$$2\int_\psi^{\psi_r} e^{2a\psi}\tan\psi\,d\psi = \int_\psi^{\psi_r} e^{2a\psi}\cos^2\psi\,d(\tan^2\psi) \sim e^{2a\psi_r}\cos^2\psi_r\int_\psi^{\psi_r} d(\tan^2\psi).$$

The contribution of the integral on the exit side is so small that it is an equally good approximation to neglect it altogether (the front tension being moderate). The calculations of Bland and Ford for zero tensions are presented in Fig. 48 in terms of r and a parameter

$$b = \mu\sqrt{\left(\frac{R'}{H}\right)} = \frac{\mu}{\alpha}\sqrt{r}.$$

The roll-force P_0 and torque G_0 are expressed non-dimensionally in terms of their values P_0^* and G_0^* if the pressure were equal to the yield stress at every point (G_0 should be corrected for the moment of the normal pressure; see (21')). The work-hardening being zero, it can be shown without

---

† W. Trinks, *Blast Furnace and Steel Plant*, 25 (1937), 617.
‡ A. Nadai, *Journ. App. Mech.* 6 (1939), A–55.
§ L. R. Underwood, *Sheet Metal Industries*, April, 1946.
‖ D. R. Bland and H. Ford, op. cit., p. 189.

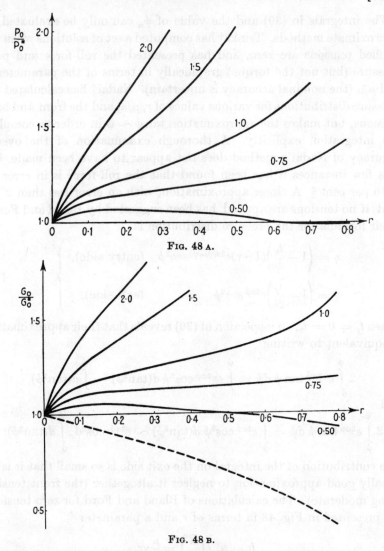

Fig. 48 a.

Fig. 48 b.

Fig. 48. Theoretical relations between (a) roll-force and reduction, and (b) torque and reduction, for various values of $b = \mu\sqrt{(R'/H)}$.

difficulty that $P_0^* = 2k\sqrt{(R'\delta)}$ and $G_0^* = kR\delta$. According to Bland and Ford's theory the least value of b for which rolling is possible at a given reduction, without tensions, is

$$b_{min} = \tfrac{1}{2}\sqrt{(1-r)}\ln\left(\frac{1}{1-r}\right)\Big/\sin^{-1}\sqrt{r}.$$

For this value of b the neutral point falls in the plane of exit. The value of b for which the angle of contact is equal to the angle of friction is $\sqrt{r}$ ($\alpha = \mu$). An empirical formula for P_0/P_0^*, correct to ± 1 per cent. in the range $0 \cdot 1 < r < 0 \cdot 6$ and $P_0/P_0^* < 1 \cdot 7$, is

$$\frac{P_0}{P_0^*} = 1 \cdot 08 + 1 \cdot 79 rb - 1 \cdot 02 r.$$

A formula for G_0/G_0^*, correct to ± 2 per cent. in the range $0 \cdot 1 < r < 0 \cdot 6$ and $G_0/G_0^* < 1 \cdot 5$, is

$$\frac{G_0}{G_0^*} = 1 \cdot 05 + (0 \cdot 07 + 1 \cdot 32 r) b - 0 \cdot 85 r.$$

From (23) the energy expended per unit volume is $2G/Rh(1+f)$ when no tensions are applied, or $2G/Rh$ with an accuracy sufficient for practical purposes; thus the ordinate in Fig. 48 b also represents

$$(1-r)W/2kr.$$

Hence the broken curve in Fig. 48 b, with equation $\{(1-r)/r\}\ln\{1/(1-r)\}$, represents the value of the torque supposing the external energy to be entirely consumed in compressing the strip uniformly. The percentage efficiency of the rolling process for given r and b is therefore $100 \times$ the ratio of the ordinate of the corresponding point on the broken curve to the value of G_0/G_0^*.

The general features of the pressure distribution are most easily seen directly from (28). The sign of dp/dx is that of $(x/R') \pm (\mu p/2k)$. Hence, on the exit side of the neutral point dp/dx is always positive, the pressure rising steadily from the value $2k - t_f$ at the point of exit. At the point of entry dp/dx is negative when $\mu > \alpha/(1 - t_b/2k)$, and positive when $\mu < \alpha/(1 - t_b/2k)$. Thus, in the first case, the pressure begins to rise along the arc of contact, and hence must continue to do so since x/R' decreases. In the second case, on the other hand, the pressure begins to fall. However, the accurate solution shows that it begins to rise again if μ is not too small, and that it becomes greater than $2k - t_f$ at a point on the arc of contact ($x \geqslant 0$) provided μ is greater than the value given by (31). The pressure peak therefore coincides with the neutral point. Typical pressure distributions, calculated from (30), are shown in Fig. 49 for $r = 0 \cdot 3$ and $b = 0 \cdot 84$ ($\mu/\alpha = 1 \cdot 53$), and for various front and back tensions.†

Work-hardening may be allowed for by the following method. Let $2k(e)$ be the compressive yield stress in plane strain, considered as a

---

† Pressure distributions for a wide range of conditions and for non-constant k have been computed by M. Cook and E. C. Larke, *Journ. Inst. Metals*, **71** (1945), 557.

function of the conventional compressive strain e (fractional reduction in height). Now the value of e at the point θ on the arc of contact is given by

$$H(r-e) = R'\theta^2.$$

Hence, if the pressure were equal to the local value of the yield stress, the roll-force would be

$$P_0^* = R' \int_0^\alpha 2k(e)\, d\theta = \sqrt{(R'H)} \int_0^r \frac{k\, de}{\sqrt{(r-e)}}.$$

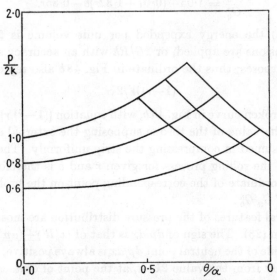

Fig. 49. Typical roll-pressure distributions according to von Karman's theory for strip-rolling with tensions (no work-hardening).

It is now assumed that the factor by which P_0^* should be multiplied to give the actual roll-force is independent of the stress-strain characteristics of the metal, and so equal to the ordinate in Fig. 48 a for a non-hardening material. Clearly this is equivalent to using a mean yield stress

$$\frac{1}{\alpha} \int_0^\alpha 2k\, d\theta.$$

Although this assumption is not adequate to reproduce the true shape of the friction hill, the pressure being overestimated on the entry side and underestimated on the exit side, the error in the calculated roll-force is found to be insignificant. However, if the same mean yield stress is used to calculate the torque, the correct value is overestimated by as much as 10 per cent. in the first pass of annealed material.† That the torque should be overestimated is evident from (21), since the quantity $\int_0^\alpha p\theta \, d\theta$ is just the moment of the pressure distribution $p(\theta)$ about the point of exit ($\theta = 0$). It is more accurate to proceed by calculating the torque G_0^* for which the pressure is equal to the yield stress, and then assume that the contribution of the friction hill is allowed for by multiplying G_0^* by the ordinate of Fig. 48 b. The value of G_0^*, uncorrected for the moment of the normal pressure about the roll centre, is

$$G_0^* = RR' \int_0^\alpha p\theta \, d\theta = RH \int_0^r k(e) \, de.$$

The theoretical prediction that there should be a pressure peak (the friction hill, as it is called) has been confirmed by experiment. A pressure-transmitting pin is held in contact with the strip through a radial bore in one of the rolls, the pressure being measured by a piezo-electric‡ or photo-elastic§ method. The distribution is found to be more rounded than the theory predicts, both at the peak and at the points of exit and entry. The rounding is probably due partly to the elastic behaviour of the strip, and partly to the distortions inherent in the method of measurement. A closer comparison is rendered uncertain by the lack of direct experimental determinations of μ. This difficulty is present, too, in the comparison of theoretical and experimental values of the roll-force or torque.‖ It seems likely that theories of rolling have not infrequently been brought into agreement with the measured roll-force by attaching spurious values to μ. For a significant test of the theory it is desirable to compare the values of both G_0 and P_0 (it should be noted that G_0/P_0 is almost independent of μ in the practical range). Fig. 50 illustrates the success with which experimental data†† is represented by the preceding theory. The material was 3 in. $\times$ 0·063 in. annealed mild steel, reduced by varying amounts between rolls of 5 in. radius with flood lubricant. The theoretical curves are based on a coefficient of friction equal to 0·08.

† D. R. Bland and H. Ford, op. cit., p. 189.
‡ E. Siebel and W. Lueg, *Mitt. Kais. Wilh. Inst. Eisenf.* **15** (1933), 1.
§ E. Orowan (to be published).
‖ See H. Ford, *Proc. Inst. Mech. Eng.* **159** (1948), 115, for experimental values of roll-force and torque for copper strip, and a detailed comparison with the values predicted by various theories.
†† This was obtained in 1948 on the experimental rolling-mill at Sheffield University, and is reproduced by the kind permission of the Director, The British Iron & Steel Research Association.

The small discrepancies may be attributed† to two factors not included in the theory: (i) the additional load due to the contact between the rolls and the elastically-recovering rolled strip, and (ii) the falling-off of the pressure towards the free edges of the strip. The former predominates at moderate reductions, and the latter at large reductions.

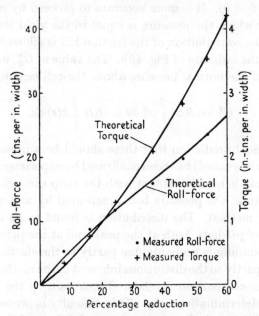

Fig. 50. Comparison of theoretical and experimental values of roll-force and torque for 0·063 in. annealed mild-steel strip. The radius of the undeformed rolls is 5 in., and the coefficient of friction is assumed to be 0·08.

(v) *Influence of applied tensions*. The effect on the pressure distribution of applying front and back tensions can be understood from (30). Except near the neutral point (Fig. 49) the pressure is reduced by an amount which, at points on the entry side of the pressure peak, is directly proportional to the back tension stress t_b, and at points on the exit side to the front tension stress t_f (this is not quite accurate since the change in roll-force slightly affects the length of the arc of contact). The reduction in pressure varies according to position on the arc of contact, being proportional to $\exp\{2a(\psi_r-\psi)\}$ on the entry side, and to $\exp(2a\psi)$ on the exit side; both factors increase towards the neutral point. In general, the position of the pressure peak is altered, being moved forward by the

† E. Orowan, *Proc. Inst. Mech. Eng.* 159 (1948), 158.

application of back tension, and backward by the application of front tension. The permissible amount of front tension is limited only by necking in the rolled strip; the amount of back tension is controlled by two factors: necking in the strip before rolling, and insufficient friction on the rolls.

The approximate solution of von Karman's equation due to Bland and Ford is equally satisfactory when moderate tensions are applied, but less so for a heavy back tension ($t_b/2k > \frac{1}{2}$, say) since the contribution of the integral in (30) is relatively more important. It would, however,

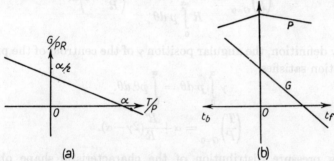

Fig. 51. Approximate linear relations between the directly measurable quantities in cold strip-rolling with tensions (diagrammatic).

be laborious to compute, and difficult to present, data sheets of the roll-force and torque covering a range of values of b, r, $t_f/2k$, and $t_b/2k$. An alternative method of approach to this problem is suggested by the consideration that the dependence of the roll-force and torque on the applied tensions is probably only slightly affected by the *precise* distribution of pressure on the rolls. This appears likely in view of the general shape of the friction hill, and the restrictions upon it imposed by the equations for the equilibrium of the strip as a whole. Theoretical considerations† indicate that there should be an approximately linear relation between the non-dimensional quantities G/PR and T/P, and that the intercept of the line on the T/P axis should be very nearly equal to the angle of contact α (Fig. 51 a). Both predictions have been demonstrated experimentally† over a wide range of conditions (Hitchcock's formula being used to calculate α). The linear relation may be expressed in the form

$$\frac{G}{PR} = \frac{G_0}{P_0 R}\left(1 - \frac{T}{P\alpha}\right), \tag{33}$$

† R. Hill, *Proc. Inst. Mech. Eng.* **163** (1950), 135.

where the subscript zero denotes no-tension values. That the intercept on the T/P axis should be approximately equal to α can be understood by reference to equation (21). When $G = 0$ (this corresponds to Steckel rolling, or drawing through idling rolls), we have

$$\left(\frac{T}{R'}\right)_{G=0} = \frac{2R'}{R}\int_0^\alpha p\theta\, d\theta - \left(\frac{R'}{R} - 1\right)\frac{\alpha P}{R'}.$$

Therefore, from (19),

$$\left(\frac{T}{P}\right)_{G=0} = \frac{2R'\int_0^\alpha p\theta\, d\theta}{R\int_0^\alpha p\, d\theta} - \left(\frac{R'}{R} - 1\right)\alpha.$$

Now, by definition, the angular position γ of the centroid of the pressure distribution satisfies

$$\gamma\int_0^\alpha p\, d\theta = \int_0^\alpha p\theta\, d\theta.$$

Hence
$$\left(\frac{T}{P}\right)_{G=0} = \alpha + \frac{R'}{R}(2\gamma - \alpha). \tag{34}$$

For any pressure distribution of the characteristic shape observed experimentally, the centroid must lie close to the mid-point of the arc of contact when the torque is zero. This follows from (20) since, when $G = 0$, the areas $\int_0^\phi p\, d\theta$ and $\int_\phi^\alpha p\, d\theta$, on either side of the pressure peak, are very nearly equal (exactly so if there were no roll-flattening). Hence

$$\left(\frac{T}{P}\right)_{G=0} \sim \alpha \quad \text{when } \gamma \sim \tfrac{1}{2}\alpha.$$

It is found that the line cuts the ordinate axis a little below the point $\tfrac{1}{2}\alpha$; the intercept is very nearly equal to the angular position of the centroid of the pressure distribution when $T = 0$.

The (algebraically) least value of T/P for which rolling is possible is found by substituting $\phi = 0$ in (20):

$$\frac{G}{RR'} = \mu\int_0^\alpha p\, d\theta + \left(\frac{R'}{R} - 1\right)\int_0^\alpha p(\theta - \tfrac{1}{2}\alpha)\, d\theta$$

$$= \left[\mu - \tfrac{1}{2}\left(\frac{R'}{R} - 1\right)\alpha\right]\frac{P}{R'} + \left(\frac{R'}{R} - 1\right)\left(\frac{T}{2R'} + \frac{\mu P}{R'}\right) \quad \text{from (18) and (19)}.$$

Hence
$$\frac{G}{PR'} = \mu - \tfrac{1}{2}\left(1 - \frac{R}{R'}\right)\left(\alpha - \frac{T}{P}\right); \quad \phi = 0.$$

Alternatively, we may use this equation to deduce μ from measurements of G and P at the greatest possible back tension. Thus

$$\mu = \frac{G}{PR} + \tfrac{1}{2}\left(1-\frac{R}{R'}\right)\left(\alpha - \frac{T}{P} - \frac{2G}{PR}\right); \qquad \phi = 0; \qquad (35)$$

or, from (33),

$$\mu = \left[1 + \left(1-\frac{R}{R'}\right)\left(\frac{RP_0\alpha}{2G_0}-1\right)\right]\left(\frac{G}{PR}\right)_{\phi=0}. \qquad (36)$$

If roll-flattening is neglected, $\mu \sim (G/PR)_{\phi=0}$; the correction may amount to some 5 per cent.

On the (very rough) assumption that the effect of tensions is to decrease the pressure by t_f on the exit side of the neutral point, and by t_b on the entry side, it may be shown from (21) that

$$G = G_0 + \tfrac{1}{2}R[(1-r)T_b - T_f] = G_0 + \tfrac{1}{2}Rh(t_b - t_f). \qquad (37)$$

This is an equation expressing the dependence of G on the tensions; it has been found to be in close accord with experiment† for annealed metals.‡ Back tension increases, and front decreases, the torque by equal amounts for the same tension *stress* (Fig. 51 b). A very simple interpretation can be given to (37). If, in the expression (23) for the work W per unit volume, we substitute for G, there results

$$(1+f)W = (1+f_0)W_0 + f(t_f - t_b),$$

where $W_0 = 2G_0/(1+f_0)Rh$. When the forward slip is small (as it usually is except when a large front tension is applied), $W \sim W_0$. Thus the energy expended per unit volume of material remains roughly constant when tensions are applied. Therefore, apart from possible savings in the drive, the efficiency cannot be significantly improved by rolling with tensions, but may be materially lowered if the back-tension energy cannot be usefully recovered.

If, now, we eliminate G between (33) and (37), we obtain a relation between P and the tensions (Fig. 51 b):

$$P = P_0 - AT_f - BT_b, \qquad (38)$$

where $\qquad A = \dfrac{RP_0}{2G_0} - \dfrac{1}{\alpha}, \qquad B = \dfrac{1}{\alpha} - (1-r)\dfrac{RP_0}{2G_0}.$

This agrees qualitatively with experiment in predicting that the roll-force is decreased more by a back tension than by the same *stress* applied as a front tension. The quantitative agreement is less good, due to the

---

† R. Hill, op. cit., p. 203.
‡ For pre-strained metals, however, G does not depend linearly on t_b, and quadratic terms in t_b^2 and $t_f t_b$ must be added to secure close agreement.

magnification of errors in (37); since $G_0/P_0 R$ is not very different from $\tfrac{1}{2}\alpha$, a small change in the coefficients of T_f and T_b in (37) produces a relatively large change in A and B. Better agreement can be secured, at the expense of some complication, by allowing for the greater decrease in pressure near the neutral point which is indicated by von Karman's theory and which is neglected in this simple analysis.

7. Machining†

In the process of machining, a surface layer of metal is removed by a wedge-shaped tool which is constrained to travel parallel to the surface

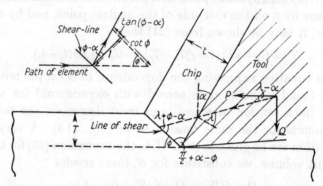

FIG. 52. Machining of a plane surface, showing the forces on the tool when the chip is continuous, with (inset) the deformation of an element crossing the line of shear.

at a chosen depth. When the lubrication is poor, a cap of dead metal accumulates around the cutting edge, from which it breaks away at intervals, leaving a rough finish; alternatively, if the metal is brittle the chip itself may rupture periodically.‡ On the other hand, if the metal is ductile and the lubrication good, the chip is a continuous coil, and the process may properly be considered one of steady motion. We shall restrict our attention to orthogonal cutting, where the tool is moved in a direction at right angles to the cutting edge. The rake α of the tool, which is the angle between the upper face and the normal to the metal surface (Fig. 52), may be either positive or negative. For simplicity, it is assumed that the state of friction over the area of contact between tool and chip can be adequately, though broadly, represented by a con-

† The following account is concerned only with the mechanics of machining; for a discussion of the metallurgical problems the reader is referred to 'Conference on Machinability', *Proc. Inst. Mech. Eng.*, **155** (1946).

‡ H. Ernst, *Machining of Metals*, Am. Soc. Metals Congress (1938). See also W. Rosenhain and A. C. Sturney, *Proc. Inst. Mech. Eng.* **1** (1925), 141.

stant coefficient μ, with a corresponding angle of friction $\lambda = \tan^{-1}\mu$. λ and μ depend, among other things, on the speed of machining. The given conditions are, then, the values of α, λ, the depth T to which the metal is machined, and its stress-strain characteristics. The problem is to calculate the distribution of stress over the face of the tool, the shape of the plastic region, and the thickness t of the chip. Since the width of the tool is generally very large compared with the depth T, the deformation is essentially plane strain.

The solution of the problem, as thus formulated, is not known, even for a non-hardening, plastic-rigid material. A useful analysis can, however, be based on the experimental observation that the deformation is mainly concentrated in a narrow zone, which springs from the edge of the tool and, moreover, is only slightly curved. The freshly machined surface does not appear to be significantly deformed nor, therefore, hardened. Following Ernst and Merchant[†] let us idealize the region of plastically deforming material, and assume it to be a single straight line (Fig. 52) inclined at some angle ϕ to the surface. The deformation then consists of a simple shear. When the work-hardening is not zero, this finite shear could not, for reasons of equilibrium, take place abruptly, but it is, nevertheless, assumed that the zone of shear has a negligibly small breadth. From geometry, the chip-thickness ratio is

$$r = \frac{T}{t} = \frac{\sin\phi}{\cos(\phi-\alpha)}. \tag{39}$$

This relation, in the form

$$\tan\phi = \frac{r\cos\alpha}{1-r\sin\alpha},$$

is used in conjunction with a measurement of r as an accurate means of determining ϕ. It will be seen from the inset in Fig. 52 that the shear strain is of amount (engineering definition)

$$\gamma = \tan(\phi-\alpha)+\cot\phi = \frac{\cos\alpha}{\sin\phi\cos(\phi-\alpha)} = \frac{r^2-2r\sin\alpha+1}{r\cos\alpha}. \tag{40}$$

For given α, γ is least when $\phi = \frac{1}{4}\pi+\frac{1}{2}\alpha$, that is, when the shear line bisects the angle between the surfaces of the metal and the tool. Values of γ as high as 5 have been observed. It is proved in Chapter XII (Sect. 4) that the direction which undergoes the greatest resultant extension during the shear is inclined, in the chip, at an angle θ to the line of shear such that

$$2\cot 2\theta = \gamma. \tag{41}$$

[†] H. Ernst and M. E. Merchant, *Trans. Am. Soc. Metals*, (1941), 299.

The elongation of individual crystals gives a fibrous appearance to the microstructure of the chip, and, if the crystals are initially oriented at random, the fibre direction is well defined and can be expected to be the direction θ.

Let P and Q be the components of the external force per unit width which is applied to the tool; P is parallel to the metal surface, and Q is normal to it and in the *downward* direction. Their resultant is inclined, by hypothesis, at the angle $\tfrac{1}{2}\pi-\lambda$ to the tool face, and hence at a downward angle $\lambda-\alpha$ to the horizontal. Thus,

$$Q = P\tan(\lambda-\alpha). \tag{42}$$

The apparent coefficient of friction can thereby be deduced from measurements of P and Q. In the experiments of Merchant[†] μ had values between about 0·5 and 1·0. These high values are understandable since the chip surface, bearing on the tool, is freshly formed and should therefore be free from adsorbed films. It is found that an increase of the speed by a factor of 5 lowers μ by about 20 per cent.; there appears also to be a scale-effect in that the apparent value of μ depends on the depth of machining.

Let k be the shear stress of the material after the strain γ. Considering the equilibrium of the chip, and resolving parallel to the line of shear, we find

$$\begin{aligned} kT/\sin\phi &= P\cos\phi - Q\sin\phi \\ &= P[\cos\phi - \sin\phi\tan(\lambda-\alpha)] \text{ from (42)} \\ &= P\cos(\phi+\lambda-\alpha)/\cos(\lambda-\alpha). \end{aligned} \tag{43}$$

The *average* value p of the compressive stress acting across the shear line is

$$p = k\tan(\lambda+\phi-\alpha). \tag{44}$$

The external work done in removing unit volume of the metal is

$$P/T = k\cos(\lambda-\alpha)/\sin\phi\cos(\phi+\lambda-\alpha). \tag{45}$$

To complete the analysis we need to relate ϕ to λ and α. Failing a full solution for the distribution of stress in the metal near the cutting edge, a further assumption must be made. Merchant[‡] postulates that ϕ is such that the work done per unit volume is a minimum. The condition for this is

$$\frac{\partial}{\partial\phi}\left(\frac{P}{T}\right) = 0,$$

[†] M. E. Merchant, *Journ. App. Mech.* 11 (1944), A-168.
[‡] M. E. Merchant, *Journ. App. Phys.* 16 (1945), 267 and 318. A similar analysis has been given by V. Piispanen, ibid. 19 (1948), 876.

or
$$\cos(2\phi+\lambda-\alpha) = \frac{1}{k}\frac{\partial k}{\partial \phi}\sin\phi\cos(\phi+\lambda-\alpha). \qquad (46)$$

In view of (40) k is a known function of ϕ, calculable from the stress-strain curve in shear at the appropriate temperature and rate of strain. If there were no work-hardening the solution† of (46) would be simply

$$\phi = \tfrac{1}{4}\pi+\tfrac{1}{2}\alpha-\tfrac{1}{2}\lambda. \qquad (47)$$

From Merchant's measurements it appears that, within experimental error, ϕ is in fact a function of the single variable $(\lambda-\alpha)$ but that this formula overestimates ϕ by some 20 to 40 per cent in the range investigated. The discrepancy cannot be due to neglecting work-hardening, because if this is included, the calculated value of ϕ is still larger (except when $\lambda = 0$, for which ϕ is always equal to $\tfrac{1}{4}\pi+\tfrac{1}{2}\alpha$). Merchant has attempted to improve the agreement by permitting the shear stress to vary with the normal pressure on the plane of shear, but the required variation is so great as to be physically out of the question, the pressure being only of the order of the yield stress. The comparative failure of the theory is almost certainly due to the inadequacy of the minimum work hypothesis.

8. Flow through a converging channel

Consider a wedge-shaped converging channel (total angle 2α) through which plastic material is being forced. Let polar coordinates (r, θ) be taken relative to the axis of symmetry and the virtual apex O of the channel (Fig. 53). The sides of the channel are rough and it is supposed that the frictional stress is constant. When the channel is very long, and the flow is steady, it is to be expected that the state of stress in material remote from the ends is effectively independent of external conditions, and that the slip-line directions depend only on θ and not on r. The corresponding flow-lines are radii through O. It is this special distribution of stress with which we shall be concerned here. The slip-line field near the ends of the channel is similar to that already described for extrusion through a wedge-shaped die (Sect. 4 (i)) if the external conditions are the same.

† It should be noticed that, if μ is regarded as constant along the tool, this formula implies that the cutting edge is a singularity for the state of stress in the chip. Thus, if the direction of maximum shear stress at a point on the chip surface adjacent to the cutting edge were parallel to the shear line, we should have $\phi = \tfrac{1}{4}\pi+\alpha-\lambda$. This means that the direction of the resultant stress acting on the tool face is a principal direction for the state of stress at the cutting edge; moreover, the other principal component of stress is zero. This formula underestimates ϕ, and correspondingly overestimates the resistance to machining; on *a priori* grounds, it is certainly invalid for large negative rakes since it implies a negative ϕ.

The yield criterion is satisfied by the introduction of a parameter ψ, depending only on θ by hypothesis, and such that

$$\tau_{r\theta} = k \sin 2\psi, \qquad \sigma_r - \sigma_\theta = 2k \cos 2\psi. \tag{48}$$

ψ is the angle between a radius and the direction of the algebraically greater principal stress. $\sigma_r - \sigma_\theta$ has been taken positive since elements on the axis are subject to a simple circumferential compression and an equal radial extension. ψ must have the same sign as θ since the friction acts to oppose the relative motion; ψ is zero on the axis, and ranges

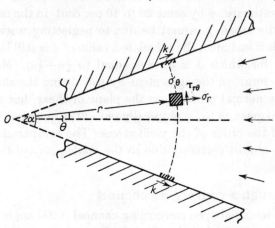

FIG. 53. Stress components for analysis of the flow of a plastic material through a converging wedge-shaped channel.

between $-\tfrac{1}{4}\pi$ and $\tfrac{1}{4}\pi$ if we suppose, for definiteness, that $|\tau_{r\theta}| = k$ on the sides of the channel (it will be shown later how the solution can be adapted for a frictional stress less than k). The equations of equilibrium are

$$r\frac{\partial \sigma_r}{\partial r} + \frac{\partial \tau_{r\theta}}{\partial \theta} + \sigma_r - \sigma_\theta = 0,$$

$$r\frac{\partial \tau_{r\theta}}{\partial r} + \frac{\partial \sigma_\theta}{\partial \theta} + 2\tau_{r\theta} = 0.$$

Combining (48) with the second of these, we find that

$$\frac{\sigma_\theta}{2k} = \frac{\sigma_r}{2k} - \cos 2\psi = -\int \sin 2\psi \, d\theta + f(r). \tag{49}$$

Substituting this in the first equation of equilibrium:

$$rf'(r) + \cos 2\psi(1 + \psi'(\theta)) = 0,$$

it follows that

$$rf'(r) = -c, \qquad \psi'(\theta) = c \sec 2\psi - 1,$$

where c is a constant. Thus

$$f(r) = -c\ln r + A, \qquad \theta = \int_0^\psi \frac{\cos 2\psi \, d\psi}{c - \cos 2\psi}, \qquad (50)$$

where A is another constant. Since we are taking ψ to have the same sign as θ, c must be positive, and moreover must be greater than unity to avoid infinities in the integral. Integrating:

$$\theta = -\psi + \frac{c}{\sqrt{(c^2-1)}} \tan^{-1}\left\{\sqrt{\left(\frac{c+1}{c-1}\right)} \tan \psi\right\} \quad (c > 1), \qquad (51)$$

where the inverse tangent is an angle between $-\tfrac{1}{2}\pi$ and $\tfrac{1}{2}\pi$. Since $\psi = \tfrac{1}{4}\pi$ when $\theta = \alpha$, c satisfies the equation

$$\frac{c}{\sqrt{(c^2-1)}} \tan^{-1} \sqrt{\left(\frac{c+1}{c-1}\right)} = \tfrac{1}{4}\pi + \alpha. \qquad (52)$$

As α varies from 0 to $\tfrac{1}{2}\pi$, c varies from ∞ to 1·192 (approx.). From (49) and (50) the stresses are, finally,

$$\left.\begin{aligned}\frac{\sigma_r}{2k} &= -c\ln r + \tfrac{1}{2}\cos 2\psi - \tfrac{1}{2}c\ln(c - \cos 2\psi) + A, \\ \frac{\sigma_\theta}{2k} &= -c\ln r - \tfrac{1}{2}\cos 2\psi - \tfrac{1}{2}c\ln(c - \cos 2\psi) + A.\end{aligned}\right\} \qquad (53)$$

These equations are due to Nadai.†

The radial velocity u (measured outwards) must be of the form

$$u = \frac{g(\theta)}{r}$$

in order to satisfy the equation of incompressibility. $g(\theta)$ is determined by the condition that the principal axes of stress and strain-rate coincide; this is

$$\frac{\dot{\epsilon}_r - \dot{\epsilon}_\theta}{2\dot{\gamma}_{r\theta}} = \frac{\sigma_r - \sigma_\theta}{2\tau_{r\theta}}, \quad \text{or} \quad \frac{-2g(\theta)/r^2}{g'(\theta)/r^2} = \cot 2\psi.$$

Hence $\quad g(\theta) = \exp\left(-2\int \tan 2\psi \, d\theta\right) = B/(c - \cos 2\psi)$

from (50), where B is a (positive) constant. Thus

$$u = \frac{B}{r(c - \cos 2\psi)}. \qquad (54)$$

Now, since the flow is radial and the shear stress is constant along any radius, the stress distribution within a sector $\pm\theta$ is the solution for a channel of angle 2θ and a frictional stress $k\sin 2\psi$. Conversely, if we require the solution corresponding to a channel of angle 2θ and a frictional

† A. Nadai, *Zeits. f. Phys.* **30** (1924), 106.

stress λk ($0 \leqslant \lambda \leqslant 1$), we simply solve (51) for the appropriate value of c, with $2\psi = \sin^{-1}\lambda$. The stress components are then given by (48), (51), and (53).

If, by some means, the shear stress on the sides of the channel were directed towards the apex, so that it assisted the motion, the state of stress would be obtained by taking ψ opposite in sign to θ. It follows from (50) that c must certainly be less than $\cos\alpha$, and that

$$\left.\begin{aligned}\theta &= -\psi - \frac{c}{\sqrt{(c^2-1)}}\tan^{-1}\left\{\sqrt{\left(\frac{c+1}{c-1}\right)}\tan\psi\right\} \quad (c \leqslant -1); \\ \theta &= -\psi - \frac{c}{\sqrt{(1-c^2)}}\tanh^{-1}\left\{\sqrt{\left(\frac{1+c}{1-c}\right)}\tan\psi\right\} \quad (-1 \leqslant c \leqslant 0).\end{aligned}\right\} \quad (55)$$

The relation between c and α is obtained by substituting $\theta = \alpha, \psi = -\tfrac{1}{4}\pi$. As c increases from $-\infty$ to -1, α increases from 0 to $\tfrac{1}{4}\pi - \tfrac{1}{2}$; as c increases still further, α continues to increase, becoming equal to $\tfrac{1}{4}\pi$ when $c = 0$. Orowan based his analysis of rolling† on the present theory. He assumed (in the notation of Sect. 6) that the distribution of stress over any circular arc, which cuts the rolls orthogonally at points where the frictional stress is $\mu p(\theta)$ and which has an angular span 2θ, would be identical with that near the centre of a long wedge-shaped channel of angle 2θ, and a constant frictional stress of amount $\mu p(\theta)$ (its direction depends on whether the arc cuts the rolls on the exit or entry side of the neutral point). When α is small, $\sin 2\psi \sim 2c\theta$, $c \sim \pm 1/2\alpha$, and

$$\tau_{r\theta} = \pm k\frac{\theta}{\alpha}, \qquad \sigma_r - \sigma_\theta = 2k\sqrt{\left(1 - \frac{\theta^2}{\alpha^2}\right)}.$$

The analogous pair of solutions for the flow of material through a *diverging* channel is obtained from the previous analysis merely by changing the sign of $\cos 2\psi$ wherever it occurs in the formulae for the stresses. The relation between θ and ψ becomes

$$\theta = \int_0^\psi \frac{\cos 2\psi \, d\psi}{c + \cos 2\psi}.$$

Finally, it is to be remarked that the present stress solution is applicable, with similar limitations, to the compression of a plastic mass between two inclined plates (see Chap. VIII, Sect. 5 (i), for a detailed discussion of certain additional assumptions). The velocity distribution is obtained by superimposing a uniform velocity $U \operatorname{cosec}\alpha$ parallel to the axis, where U is the inward speed of the plates normal to their lengths.

† E. Orowan, op. cit., p. 189.

VIII
NON-STEADY MOTION PROBLEMS IN TWO DIMENSIONS. I

1. Geometric similarity and the unit diagram

WE now consider problems in which the stress and velocity at any fixed point are varying from moment to moment. To begin with, we restrict our attention to problems where the plastic region develops in such a way that the entire configuration remains geometrically similar. We have already encountered two examples, namely the expansion of a cylindrical or a spherical cavity from zero radius in an infinite medium; when the configuration at any moment is scaled down to a constant cavity radius the same distribution of stress is obtained. Other examples are the indenting of the plane surface of a semi-infinite block by a conical or pyramidal punch. For geometrical similarity it is necessary that the distortion should be initiated at a point or along a line, and that the specimen should be infinite in at least one dimension. Although either condition precludes the exact realization of similarity in practice, it is easily attained within experimental error. For example, Vickers pyramid hardness is observed[†] to be effectively independent of the applied load, provided the specimen is large compared with the diameter and depth of the impression, and provided the impression is large compared with the grain size and the radius of curvature of the tip of the cone.

To formulate the idea of similarity more precisely, let $\mathbf{r}$ be the position vector of an element referred to the origin of distortion, and let c be a characteristic length defining a stage of the deformation; for example, c might be the radius of the plastic region round an expanding spherical cavity, or the depth of penetration by a conical indenter. Now in a general problem of non-steady motion the stress and velocity are certain functions of $\mathbf{r}$ and c. When geometric similarity is preserved, the stress and velocity are functions only of the single variable $\mathbf{r}/c$. Hence, if the configuration at any stage is scaled down in the ratio $c:1$, we always obtain exactly the same geometrical figure, at a fixed point of which the corresponding stress and velocity are unvarying; in particular, the region representing the plastic material does not alter. This figure, in

[†] P. Field Foster, *The Mechanical Testing of Metals and Alloys*, 4th edition, p. 161 (Pitman & Sons, 1948).

which the characteristic length has been made unity, will be called the *unit diagram*.† An element, whose position vector is **r** in actual space, is represented in the unit diagram by a point whose position vector is $\boldsymbol{\rho} = \mathbf{r}/c$.

There is a close resemblance between problems where geometric similarity is maintained and problems of steady motion. When the external conditions are completely specified, there is no development of the plastic region to be traced and, in both, a system of coordinates can be found such that the corresponding distribution of stress and velocity is constant. Both types of problem are statically undetermined and the same trial-and-error procedure must be followed to construct a slip-line field satisfying the stress and velocity boundary conditions. In both, the fields are finally justified by showing that positive work is done on all plastic elements, and that in the rigid material there is an associated state of stress such that the yield limit is nowhere exceeded. The only feature of difference in the method of solution is the form taken by the velocity boundary conditions expressing the maintenance of similarity. It is more advantageous to phrase these in terms of the movement of elements in the unit diagram, rather than in actual space. If the velocity of an element in the actual space is $\mathbf{v} = d\mathbf{r}/dc$ (referred to c as the scale of 'time'), and the velocity of the corresponding point in the unit diagram is $d\boldsymbol{\rho}/dc$, then

$$\mathbf{v} = \frac{d}{dc}(c\boldsymbol{\rho}) = c\frac{d\boldsymbol{\rho}}{dc} + \boldsymbol{\rho}, \quad \text{or} \quad c\frac{d\boldsymbol{\rho}}{dc} = \mathbf{v} - \boldsymbol{\rho}. \tag{1}$$

This equation states that the corresponding point in the unit diagram moves, at each moment, towards a focus whose position vector is **v**, and that its speed is equal to the quotient of its focal distance and the parameter c. The path of an element in actual space is represented by some trajectory in the unit diagram which is calculable from (1); in particular, the trajectory of an element which is at rest in actual space (for example, an element not yet overtaken by the spreading plastic region) is a straight line directed towards the origin. After the element becomes plastic the corresponding point in the unit diagram describes, in general, a curved trajectory. It is evident that all elements initially situated on the same radius through the origin describe the same trajectory in the unit diagram; this is a complete expression of the continuing geometric similarity. We shall see later how this property of the unit diagram facilitates a calculation of the distortion of a square grid.

† R. Hill, E. H. Lee, and S. J. Tupper, *Proc. Roy. Soc.* A, **188** (1947), 273.

2. Wedge-indentation†

(i) *Method of solution.* The plane surface of a semi-infinite block of plastic-rigid material is penetrated normally by a smooth rigid wedge of total angle 2θ. In Fig. 54 (right-hand half) $ABDEC$ is the region of plastically deforming material; AC is the displaced surface (whose shape is to be determined); AB is the line of contact with the wedge, and $BDEC$ is a slip-line. The most convenient starting slip-line is BD. When its position has been assumed, the condition that slip-lines meet the wedge at 45° defines the field ABD uniquely (third boundary-value

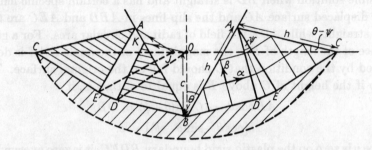

Fig. 54. Indentation of a plane surface by a smooth wedge, showing the slip-line field on the right and the main features of the distortion on the left.

problem). Since the free surface will not necessarily meet the wedge orthogonally, the point A must be a stress singularity. This, with the slip-line AD, defines the field ADE, which may be continued round A through any desired angle (first boundary-value problem, special case). The slip-line AE, together with the requirement that AC must be a free surface, defines the field AEC and, incidentally, the shape of AC (converse of second boundary-value problem). Now the point C must lie on the original plane surface; this determines the angular span ψ of the field ADE. We have next to examine whether, with our initial choice of BD, the velocity boundary conditions are satisfied. Along AB the component of velocity normal to the wedge is equal to the normal component of the speed of penetration; along $BDEC$ the normal component of velocity is zero since the material underneath is rigid. The velocity solution may therefore be begun in ABD (third boundary-value problem), and extended successively to ADE and AEC (first boundary-value problem). The calculated velocities of elements on the free surface must

† R. Hill, E. H. Lee, and S. J. Tupper, op. cit., p. 214. The problem of oblique penetration by a wedge has been treated by R. Hill and E. H. Lee, Ministry of Supply, Armament Research Department, Theoretical Research Rep. 1/46.

be such that the surface is continually displaced in such a way that geometric similarity is preserved. This is the condition which controls the shape of the starting slip-line BD. In the unit diagram the curve corresponding to the free surface must be the trajectory for surface elements. Hence, according to the interpretation of (1), the tangent at any point on this curve must pass through the associated focus with position vector $\mathbf{v}$. If the tentative solution has this property, similarity is maintained.

(ii) *Position of the displaced surface.* We now verify that there is a possible solution when BD is straight and has a certain specific length. The displaced surface AC and the slip-lines in ABD and AEC are then also straight, while ADE is a field of radii and circular arcs. For a given choice of the length of BD, the magnitude ψ of the angle DAE is determined by the condition that C should fall on the original surface. This is so if the height of C above B is equal to c; that is, if

$$AB\cos\theta - AC\sin(\theta-\psi) = OB,$$

or
$$h[\cos\theta - \sin(\theta-\psi)] = c. \qquad (2)$$

Since v is zero on the plastic-rigid boundary $BDEC$, it is zero everywhere by Geiringer's equation for the variation of v along the straight β-lines. It follows that u is constant on each α-line, and hence, by the boundary condition on AB, it is universally equal to $\sqrt{2}\sin\theta$ (the downward speed of the wedge is unity on the scale c). Thus, at any moment, all elements are moving with the same speed along the α-lines. The surface AC is therefore displaced to a parallel position, and the new configuration can be made geometrically similar by a suitable choice of the length of BD or, equivalently, the position of A. Referring, now, to the unit diagram (Fig. 55), the foci all lie on a circular arc HK of span ψ and radius $|\mathbf{v}| = \sqrt{2}\sin\theta$, where OH and OK are parallel respectively to EC and BD. H is the focus for elements in AEC, and K is the focus for elements in ABD. Since OK is of length $\sqrt{2}\sin\theta$ and is inclined at $45°$ to AB, K must lie on AB. In ADE the focus for an element at P is the point F where the arc HK is intersected by the perpendicular from O to AP. Now we have seen that, for similarity, the tangent at any point on AC must be directed towards its associated focus; hence AC must pass through H. The condition for this is that the projection of AB perpendicular to CA should be equal to the sum of the projections of OH and OB; that is,

$$AB\cos\psi = OH\sin\tfrac{1}{4}\pi + OB\cos(\theta-\psi),$$

or
$$h\cos\psi = c[\sin\theta + \cos(\theta-\psi)]. \qquad (3)$$

The elimination of h/c from (2) and (3) gives the relation between ψ and θ:

$$\cos(2\theta-\psi) = \frac{\cos\psi}{1+\sin\psi}. \quad (4)$$

Since similarity is maintained by this solution, and since the velocity equations express incompressibility, it necessarily follows that the volume of material displaced above the original surface is equal to the

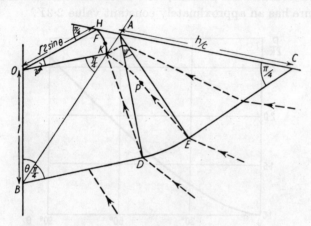

FIG. 55. Unit diagram and trajectories of elements for wedge-indentation.

volume of the impression below it. This may be immediately verified from the observation that triangles AOC and AOB have an equal side ($AC = AB$) and equal perpendiculars (length $\sin\theta$) to these sides from O; their areas are therefore equal. The result follows on subtracting triangle AOL from both, where L is the meet of OC and AB.

In this instance it is easy to show that the rate at which work is done is nowhere negative. This is so for the finite shearing of elements crossing the plastic-rigid boundary $BDEC$, if the boundary is an α-line. Since the velocity is uniform in ABD and ACE, zero rate of work is done on elements traversing these areas. Finally, for elements in ADE the rate of work per unit volume is equal to $\sqrt{2}k\sin\theta$ divided by the distance from A (see the formula for the shear strain-rate in polar coordinates, Appendix II).

The mean compressive stress has the value k on a free surface in compression, and hence, by Hencky's theorem, its value on the wedge face AB is $k(1+2\psi)$. The pressure P on the wedge is therefore distributed uniformly, and is of amount

$$P = 2k(1+\psi). \quad (5)$$

The load per unit width is $2Ph\sin\theta$, and the work expended per unit volume of the impression below OC is $Ph\cos\theta/c$. The relation between P and θ is shown in Fig. 56; P rises steadily from $2k$ to $2k(1+\tfrac{1}{2}\pi)$ as the angle increases. This should be contrasted with the experimental observation by Bishop, Hill, and Mott[†] that, when cold-worked copper is indented by a lubricated *cone*, the mean resistive pressure decreases as the cone becomes less pointed; for $\theta > 30°$ the decrease is slight and the pressure has an approximately constant value $2·3Y$.

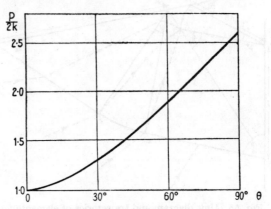

Fig. 56. Relation between the pressure and the semi-angle in wedge-indentation.

The distribution of stress in the rigid material is not known, but there is no reason to suppose that the material is incapable of supporting the calculated stresses along $BDEC$. It is observed in the indentation of hard materials by a smooth wedge that the plastic region extends a little way below the tip (more if the wedge is rough or the material is annealed), but that the strains are small; this corresponds to the rigid part of the plastic region (sketched diagrammatically in Fig. 54) for our hypothetical plastic-rigid body. The present solution would continue to hold even for a block of finite dimensions, provided it could be associated with a non-plastic state of stress in the rigid material. In other words, to the approximation achieved by the hypothetical material, the state of stress in the plastically deforming region can remain similar even if the block is finite, though the non-plastic stress distribution, of course, can not. As the penetration increased, however, a stage would be reached where a possible state of stress in the rigid material could not

[†] R. F. Bishop, R. Hill, and N. F. Mott, *Proc. Phys. Soc.* **57** (1945), 147. See also R. L'Hermite, *Proc. 7th Int. Cong. App. Mech.*, London (1948).

be found; this would imply that plastic *deformation* had begun elsewhere.†

(iii) *Distortion round the impression.* An inspection of the trajectories in Fig. 55 reveals that an element initially between OD and OB remains in KBD and so always moves parallel to BD after becoming plastic. Hence the material initially situated in OBD finally occupies KBD, and its deformation is equivalent to a simple shear parallel to BD. Similarly, an element which is finally within AEC has always moved parallel to EC after becoming plastic. The material finally occupying AEC has, in effect, been sheared parallel to EC from the initial position JEC, where J is the meet of OC with the parallel to CE through A. The deformation of material initially in $ODEJ$ has no simple properties, except that surface elements initially on OJ are finally in contact with the wedge along KA. Thus, part of the original surface is drawn down the side of the wedge; this phenomenon is accentuated if the wedge is not free from friction.

The calculation of the distortion of a square grid is most conveniently carried out in the unit diagram. The problem is to find the final position, when the penetration is c, of the corner of a square whose initial position is $\mathbf{r}_0$. Suppose that the plastic boundary first reaches the corner when the penetration is c_0 ($< c$), and let $\boldsymbol{\rho}_0 = \mathbf{r}_0/c_0$ be the position of the corresponding point in the unit diagram. This point afterwards moves along the trajectory through $\boldsymbol{\rho}_0$. The trajectory having been calculated, let s be the further distance traversed by the point when the penetration has increased to c, and let $f(s)$ be the corresponding focal distance. Then, from (1),

$$\ln \frac{c}{c_0} = \int_0^s \frac{ds}{f(s)}. \qquad (6)$$

This determines s in terms of c, and hence the position vector $\mathbf{r} = c\boldsymbol{\rho}$. The integral must be evaluated numerically in ADE, but in KBD and AEC (6) simplifies to

$$\frac{c}{c_0} = \frac{d}{d-s},$$

where $f(s) = d-s$, and d is the distance from $\boldsymbol{\rho}_0$ of the focus K or H respectively. The calculated distortion for a wedge of total angle 60°

† The critical penetrations into rectangular blocks of finite breadth or depth have been calculated for wedges and flat indenters: R. Hill, *Phil. Mag.* **41** (1950), 745. The results are of significance in deciding how large a specimen should be for a valid hardness test.

is shown in Fig. 57. The agreement with experiment is remarkably close.† Hodge‡ has shown that a good approximation to the distortion is obtained by formally replacing the region ADE by an arbitrary small number of stress discontinuities across certain radii through A; the material between two neighbouring discontinuities is taken to move as a rigid body. As might be expected, this method is not useful where most of the plastic material is actually deforming, and where, apart from the plastic-rigid boundary, there are no zones of severe shear (for example, in the compression of a block between rough plates).

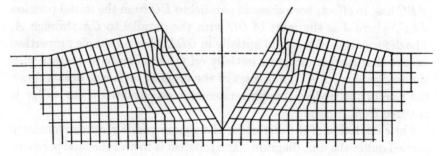

Fig. 57. Distortion of a square grid in wedge-indentation.

(iv) *Influence of friction and strain-hardening.* The solution when the wedge is rough (sliding friction) differs only in that the slip-lines do not meet the face at 45°. The displaced surface AC is still straight, though inclined at a smaller angle, and the pressure on the wedge is uniformly distributed. For a sufficiently blunt wedge and a large enough coefficient of friction, the theory indicates that the indenter is covered by a 90° wedge-shaped cap of dead metal.

The effect of work-hardening and elastic compressibility can only be assessed qualitatively. It is helpful to visualize the actual mode of deformation as a compromise between two extreme possibilities. If the displaced material were accommodated by the elastic resilience of the bulk of the specimen, the plastic strains would be relatively small and diffused over a wide area. If, on the other hand, the material were displaced sideways the strains would be relatively large. Now for an annealed metal the work-hardening which would accompany the second mode of deformation is so great that the first mode is initiated at a smaller load, and the familiar sinking-in impression is obtained. For a

† See the photographs of distorted grids on lead in the paper by R. Hill, E. H. Lee, and S. J. Tupper, op. cit., p. 214.

‡ P. G. Hodge, Jr., *Graduate Division of App. Math.*, Brown University, Tech. Rep. 30 (1949).

heavily pre-strained metal the further work-hardening is slight, and the load needed to effect the second mode of deformation is reached before that needed to overcome the elastic resistance in the first. In *lubricated* indentation of a pre-strained metal a ridge or coronet is always thrown up.

3. Compression of a wedge by a flat die†

The apex O of an infinite wedge (total angle 2θ) is compressed symmetrically by a smooth flat die (Fig. 58). Let AA be the contact at any

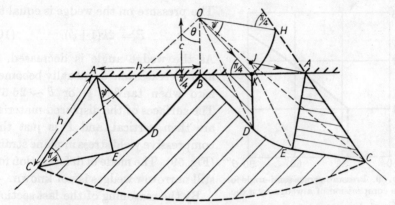

FIG. 58. Slip-line field and deformation in a wedge compressed by a smooth flat die.

moment between the die and the flattened wedge, and let the length OB be taken as the time-scale c. Proceeding as in the last section, we can show that there is a possible plastic region preserving similarity when the displaced surfaces AC are straight. The slip-line field shown in Fig. 58 is self-explanatory, and it is evident that elements move with uniform speed $\sqrt{2}$ along the slip-lines parallel to the plastic-rigid boundary $BDEC$. If $AB = AC = h$, and angle $DAE = \psi$, the requirement that C should lie on the original surface is satisfied when

$$\frac{AB + AC\sin\psi}{OB + AC\cos\psi} = \tan\theta,$$

or $\qquad h(1+\sin\psi) = \tan\theta(c+h\cos\psi).$ (7)

The foci lie on a circular arc HK of span ψ and radius $\sqrt{2}$, where OH and OK are parallel respectively to EC and BD (we may, without confusion, regard the right-hand half of Fig. 58 either as the unit diagram or the actual configuration). Since OK is inclined at $45°$ to BA, K must be situated on the die face. For geometric similarity H must lie on CA

† R. Hill, *Proc. 7th Int. Cong. App. Mech.*, London, (1948).

produced; this requires that the projection of AB perpendicular to CA should be equal to the sum of the projections of OH and OB, that is,

$$AB\cos\psi = OH\sin\tfrac{1}{4}\pi + OB\sin\psi,$$

or
$$h\cos\psi = c(1+\sin\psi). \qquad (8)$$

Solving (7) and (8), we find

$$\tan\theta = \frac{(1+\sin\psi)^2}{\cos\psi(2+\sin\psi)}; \quad \frac{h}{c} = \frac{1+\sin\psi}{\cos\psi}, \qquad (9)$$

The pressure on the wedge is equal to

$$P = 2k(1+\psi). \qquad (10)$$

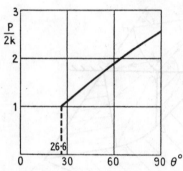

Fig. 59. Pressure versus semi-angle for the compression of a wedge by a die.

As the wedge angle is decreased, ψ also decreases and eventually becomes zero when $\tan\theta = \tfrac{1}{2}$, or $\theta \sim 26\cdot 6°$. The surfaces of the displaced material are then vertical, and P is just the compressive yield stress in plane strain (Fig. 59). The mode of deformation for still narrower wedges is not known.

By the reasoning of the last section it may be proved that the triangle KBD has, in effect, been sheared parallel to BD from its initial position OBD, and that triangle AEC has been sheared from JEC, where J is the meet of OC and the parallel to CE through A. The section OJ of the original surface is finally in contact with the die along KA.

There is another configuration which is analytically possible, all stress and velocity boundary conditions being satisfied. The displaced surface is still straight but the slip-line from C passes, not through the mid-point B of the die, but through the opposite corner A of the surface of contact. A steadily increasing 90° wedge-shaped cap of dead metal is carried down with the die. The pressure on the die in this second configuration is greater than in the first. The reader should have no difficulty in proving that

$$\tan\theta = \frac{(1+2\sin\psi)^2}{4\cos\psi(1+\sin\psi)}; \quad \frac{h}{c} = \frac{1+2\sin\psi}{2\cos\psi}. \qquad (11)$$

It is to be presumed that both configurations could be obtained in practice by suitably modifying the shape of the wedge tip. This problem exemplifies a previous statement that the final steady state may be dependent on the initial conditions.

4. Expansion of a semi-cylindrical cavity in a surface†

A semi-cylindrical cavity of current radius c is expanded by suitably distributed normal pressure from a point on the plane surface of an infinite block of material. In Fig. 60 BEB is the surface of the cavity, and ABC is the coronet. The impression is supposed to have been initiated at the point O, and the pressure is applied only over BEB and not on AB. Since geometric similarity is preserved during the formation

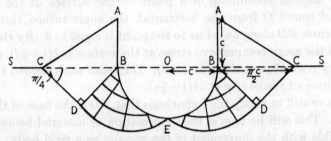

Fig. 60. Slip-line field and deformation round a semi-cylindrical cavity expanded in a plane surface.

of the cavity, the process is not the two-dimensional analogue of a ball hardness test.

Let us examine whether there is a possible mode of deformation in which the plastic-rigid boundary EDC passes through the deepest point E of the impression. Since, by hypothesis, there are no tangential stresses over the cavity surface, slip-lines meet it at 45°. BD is the most convenient starting slip-line; when its position has been chosen, the field BED is uniquely determined (third boundary-value problem). Now since the surface of the coronet is free from applied forces, the elements in a part, at least, of the coronet must have unloaded from their former plastic state. This part of the coronet moves as a rigid body, and is carried outwards on the plastic layer beneath; furthermore, it must be separated from the plastically deforming material by a slip-line.

The natural trial hypothesis is that BD is this slip-line. It then follows that the slip-line CD is straight, since the coronet moves as a rigid body. Consider, now, the equilibrium of $ABCD$ under the stresses acting on BD and CD. The normal pressure along CD is constant by Hencky's theorem, and equal to p_0, say. The distribution of normal pressure along BD is fixed in terms of p_0 and the shape of BD by the same theorem. The shear stress at all points on BD and CD is of course equal to k; its sign is such that CD is an α-line. By a little experimenting the reader

† R. Hill, op. cit., p. 221.

will soon convince himself that $ABCD$ can only be in equilibrium under this distribution of forces when BD is straight and $p_0 = k$. All β-lines in BDE are then straight, the α-lines being the involutes of the circle with centre O and radius $c/\sqrt{2}$ (cf. Fig. 23). Using the property that the β-lines may be generated by unwinding a taut thread from the circle-evolute, we see that BD is of length $\pi c/2\sqrt{2}$ since the β-lines between E and B are tangential to the evolute over an angular span of $\tfrac{1}{2}\pi$. If θ is the angular coordinate of a point on the surface of the cavity, measured round O from the horizontal, the angle turned through in passing from BD along an α-line to the point is equal to θ. By Hencky's theorem the mean compressive stress at the surface is $k(1+2\theta)$, and the necessary normal pressure is $2k(1+\theta)$. The mean work needed to make unit volume of the cavity is $2k(1+\tfrac{1}{4}\pi)$.

We have still to verify the hypothesis that BD is the base of the rigid coronet. This will be true if the deformation of material below BD is compatible with the movement of the coronet as a rigid body; that is, if the normal component of velocity on BD is uniform. Now, by one Geiringer equation, the velocity component v is constant on each straight β-line, and is therefore universally zero since it is zero on the plastic-rigid boundary DE. Hence, by the second Geiringer equation, the component u is constant along each α-line. But, since the cavity expands radially at unit rate, u is equal to $\sqrt{2}$ on BE. The speed of all elements in BDE is therefore the same, and the coronet slides outwards along DC with speed $\sqrt{2}$. In passing, we note that the rate of work per unit volume is positive, being equal to $\sqrt{2}k$ times the local curvature of the α-line.

Finally, in order that similarity is maintained, we must suitably determine the shape of the coronet. In the unit diagram (Fig. 61) the foci are situated on the circular arc AZ of radius $\sqrt{2}$ and span $\tfrac{1}{2}\pi$; the focus of elements in $ABCD$ is a point A of this arc such that OA is parallel to DC, and so inclined at 45° to the horizontal. The requirement that the trajectories of elements in the coronet are directed towards their common focus A demands that the surfaces of the coronet are plane and that A is the apex. The length of AB is equal to c, and the length of BC to $\tfrac{1}{2}\pi c$ (since $BD = \pi c/2\sqrt{2}$). Hence the area of triangle ABC is $\tfrac{1}{4}\pi c^2$; this is equal to half the sectional area of the cavity, as it naturally must be since the continuing similarity implies that the cavity was expanded from a point.

To construct the focus F of a point P in BDE, the slip-line PQ is produced to touch the circle-evolute at T; the radius OT produced meets the focal circle in the required point F. It is evident that QF is tangential to the cavity surface and that it is of unit length. F is the focus for all points on the straight slip-line through Q. If (ρ, θ) are polar coordinates with respect to O and the horizontal, the equation of a trajectory in BDE is

$$\frac{d\theta}{d\rho} = \frac{\sqrt{(2\rho^2-1)}}{\rho(\rho^2-1)}. \tag{12}$$

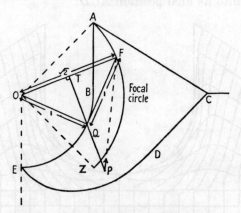

Fig. 61. Unit diagram and trajectories of elements for a semi-cylindrical cavity expanded in a plane surface.

If s is the arc-length of a trajectory measured from its intersection with DE,

$$ds = \sqrt{(d\rho^2 + \rho^2 d\theta^2)} = -\frac{\rho^2 d\rho}{(\rho^2-1)}.$$

It will be seen from Fig. 61 that the focal distance $f(s)$ is equal to ρ, and so, from (6),

$$\ln\frac{c}{c_0} = \int_0^s \frac{ds}{f(s)} = -\int_{\rho_0}^\rho \frac{\rho \, d\rho}{\rho^2-1} = \tfrac{1}{2}\ln\left(\frac{\rho_0^2-1}{\rho^2-1}\right),$$

where ρ_0 is the length of the radius to the point where the trajectory meets DE. Therefore,

$$c^2(\rho^2-1) = c_0^2(\rho_0^2-1) = r_0^2 - c_0^2. \tag{13}$$

To find the final position (ρ, θ), when the cavity radius is c, of a point overtaken by the plastic boundary when its position was (ρ_0, θ_0), we first calculate the corresponding trajectory $\theta(\rho)$ from (12). From (13) we obtain ρ, and hence θ. The final position in actual space is (r, θ), or $(c\rho, \theta)$. In this way the deformation of a square grid (Fig. 62) has been computed. The surface of the cavity is formed from elements initially situated vertically below O, and it will be seen that these elements have been subjected to particularly heavy distortion. It will be noticed, too, that part of the coronet has been strained in simple shear. Reference to the unit diagram shows that elements which cross CD always remain in the rigid area ACD, and hence that the finite shear received while crossing CD is retained unchanged. The triangle OCD is therefore sheared parallel to CD into its final position ACD.

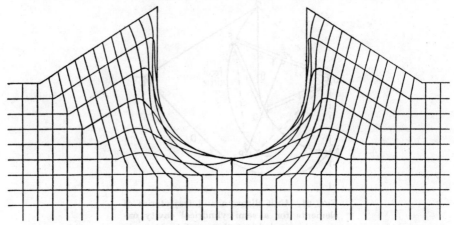

Fig. 62. Distortion of a square grid near a semi-cylindrical cavity expanded in a plane surface.

5. Compression of a block between rough plates

(i) *Fundamental assumptions.* A rectangular block of plastic-rigid material is compressed between rigid parallel plates which are assumed to be so rough that the greatest possible frictional stress, namely the yield stress in shear, is induced wherever the relative displacement exceeds an elastic order of magnitude. It is supposed, to begin with, that the block is wider than the plates (Fig. 63 a). In this problem geometric similarity is not maintained, and we have therefore to follow, from the very beginning, the progressive changes in the configuration.

Plastic zones are initiated at the sharp edges of the plates by the first application of load, and spread inwards. In a plastic-elastic material the direction of spread, and the shape of the plastic boundary, are controlled by the frictional stresses induced by relative displacements of an elastic order of magnitude. The relation between the friction and the amount of slip in this range must be very dependent on the particular metals used for the block and the plates, and on the precise state of the surfaces of contact. However, experiments with artificially roughened plates, and a variety of materials and frictional properties, indicate that the plastic zones spreading inward from left and right invariably meet first in the geometric centre of the block. No theoretical investigation has been carried out, nor indeed is it essential for a useful analysis of the subsequent distortion. We take the experimental observations as our starting-point, and *assume* that the plastic region develops in such a way that a non-plastic area is left in contact with each plate over a certain length near the centre.

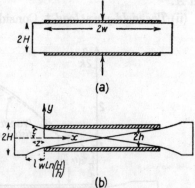

Fig. 63. Compression of a rectangular block between parallel plates, showing the configuration, (a) initially, and (b) during the compression.

Our idealized material being rigid when stressed below the yield limit, the plates cannot move together so long as a remaining non-plastic strip of finite width spans the block. Even when the plastic zones have fused, compression may still not be possible. It is necessary, in addition, that the two slip-lines through the centre should be entirely contained within the plastic zone up to their junctions with the plates. So long as these slip-lines intersect the non-plastic areas, no deformation consistent with Geiringer's equations is possible, and the block remains rigid. We may legitimately, and conveniently, refer to the moment when the plates first approach each other as the yield-point of the block as a whole, and to the corresponding load on the plates as the yield-point load.

A plastic-elastic block is necessarily compressed by any load, however slight, but, when work-hardening is absent, there is a pronounced bend in the load-compression curve marking the initiation of large plastic strains. The loading interval corresponding to the bend is narrow and well defined, and the mean load in this interval should approximate closely to the yield-point of the plastic-rigid block. At lower loads the amount of compression is restricted by the elastic resistance of the non-plastic strip, and all plastic strains are of an elastic order of magnitude. The bend is the more rounded and less definite, the more rapidly the material work-hardens. If the metal is fully annealed, it is altogether impossible to define a load which may be compared with the value calculated by the theory.

Plastic deformation is also restricted by the overhang, part of which must be non-plastic since the surfaces are stress-free. Since we have made a hypothesis about the development of the plastic region in one direction, we cannot avoid making another about its development in the opposite direction. Experiments† suggest that the overhang, even if partly plastic, is thrust sideways as a rigid whole. It follows that the external plastic-rigid boundary must pass through the edges O of the plates (Fig. 64), intersecting the horizontal axis of symmetry at 45° in A.

(ii) *The yield-point load.*‡ Consider, now, the slip-line field in the upper

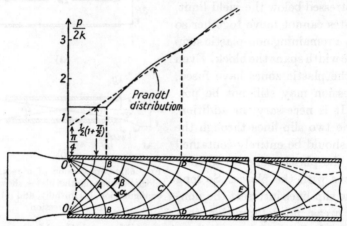

Fig. 64. Slip-line field and plastic region in a block compressed between perfectly rough plates, together with a comparison of the actual pressure distribution with Prandtl's.

left-hand quadrant (Fig. 64). OA is a convenient starting slip-line from which to build up the field; let us suppose, as the simplest trial assumption, that it is straight. Since O must be a stress singularity, a field of radii and concentric circular arcs can be extended round O. Now one family of slip-lines meets the plate orthogonally, since the frictional stress is equal to k; hence (third boundary-value problem) this radial field may be continued up to the horizontal through O, the final slip-line OB being completely coincident with the surface. The arc AB and its reflection in the axis of symmetry define the field ABC (as in Fig. 30). The field BCD is defined by the slip-line BC and the boundary condition on the plate; slip-lines of one family meet the plate tangentially, which

---

† J. F. Nye, Ministry of Supply, Armament Research Dept., Rep. 39/47.
‡ R. Hill, E. H. Lee, and S. J. Tupper, Ministry of Supply, Armament Research Dept., Theoretical Research Rep. 28/45. For a shorter account see R. Hill and E. H. Lee, *Proc. 6th Int. Cong. App. Mech.*, Paris, (1946); *Journ. App. Mech.* (in press).

therefore happens to be a natural boundary of the solution (Chap. VI, Sect. 4 (ii)). The construction of the field continues in similar fashion up to the slip-line through the centre of the block. The wedge-shaped area included between this slip-line and its counterpart in the right-hand quadrant is, by hypothesis, rigid; the boundary of the plastic region is indicated diagrammatically by the broken curve. The rigid wedge is borne down with the plate, losing material to the plastic region so that it continues to make point contact with its fellow.

Two boundary conditions are available for the velocity solution in the part of the field adjacent to the wedge; these are that the component of velocity normal to the horizontal axis is zero, and that the component normal to the wedge boundary is equal to the component speed of the wedge in that direction. If the plate is moved with unit speed, there is obviously a tangential velocity discontinuity of amount $\sqrt{2}$ along the wedge boundary; its sense requires that the boundary is an α-line. The discontinuity is not propagated farther since the boundary meets the plate tangentially. The solution may be systematically continued up to the circular arc AB. Since the normal component of velocity on the slip-line OB is constant, it is constant on OA, whatever the distribution of velocity on AB. The distortion is therefore compatible with the rigid-body displacement of the overhang; this verifies the initial choice of the shape of OA. There is no velocity discontinuity across OA.

Considerations of the equilibrium of the overhang show that the mean compressive stress must have the value k along OA. By Hencky's theorem the mean compressive stress on OB is equal to $k(1+\frac{1}{2}\pi)$; this is also the value of the normal pressure acting on the plate. The slip-line field has been calculated by the (R, S) method (Chap. VI, Sect. 5 (i)), up to the slip-line DE. E coincides with the centre of the block when the ratio of width to height is 6·72, approximately. The corresponding pressure distribution on the plate is represented by the solid curve in Fig. 64. Since we do not know the state of stress within the rigid wedge, we can only calculate the *average* pressure between the plate and the wedge; this is obtained by integrating the resolved components of the stresses acting on DE. (The load on this section of the plate is *less* than it would be if the plastic region extended farther to the right, since there is then an additional upward thrust from the shear stresses acting over the vertical section through E.) By cutting off the field at various points between C and E we may obtain, from the same solution, the pressure distribution for any block with a width/height ratio between 3·64 and 6·72. The average pressure P over the plate at the yield-point is found to depend on this ratio very nearly according to the equation

$$\frac{P}{2k} = \frac{3}{4} + \frac{w}{4h} \quad \left(\frac{w}{h} \geqslant 1\right), \tag{14}$$

where $2w$ is the width of the block, and $2h$ is the height. Experiments[†] with tellurium lead (0·05 per cent. tellurium) have confirmed this relation in the range of strain for which the rate of work-hardening is small ($k \sim 1·3 \times 10^8$ dynes/cm.² after about 30 per cent. pre-strain in compression).

When w/h is less than 3·64, but greater than unity, there is a solution in which the rigid wedge extends over the whole plate (Fig. 65); the plastic

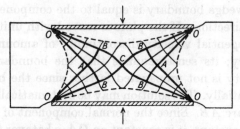

Fig. 65. Slip-line field and plastic region in a block compressed between rough plates, when the width of the plates is between 1 and 3·64 times the height of the block.

region is compatible with the rigid-body displacement of the overhang since the velocity component normal to OB is still constant. It is not known with certainty for what range of frictional conditions this solution is appropriate, though similar zones of intense shear, radiating from the edges of the plates, have often been observed.[‡] For the slip-line field and associated velocity distribution when w/h is less than unity see Chapter IX (Sect. 5 (iii)).

(iii) *Distortion of a block after a finite compression.* It is obvious that the present solution is valid no matter what the amount of compression, since the overhang remains rigid. As the compression continues, we have to deal, in effect, with a series of blocks of increasing width/height ratio. No calculation has been made of the distortion of a square grid after a finite compression, but only of the displacements during an infinitesimal compression following the yield-point.[§] It is found that the horizontal component of velocity increases steadily, over each

[†] J. F. Nye, op. cit., p. 228.
[‡] See, for example, A. Nadai, *Plasticity*, chap. 17 (McGraw-Hill Book Co., 1931).
[§] R. Hill, E. H. Lee, and S. J. Tupper, op. cit., p. 228.

vertical section, from the plate to the axis of symmetry. Hence a vertical line is deformed into a curve which is convex to the edge of the block. The final quantity obtained in a calculation of the velocity solution is the outward speed of the overhang; this may be compared with the accurate value, which is w/h times the speed of the plates, since the material is incompressible. It has been found† that the resultant error, after first computing a 5° net as far as E by the (R, S) method, and then computing the velocity components from E outwards by the equations of Section 6 (i) in Chapter VI, is only 0·2 per cent.

The equation of the contour of the material squeezed from between the plates is easily found. Let (x, y) axes of reference be taken such that the axis of x coincides with the horizontal axis of symmetry, and the axis of y passes through the left-hand edges of the plates (Fig. 63 b). Let $2t$ be the thickness of the overhang at a distance z from the plane $x = 0$. During a further increment of compression the overhang moves out through a distance $-w\, dh/h$. The tangent to the surface at the edge of the plate is therefore in the direction of the block diagonal. This is true at all moments of the compression, and hence

$$\frac{dt}{dz} = \frac{t}{w}, \quad \text{or} \quad t = he^{z/w}. \tag{15}$$

The total distance through which the overhang has been displaced is therefore $w\ln(H/h)$, where $2H$ is the initial height of the block.

(iv) *Compression between partially rough plates.* For simplicity, suppose that the frictional conditions are such that the shear stress on the plates has some constant value, less than k. The slip-lines then intersect the plates at constant angles (if the coefficient of friction were constant, the angle of intersection would vary with the normal pressure along the plate). The reader should have no difficulty in constructing the field, analogous to Fig. 64, assuming that the slip-lines through the edges of the plates are straight. This, however, cannot be the correct solution for all width/height ratios since the velocity discontinuities, initiated at the block centre, are now propagated by successive reflection from the plates down the whole length of the block. It is easy to show that the amount of the discontinuity is multiplied at each reflection by the factor $\tan\theta$, where θ ($\leqslant \tfrac{1}{4}\pi$) is the angle at which the α-lines meet the plate; the discontinuity is therefore progressively diminished (unless the plate is smooth) but never becomes zero. When the dimensions of the block are such that the discontinuities terminate at the edges

† Ibid. p. 228.

of the plates, the field is valid. For all other dimensions the discontinuities terminate on the exit slip-lines; this is incompatible with the rigid-body movement of the overhang. The correct solution is not known; the situation is analogous to that noted in sheet-drawing (Chap. VII, Sect. 2 (i)). In particular, when there is *no* friction on the plates, the solution fails unless the width/height ratio is integral; the block then deforms, momentarily, as a number of independent rigid units, which slide along a criss-cross of slip-lines. When there is no overhang, a block compressed between frictionless plates deforms uniformly.

(v) *Prandtl's cycloid solution.* When the block is very wide compared with its height, it might be expected that the field of slip-lines becomes more uniform with increasing distance from an edge, and that, in the limit, the slip-lines of each family are parallel curves. It has in fact been shown† that such a limiting configuration is approached, not steadily, however, but in a quasi-oscillatory manner typical of hyperbolic differential equations.

In order to find this limiting field, it is simplest to return to the Cartesian form of the basic equations (Chap. VI, Sect. 2). With the axes of Fig. 63 b, if the slope of the slip-lines is independent of x, so also are $\sigma_x - \sigma_y$ and τ_{xy} (cf. equation (11) of Chap. VI). Thus, we may write

$$\tau_{xy} = kf(y), \qquad \sigma_x - \sigma_y = 2k\sqrt{(1-f^2)},$$

where $f(y)$ is to be determined. Substitution for τ_{xy} and σ_x in the equations of equilibrium gives

$$\frac{\partial \sigma_y}{\partial x} + kf'(y) = 0, \qquad \frac{\partial \sigma_y}{\partial y} = 0.$$

These are compatible if and only if

$$f(y) = ay+b, \qquad \sigma_y = -k(ax+c),$$

where a, b, and c are constants. The boundary conditions are $\tau_{xy} = 0$ when $y = 0$, and $\tau_{xy} = mk$ when $y = h$, where $0 \leqslant m \leqslant 1$ ($m = 1$ for perfectly rough plates). Hence $a = m/h$, $b = 0$, and

$$\left. \begin{array}{l} \dfrac{\sigma_x}{k} = -c - \dfrac{mx}{h} + 2\sqrt{\left(1 - \dfrac{m^2 y^2}{h^2}\right)}, \\[1em] \dfrac{\sigma_y}{k} = -c - \dfrac{mx}{h}, \\[1em] \dfrac{\tau_{xy}}{k} = \dfrac{my}{h}. \end{array} \right\} \qquad (16)$$

† R. Hill, E. H. Lee, and S. J. Tupper, op. cit., p. 228.

COMPRESSION OF BLOCK BETWEEN ROUGH PLATES

These equations are due to Prandtl.† The slip-lines may be shown to be cycloids with slopes

$$\sqrt{\left(\frac{1\pm(my/h)}{1\mp(my/h)}\right)},$$

and radii of curvature

$$\frac{2h}{m}\sqrt{\left\{2\left(1\mp\frac{my}{h}\right)\right\}}.$$

When $m = 1$ the projection on the x-axis of their length from plate to plate is πh; they are divided by the x-axis in the ratio $(\tfrac{1}{2}\pi+1):(\tfrac{1}{2}\pi-1)$.

Since the left-hand edge of the block is free from stress, the resultant horizontal thrust on a vertical section must balance the frictional resistance exerted by the plates. Hence

$$\int_0^h \sigma_x\,dy = -mkx.$$

This requires
$$c = \frac{\sin^{-1} m}{m} + \sqrt{(1-m^2)}. \tag{17}$$

Note that ck would be the value of the pressure on the plate at the edge ($x = 0$) *if* this solution extended so far; that is, if the distribution of external stresses required by (16) were applied over the edge of the block. Prandtl's solution for $m = 1$ ($c = \tfrac{1}{2}\pi$) is compared with the accurate solution in Fig. 64. It will be seen that Prandtl's solution is a very good approximation, even up to a distance h from the edge; the correct pressure distribution oscillates about the Prandtl distribution, the amplitude presumably decreasing steadily. This may be regarded as the plastic analogue of Saint-Venant's principle in elasticity.

If we regard Prandtl's field as applicable (with negligible error) in the central part of a sufficiently wide block, the rigid wedge is bounded by cycloids. The corresponding velocity distribution, constructed as explained in (ii), has been calculated when $m = 1$ by Geiringer,‡ in a region near the centre, using the Green's function method (Chap. VI, equation (30)). It might be expected that, at a sufficiently great distance from the centre (the velocity discontinuity having diminished to negligible proportions), a limiting distribution of velocity would be approached, such that the strain-rate is independent of x. This has not been investigated, but there is a possible limiting distribution with Cartesian components

$$\frac{u}{U} = C + \frac{x}{h} - \frac{2}{m}\sqrt{\left(1 - \frac{m^2y^2}{h^2}\right)}, \qquad \frac{v}{U} = -\frac{y}{h}, \tag{18}$$

† L. Prandtl, *Zeits. ang. Math. Mech.* **3** (1923), 401.
‡ H. Geiringer, *Proc. 3rd Int. Cong. App. Mech.*, Stockholm, **2** (1930), 185; *Mém. Sci. Math.* **86** (1937).

where U is the speed of the plates. These expressions are due to Nadai;† it may be verified that they satisfy equations (4) and (7) of Chapter VI. They are not valid near the centre, since they are incompatible with the movement of the wedge as a rigid body. For the rate of work to be positive, the radical in (16) and (18) must be given the positive sign. The parameter C must be determined so that the horizontal flow across a vertical section is equal to the rate at which material to the right is displaced by the plates. This requires

$$-\int_0^h u\, dy = (w-x)U,$$

or
$$C = -\frac{w}{h} + \frac{\sin^{-1} m}{m^2} + \frac{1}{m}\sqrt{(1-m^2)}. \tag{19}$$

It is evident from (18) that an element always remains at the same relative distances from the plate and the axis, since

$$\frac{d}{dt}\left(\frac{y}{h}\right) = \frac{1}{h}\frac{dy}{dt} - \frac{y}{h^2}\frac{dh}{dt} = \frac{v}{h} + \frac{yU}{h^2} = 0.$$

Thus, equally spaced horizontal lines remain equally spaced. Consider, now, the horizontal displacement of an element for which y/h has the constant value η. At any moment, let ξ be the distance by which it is in advance of the surface element which was originally situated in the same vertical section. Then, from (18),

$$-\frac{d\xi}{dh} = \frac{1}{U}\frac{d\xi}{dt} = \frac{\xi}{h} + \frac{2}{m}\sqrt{(1-m^2\eta^2)} - \frac{2}{m}\sqrt{(1-m^2)}.$$

Integrating:
$$h\xi = \frac{(H^2-h^2)}{m}[\sqrt{(1-m^2\eta^2)} - \sqrt{(1-m^2)}],$$

or
$$\left[\frac{mh\xi}{H^2-h^2} + \sqrt{(1-m^2)}\right]^2 + \frac{m^2 y^2}{h^2} = 1.$$

This is the equation of the curve into which an original vertical line is distorted. It is a section of an ellipse with semi-major axis h/m and semi-minor axis $(H^2-h^2)/mh$. When the plates are perfectly rough ($m = 1$), the ellipse is tangential to the surface at the extremities of its major axis. Under these conditions the calculated curve has been found to be in good agreement with that observed when the rate of work-hardening is small.‡ When the plates are partially rough, it is evident that the deformation can be visualized as though the block were part of a larger block whose initial height is H/m.

---

† A. Nadai (unpublished work).   ‡ J. F. Nye, op. cit., p. 228.

(vi) *Solution when the plates overlap the block.* Subject to similar provisos concerning the prior development of the plastic region, the field of Fig. 64 is also valid at the yield-point of a rectangular block compressed between perfectly rough, overlapping plates. The right-angled triangular region to the left of *OA* is stressed in pure compression, and, although entirely plastic, is displaced outwards as a rigid whole. The solution therefore continues to hold throughout the subsequent compression,† so that, rather surprisingly, there is no barrelling. It may be shown that elements on the free surface gradually move round the corners to the surfaces in contact with the plates. The slip-line field for this problem was first given (qualitatively) by Prandtl;‡ however, he was not aware that the associated velocity solution is such that the plane edges remain plane.

If the edge is initially *concave*, the field is different. Assuming that the edge is plastic, the solution is defined in the region bounded by the two slip-lines from the corners, making 45° with the edge (second boundary-value problem). Each of these slip-lines and the singularity at the corresponding corner define a region up to the (curved) slip-line which is tangential to the plate. The construction of the field continues in the usual way up to the boundary of the central rigid wedge. The velocity boundary conditions impose no restrictions on the field since the edge of the block is plastic (it is necessary, of course, that the rate of work should everywhere be positive). During the ensuing compression the shape of the edge, and with it the slip-line field, progressively change. If, on the other hand, the edge is initially *convex*, the adjacent part of the block is presumably displaced as a rigid whole, so that the starting slip-line is straight, as in Fig. 64. The same field is then valid throughout the compression.

Suppose, now, that the plates are only partially rough. If the edge is initially concave or straight, the 'natural' field is valid. When the velocity discontinuities do not terminate at the corners, they are automatically accommodated by the *plastic* edge, whose shape is thereby altered. An analogous field continues to hold during the subsequent compression if the edge remains plastic. If the edge is initially convex, the solution is valid only if the block dimensions are such that the discontinuities end at the corners. Sokolovsky§ has computed the pressure distribution on the plates when the block is rectangular,

† R. Hill, *Dissertation*, p. 138 (Cambridge, 1948), issued by Ministry of Supply, Armament Research Establishment, as Survey 1/48.
‡ L. Prandtl, *Zeits. ang. Math. Mech.* **3** (1923), 401.
§ W. W. Sokolovsky, *Theory of Plasticity*, p. 180 (Moscow, 1946).

without, however, recognizing the limits to the range of validity of the solution.

If the frictional conditions are such that there exists a coefficient of friction μ, we can easily obtain an approximate estimate of the yield-point load, which should not be too inaccurate when μ is small. The assumptions of the analysis are analogous to those of Sachs' theory of drawing, or von Karman's theory of rolling. Let p be the pressure on the plates at a distance x from the left-hand edge of the block. Let $2qh$ be the total horizontal thrust on a vertical section; q is then the mean *compressive* stress acting over the section. For equilibrium of a slice of the block contained between the vertical planes x and $x+dx$,

$$(q+dq)h - qh = \mu p\, dx,$$

or
$$\frac{dq}{dx} = \frac{\mu p}{h}. \tag{20}$$

It is now assumed (cf. equation (26) of Chap. VII) that the yield condition may be written approximately as

$$p - q = 2k. \tag{21}$$

Eliminating q between these equations:

$$\frac{dp}{dx} = \frac{\mu p}{h}, \quad \text{or} \quad p = 2ke^{\mu x/h}, \tag{22}$$

after using the fact that $q = 0$ and $p = 2k$ when $x = 0$. The mean pressure is

$$\bar{p} = \frac{1}{w}\int_0^w p\, dx = \frac{2kh}{\mu w}(e^{\mu w/h} - 1),$$

or since μ is, by hypothesis, small,

$$\bar{p} \sim 2k\left(1 + \frac{\mu w}{2h}\right). \tag{23}$$

If the compression plates are not carefully lubricated, the apparent yield stress $\bar{p}$ may significantly exceed the true yield stress $2k$, particularly when the block is wide compared with its height. For the corresponding analysis for a cylindrical compression specimen, see Chapter X (Sect. 7).

IX
NON-STEADY MOTION PROBLEMS IN TWO DIMENSIONS. II

1. Introduction

WE turn now to two-dimensional problems in which, for simplicity, it is necessary to restrict the analysis of the state of stress to the initial part of the loading path, when the total strain is still small. It will be assumed, in fact, that *all changes in the external dimensions of the body are negligible*, so far as the boundary conditions are concerned. The plastic strains being small, it is generally inaccurate, even for a calculation of the stress distribution, to suppose that the material is plastic-rigid; however, it will be shown that in certain circumstances the stress is independent of the value of Young's modulus. We shall therefore carry out the analysis for a plastic-elastic material, and assume only that it has been pre-strained to a degree such that the work-hardening is negligible in the range of strain under consideration. When such a body is continuously loaded from a stress-free state the plastic and elastic components of the strain are at first comparable. The non-plastic part of the body constrains the displacement of the remainder, and the overall distortion of the body is of order $1/E \times$ the mean stress. As the loads are raised, the plastic region expands to a size where this constraint becomes locally ineffective; large plastic strains are then possible, and the overall distortion increases at a rate (relatively to the applied loads) controlled only by the changing shape of the specimen. In a non-hardening material the loading interval during which this transition is effected is well defined when there is a sufficient freedom of flow; a curve of load plotted against some measure of the overall distortion would then have a rapid bend corresponding to the transition. In a plastic-rigid body, on the other hand, no deformation at all is possible until the plastic region attains a certain critical size; the load-distortion curve has a discontinuity in slope at the load under which distortion begins. The bend, whenever it is sufficiently well defined, will be described as the *yield-point of the body* for the particular loading path; this is a term already introduced in the discussion of the compression of a mass between rough plates (Chap. VIII, Sect. 5 (i)). In many problems the yield-point load is the quantity of greatest interest, and a main object of the analysis will be to calculate it. In view of the initial assumption, the yield-point load represents the upper limit to the range of validity of the present analysis.

2. Formulation of the problem

(i) *Assumption of incompressibility.* In a plastic-elastic body, subjected to plane strain, the stress σ_z in a plastic element depends on the strain-history. Since the displacements in the planes of flow depend on σ_z, their calculation is a matter of extreme difficulty. We have seen, however, that when Poisson's ratio is $\frac{1}{2}$ this difficulty disappears, since σ_z is always equal to $\frac{1}{2}(\sigma_x+\sigma_y)$. Throughout this chapter we shall take advantage of this and set $\nu = \frac{1}{2}$. The state of stress in any element is then a pure shear combined with a hydrostatic pressure, while the maximum shear stress is a constant k throughout the plastic region.

Although this simplification is not adequate to afford even a tolerably accurate value of σ_z for a compressible material (the plastic strain being small), this does not greatly matter since a knowledge of σ_z is not usually of much interest. On the other hand, the accompanying error in the yield criterion is negligible for most purposes (cf. the expansion of a cylinder by internal pressure, discussed in Chap. V). Thus, when the boundary conditions are such that the problem is statically determined, the calculated stresses σ_x, σ_y, and τ_{xy} should be very accurate. Even when the boundary conditions involve prescribed external displacements the errors in σ_x, σ_y, and τ_{xy} can be expected to be fairly small.

(ii) *The basic equations.* In a problem of non-steady motion the configuration of stress and strain is continually changing. The problem is properly posed only when the entire loading (or displacement) path is specified; that is, when the loads (or displacements) applied to the surface are given at all times from the moment when the body was originally stress-free or in a known state of initial stress. In a body liable to plastic yielding the state of stress does not depend only on the current loads. It is apparent that the intermediate development of the plastic region, and hence its final extent, depends also on the route by which the current loading has been reached. For example, the individual loads might be increased from zero in proportion; some might be raised to their final values before the others; certain loads might be applied and subsequently removed so that they do not appear in the final system, and so on. The following typical boundary-value problem therefore presents itself at each stage of the loading-path: given the previously calculated state of stress throughout the body at one moment, what are the (infinitesimal) increments in stress and strain produced by further increments in the loads and displacements applied to the surface.

Instead of the differentials $d\sigma_x$, $d\sigma_y$, and $d\tau_{xy}$, we may equally well use the stress-rates $\dot\sigma_x$, $\dot\sigma_y$, and $\dot\tau_{xy}$, obtained by dividing by the differential dt, where t denotes time or any other monotonically varying quantity (such as an applied load). To the approximation involved in neglecting changes in the external dimensions of the body we may disregard the movement of any element and interpret the time-derivative as referring to the variation at a fixed point.

The stress-rates must satisfy the equations of equilibrium, which, to the same order of approximation, are

$$\frac{\partial \dot\sigma_x}{\partial x}+\frac{\partial \dot\tau_{xy}}{\partial y}=0, \qquad \frac{\partial \dot\tau_{xy}}{\partial x}+\frac{\partial \dot\sigma_y}{\partial y}=0. \tag{1}$$

In an element which undergoes continued plastic deformation, so remaining stressed to the yield limit, we must have

$$\partial/\partial t\{\tfrac{1}{4}(\sigma_x-\sigma_y)^2+\tau_{xy}^2\}=0,$$

or
$$\tfrac{1}{4}(\sigma_x-\sigma_y)(\dot\sigma_x-\dot\sigma_y)+\tau_{xy}\dot\tau_{xy}=0. \tag{2}$$

On the other hand, in a plastic element which begins to unload, or in any element in the elastic region, the stress-rates satisfy the compatibility equation

$$\left(\frac{\partial^2}{\partial x^2}+\frac{\partial^2}{\partial y^2}\right)(\dot\sigma_x+\dot\sigma_y)=0, \tag{3}$$

expressing the condition for the existence of a continuous velocity satisfying the elastic stress-strain equations. Across the existing plastic-elastic boundary (or, indeed, any curve) the normal and shear components of the stress-rate must be continuous for equilibrium; the normal component acting parallel to the boundary may, however, be discontinuous (this happens, for instance, in a bent or twisted bar). It may be shown without difficulty that equations (1) and (2) are hyperbolic, with the slip-lines of the existing state of stress as characteristics. (1) and (2) may therefore be transformed into relations giving the variations $d(\dot p)$ and $d(\dot \phi)$ along the slip-lines in the loading part of the plastic region. In principle, these relations can be integrated by the usual small-arc process, starting from values of $\dot p$ and $\dot \phi$ known (or assumed) on the plastic-elastic boundary, or on a plastic section of the surface.

However, the differential form of the stress equations is best suited for general investigations, such as the discussion of uniqueness (see below), and it is usually simpler in approximate computation to work with the equations in their integrated form. For this purpose, let σ_x, σ_y, τ_{xy} now refer to the unknown stresses whose values are required after a further small, but finite, change in the external loads or displacements. The equations of equilibrium are

$$\frac{\partial \sigma_x}{\partial x} + \frac{\partial \tau_{xy}}{\partial y} = 0, \qquad \frac{\partial \tau_{xy}}{\partial x} + \frac{\partial \sigma_y}{\partial y} = 0. \tag{1'}$$

The plastic-elastic boundary alters slightly during this interval, and its new position has to be determined. This additional unknown quantity is balanced by the condition that *all* components of stress must be made continuous across the boundary, since elements just on the elastic side must be on the point of yielding. (This should be carefully contrasted with the formulation of the stress-rate problem, where it was mentioned that one component of the stress-rate is not necessarily continuous. There the value of the stress in a given element after a small interval is obtained, in principle, by integration of the stress-rate; since the latter is only momentarily discontinuous as the element is traversed by the moving plastic boundary, the stress components themselves are all continuous.) Throughout the new plastic region the stress satisfies the yield criterion

$$\tfrac{1}{4}(\sigma_x - \sigma_y)^2 + \tau_{xy}^2 = k^2. \tag{2'}$$

All the theorems proved in Chapter VI, relating to the slip-line field, are applicable. In an element which has been stressed elastically ever since it was stress-free,

$$\left(\frac{\partial^2}{\partial x^2} + \frac{\partial^2}{\partial y^2}\right)(\sigma_x + \sigma_y) = 0. \tag{3'}$$

In an element which has unloaded from a plastic state during the interval under consideration, or at any previous time,

$$\left(\frac{\partial^2}{\partial x^2} + \frac{\partial^2}{\partial y^2}\right)(\sigma_x + \sigma_y - \sigma_{x0} - \sigma_{y0}) = 0, \tag{3''}$$

where the subscript zero refers to the moment when unloading began.

Coming, now, to the calculation of the increments of displacement, let u_x and v_y be the components of velocity. The equation of incompressibility is

$$\frac{\partial u_x}{\partial x} + \frac{\partial v_y}{\partial y} = 0. \tag{4}$$

The components of the plastic strain-rate are

$$\frac{\partial u_x}{\partial x} - \frac{\dot\sigma'_x}{2G}, \qquad \frac{\partial v_y}{\partial y} - \frac{\dot\sigma'_y}{2G}, \qquad \frac{1}{2}\left(\frac{\partial u_x}{\partial y} + \frac{\partial v_y}{\partial x}\right) - \frac{\dot\tau_{xy}}{2G},$$

where the subtracted terms constitute the elastic component of the strain-rate. Now, in an element of ideal material undergoing plastic deformation, the principal axes of the plastic strain-rate coincide with the principal axes of the stress. The condition for this is

$$\frac{2\tau_{xy}}{\sigma_x - \sigma_y} = \frac{\dfrac{\partial u_x}{\partial y} + \dfrac{\partial v_y}{\partial x} - \dfrac{\dot{\tau}_{xy}}{G}}{\dfrac{\partial u_x}{\partial x} - \dfrac{\partial v_y}{\partial y} - \dfrac{(\dot{\sigma}_x - \dot{\sigma}_y)}{2G}}, \tag{5}$$

in view of the identity $\sigma'_x - \sigma'_y \equiv \sigma_x - \sigma_y$.

When the stress-rates have been calculated, (4) and (5), together with suitable boundary conditions, serve to determine the velocities in the loading part of the plastic region. The characteristics are, of course, still the slip-lines, and the relations along them are found by the method of Chapter VI. Let the (x, y) axes be taken coincident with the (α, β) directions at some point P. Since $\sigma_x = \sigma_y$, (5) reduces to

$$\frac{\partial u_x}{\partial x} - \frac{\partial v_y}{\partial y} = \frac{1}{2G}(\dot{\sigma}_x - \dot{\sigma}_y)$$

at P. When combined with (4), this leads to

$$\frac{\partial u_x}{\partial x} = \frac{1}{4G}(\dot{\sigma}_x - \dot{\sigma}_y) = -\frac{\partial v_y}{\partial y}. \tag{6}$$

Now, in the usual notation,

$$u_x = u\cos\phi - v\sin\phi,$$
$$v_y = u\sin\phi + v\cos\phi,$$
$$\sigma_x - \sigma_y = -2k\sin 2\phi.$$

Substituting in equations (6):

$$\left\{\frac{\partial}{\partial x}(u\cos\phi - v\sin\phi)\right\}_{\phi=0} = -\frac{k}{2G}\left\{\frac{\partial}{\partial t}(\sin 2\phi)\right\}_{\phi=0},$$

$$\left\{\frac{\partial}{\partial y}(u\sin\phi + v\cos\phi)\right\}_{\phi=0} = \frac{k}{2G}\left\{\frac{\partial}{\partial t}(\sin 2\phi)\right\}_{\phi=0};$$

whence

$$\left.\begin{array}{l} du - v\,d\phi = -\dfrac{k}{G}\dot{\phi}\,ds_\alpha \quad \text{on an } \alpha\text{-line,} \\[6pt] dv + u\,d\phi = \dfrac{k}{G}\dot{\phi}\,ds_\beta \quad \text{on a } \beta\text{-line.} \end{array}\right\} \tag{7}$$

These are the analogues of Geiringer's equations ((14) of Chap. VI), to which they reduce when G is infinite (u and v being non-zero) or when the slip-line field does not change in time (for example, in a certain area near a free plastic surface). For approximate computation these equations can be rewritten in terms of small increments of displacement; the total displacement is given by a summation over the whole strain-path. Equations (7) are solved, in the usual way, by integrating along the slip-lines from the plastic-elastic boundary or from the surface of the body, using as boundary conditions the values of u and v calculated in the elastic region or prescribed on the surface.

In the elastic region and in the unloading part of the plastic region the stress-strain relations (in differential form) are

$$\left. \begin{aligned} 4G\frac{\partial u_x}{\partial x} = -4G\frac{\partial v_y}{\partial y} &= \dot{\sigma}_x - \dot{\sigma}_y, \\ G\left(\frac{\partial u_x}{\partial y} + \frac{\partial v_y}{\partial x}\right) &= \dot{\tau}_{xy}, \end{aligned} \right\} \quad (8)$$

where use has been made of the identity $E = 3G$ for an incompressible material. Equation (3) is, of course, obtained by the elimination of u_x and v_y from (8); conversely, when the stress-rates have been found, satisfying (3), there exist velocities u_x and v_y which satisfy (8).

(iii) *Statically determined problems*.[†] Since we are considering small strains only, the uniqueness theorem of Chapter III (Sect. 2 (ii)) assures us that if a distribution of stress-rate is found such that all the stress and velocity equations, together with the continuity and boundary conditions, are satisfied, then it is unique. We must now inquire when the problem is statically determined: that is, when the boundary conditions are such that the stress-rate can be calculated uniquely from the stress equations alone, without reference to the associated velocity solution.[‡] Now it is obvious that when displacements are prescribed over a part of the boundary (for example, in a body indented by a rigid die of given shape) the state of stress and the extent of the plastic region cannot, in general, be found independently of the associated velocities. (There are certain apparent exceptions, to be noted later.) However, even when all the boundary conditions relate to stresses only, the problem is still not necessarily statically determined. Taking, for definiteness, the case when no plastic element unloads, it is necessary in addition that no slip-line should cut the plastic-elastic interface more than once;[§] in other

---

[†] The following discussion and theorems are taken from unpublished work of the writer (1949).

[‡] Apart from verifying that the rate of plastic work is positive.

[§] That this is not universally necessary can be seen by reference to the case when the entire body unloads; the stress-rates then satisfy the elastic equations everywhere and are uniquely determined by them when the applied loads are given.

words, every plastic element must be linked with the surface by two slip-lines lying entirely within the plastic region. If this condition is not fulfilled, the stress equations alone are not sufficient to determine the stresses and the plastic boundary uniquely.

To prove this, consider Fig. 66a where the shaded area P represents the plastic region at some moment, and the unshaded area E the elastic region. By hypothesis, certain slip-lines in P cut the plastic-elastic interface Σ twice. Let AB be one of these slip-lines, and let CD be the particular slip-line of the other family which crosses AB and intersects

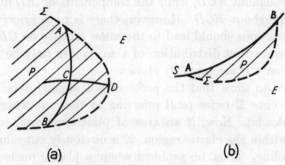

Fig. 66. Constructions used in examining when a plastic-elastic state of stress is uniquely defined by the stress boundary conditions alone.

Σ at right angles. Suppose that one stress-rate distribution has been found, satisfying all equations and boundary conditions throughout the body. This solution defines certain values of $\dot{p}$ and $\dot{\phi}$ along AB. Now let an arbitrary distribution of $\dot{\phi}$ be chosen on CD, subject only to the limitation that it is continuous along CD and at C; the differential relation between $\dot{p}$ and $\dot{\phi}$ along CD then gives the corresponding values of $\dot{p}$. Since the slip-lines are characteristics for the stress-rates, the values of $\dot{p}$ and $\dot{\phi}$ along AC and CD define a solution of the plastic stress-rate equations throughout the whole of ACD. Similarly, a solution is defined in BCD by the values along BC and CD. By combining these with the part of the known solution which refers to plastic elements outside ACB, we have evidently constructed another stress-rate distribution in P. The corresponding values of $\dot{p}$ and $\dot{\phi}$ on Σ determine values of the normal and shear components of the stress-rate acting across Σ. We draw, now, on a well-known theorem in elasticity to the effect that for any applied external stresses (in equilibrium) there exists a corresponding distribution of stress within the body satisfying the elastic equations; this is obviously true also for the stress-rate. By applying the theorem to E, we see that a solution of the stress-rate equations, satisfying all boundary conditions, can be associated with an arbitrary choice of $\dot\phi$ on CD. This completes the proof. Now the uniqueness theorem of Chapter III indicates that the actual distribution of stress-rate is that with which there can be associated a solution of the velocity equations. To understand how this requirement imposes restrictions on the stress-rate, note first that there is a unique velocity solution in E associated with any distribution of stress-rate. By the theorem of Section 6 (ii) in Chapter VI, the corresponding velocity components on AD determine the velocity throughout ACD, while the components on BD determine the velocity throughout BCD. However, there is no *a priori* reason why these two solutions should lead to the same velocity on CD. It follows that a self-consistent distribution of velocity can only be found if the values of $\dot\phi$ on CD are properly chosen.

It remains to show that the problem is statically determined when no slip-line cuts Σ twice (still referring to the case where no plastic element unloads). Now, if an area of plastic material is completely contained within the elastic region, Σ is obviously cut more than once by any slip-line. Thus, no problem where a plastic nucleus originates within the body is statically determined. Consider, therefore, a plastic region comprising one or more zones intersecting the surface S; Fig. 66 b shows a typical zone which includes a section AB of S. Since, by hypothesis, any slip-line crossing Σ also intersects AB, P lies within the triangular area bounded by AB and the intersecting slip-lines through A and B. Hence the prescribed external stress-rates on AB uniquely determine the stress-rates throughout P. The corresponding normal and shear components of the stress-rates acting across Σ, together with the other boundary conditions for E, then uniquely define the stress-rate distribution in E (the normal component acting parallel to Σ will therefore usually be discontinuous across Σ, as was mentioned previously). The problem is thus statically determined. Knowing the stress-rates in E, we can evaluate the corresponding velocities uniquely, and in particular their values on Σ. The latter define the velocity distribution in P, which is directly calculable from (7); no inconsistency can emerge since no slip-line cuts Σ more than once.

(iv) *Influence of the value of G.* If the velocity equations (7) and (8) are re-written with quantities $u' = Gu$ and $v' = Gv$ as dependent variables, no elastic constants appear explicitly. It follows that, *when the boundary conditions do not specify the absolute values of any displacements*, the quantities u' and v' and the distribution of stress are independent of G to the approximation involved in neglecting changes in the shape of the body; thus, at loads below the yield-point, the value of G is effectively without influence on the stresses and the quantities u' and v'. The displacement of an element at a given moment is therefore directly proportional to $1/G$.

In a plastic-rigid body the stress distribution is the same, but the displacements are, of course, zero prior to the yield-point. Nevertheless, the quantities u' and v' have still to be introduced when the problem is *not* statically determined (even though the boundary conditions relate to stresses only). Moreover, since the stress-rate terms in equations (5) and (7), re-written with u' and v', are comparable with the other terms, they cannot usually be omitted in the rigid part of the plastic region without introducing an appreciable error in the calculated *stresses*. This shows, in a particularly vivid way, why a plastic-rigid material must be regarded as a plastic-elastic material in which E is made to increase without limit. If $1/E$ were naïvely set equal to zero at the outset, thereby forestalling the introduction of u' and v', there would be too few conditions to define the extent of the plastic region when the problem is not statically determined; the result would be an absence of uniqueness.

3. Yielding of notched bars under tension

In illustration of the foregoing principles consider a long bar, notched symmetrically on opposite sides, and pulled in tension by forces which are directed along the longitudinal axis and which are steadily increased from zero. The surfaces of the bar perpendicular to the planes of flow are stress-free. The bar is assumed so long that the state of stress in the neighbourhood of the notch is, to any desired approximation, independent of the precise distribution of the end load. All boundary conditions refer to loads, and it follows accordingly from Section 2 (iv) that the distribution of stress is not dependent on the elastic constants.

Southwell and Allen[†] have shown how the stress equations can be solved numerically by the application of relaxation methods.[‡] The equilibrium equations are satisfied by the introduction of a stress function ψ such that

$$\frac{\sigma_x}{k} = \psi_{yy}, \qquad \frac{\sigma_y}{k} = \psi_{xx}, \qquad \frac{\tau_{xy}}{k} = -\psi_{xy},$$

where the subscripts attached to ψ refer to partial derivatives. The

[†] R. V. Southwell and D. N. de G. Allen, *Phil. Trans. Roy. Soc.* A **242** (1950), 379.
[‡] R. V. Southwell, *Relaxation Methods in Theoretical Physics* (Clarendon Press, Oxford, 1946).

(x, y) plane is subdivided into a square network of arbitrarily fine mesh, and equations (2′) and (3′), in the forms

$$\tfrac{1}{4}(\psi_{yy}-\psi_{xx})^2+\psi_{xy}^2 = 1$$

and
$$\nabla^4\psi \equiv \left(\frac{\partial^2}{\partial x^2}+\frac{\partial^2}{\partial y^2}\right)^2\psi = 0,$$

are replaced by finite difference equations across the nodal points of the net. Values of ψ are tentatively assigned to the nodal points and modified until, after a process of trial and error, the difference equations and boundary conditions are satisfied to an accuracy warranted by the fineness of the mesh. Only one such solution is needed while the bar is still wholly elastic, since the stress at any point is proportional to the applied tension. After a plastic nucleus has formed (normally at the root of the notch) separate solutions have to be found for each of a succession of small increments of the tension. It is reasonably assumed that no element unloads, so that at each stage one of the conditions helping to determine the new plastic region is that it should completely enclose the existing one; it is in this way that the previous history continues to influence the course of events. Now the increments of stress are capable of being evaluated uniquely only when the steps in the applied tension are infinitesimal; the steps should therefore be taken so small that the uncertainty in placing the plastic boundary is comparable with the errors inherent in the finite difference formulae.

(i) Consider, first, a *semicircular notch* (Fig. 67), whose radius r is equal to one quarter of the width w of the bar; the width $2a$ of the minimum section is therefore equal to $2r$ or $\tfrac{1}{2}w$. This problem has been investigated by Southwell and Allen.† It is found that the bar deforms elastically up to a mean stress of approximately $0.33 \times 2k$ distributed over the ends (the mean longitudinal stress across the minimum section is twice this). Yielding now occurs at the roots of the notches, and it appears that the shape of the plastic region is initially such that the problem is statically determined; since the surface is circular and stress-free the slip-lines are logarithmic spirals, as in a tube expanded symmetrically. The positions of the boundary for various values of the applied tension are indicated roughly in the figure. At tensions greater than about $0.45 \times 2k$, however, the plastic regions obtained by Southwell and Allen are such that the plastic-elastic interface is cut twice by certain slip-lines. According to the theorem proved in Section 2 (iii) the problem is no longer statically determined, and the position of the

† R. V. Southwell and D. N. de G. Allen, op. cit., p. 245.

interface can only be found by satisfying the condition for the existence of a velocity solution. Since Southwell and Allen did not investigate this point their results become progressively less reliable as the load is raised. Nevertheless, the general extent of the plastic region and its direction of spread are probably not far wrong, since these are mainly a matter of equilibrium. The results indicate that the plastic boundary

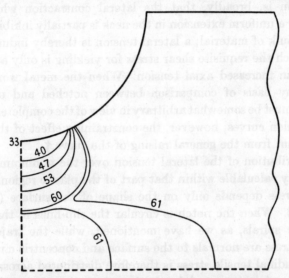

Fig. 67. Successive positions of the plastic boundary in a tensioned bar with two semicircular notches (after Southwell and Allen). The numbers attached to the curves denote the applied mean stress (the uniaxial tensile yield stress is taken as 100). Plane strain.

moves inwards from the notch at a rate roughly proportional to the increase in load until, at a tension of about $0 \cdot 60 \times 2k$, two plastic nuclei appear on the longitudinal axis about a distance a from the mid-point of the neck. Plastic zones now develop very rapidly from the nuclei, and fuse with the primary zones by the time the tension has been increased only to $0 \cdot 61 \times 2k$. Before this, the overall extension of the bar is of an elastic order of magnitude because of the residual core of non-plastic material running the whole length of the bar. Following the fusion of the primary and secondary zones the plastic region must continue to spread rapidly, and large distortions soon become possible. The yield-point load is therefore approximately equal to $0 \cdot 61 \times 2kw$ (per unit width normal to planes of flow).

It follows that the average longitudinal stress acting across the minimum section at the yield-point is $1 \cdot 22 \times 2k$. This is greater than the plane strain yield stress $2k$ in uniaxial tension, despite the fact that the notched bar yields locally much earlier. This is in accord with the familiar experimental observation that in a notched-bar test the mean stress in the neck, measured at the yield-point, is greater than the true yield stress measured with an un-notched specimen. The reason for this phenomenon is, broadly, that the lateral contraction which would accompany a uniform extension in the neck is partially inhibited by the adjoining bulk of material; a lateral tension is thereby induced, in the face of which the requisite shear stress for yielding is only attained by means of an increased axial tension. When the metal work-hardens rapidly, any basis of comparison between notched and un-notched specimens must be somewhat arbitrary in view of the completely rounded load-extension curves; however, the constraining effect of the notch is still apparent from the general raising of the curve.†

The distribution of the lateral tension over the minimum section is immediately calculable within that part of the plastic region where the state of stress depends only on the shape of the surface (Chap. VI, Sect. 5 (ii)). When the notch is circular the slip-lines in this part are logarithmic spirals, as we have mentioned, while the trajectories of principal stress are normals to the surface and concentric circular arcs. The longitudinal tensile stress is therefore distributed across the minimum section, *in this part of the plastic region*, according to the formula

$$\sigma = 2k\left[1 + \ln\left(1 + \frac{x}{r}\right)\right], \tag{9}$$

where x is distance measured from the root. σ rises steadily from the value $2k$ at the root. The lateral stress is a tension of amount $\sigma - 2k$, which increases from zero as we go inwards to the plastic boundary. Plastic yielding thus disperses the concentration of stress near the root, present while the bar is still elastic, and reverses the gradient of the distribution of stress across the neck.

(ii) Consider, next, a notch whose root is a circular arc of radius r, and for the present leave the relative values of w, r, and a unspecified. The state of stress depends on the two parameters w/a and r/a, which define the shape of the notch. No detailed investigation has been carried out, but we can make what should be a tolerably accurate estimate‡

---

† See, for example, M. L. Fried and G. Sachs, *Am. Soc. Test. Mat.*, Spec. Tech. Pub. No. 87 (1949), 83.
‡ R. Hill, *Quart. Journ. Mech. App. Math.* 2 (1949), 40.

of the yield-point load when the shape of the notch is such that the plastic region spreads directly across the neck, somewhat as in Fig. 68. For a given value of r/a we should naturally expect this to happen when w/a is sufficiently large. It is convenient, for the present purpose, to re-introduce the plastic-rigid material in view of the precision with which the yield-point can be located. When the plastic region develops as in Fig. 68, a little consideration shows that the yield-point corresponds to the moment when the slip-lines from the points S, where the plastic boundary Σ meets the surface, fall within Σ and intersect at the geometric centre O of the bar. No extension is previously possible since, by the theorem of Section 6 (ii) (Chap. VI), the whole of the plastic region is rigidly constrained.† Following this moment, however, the ends are free to move apart, and only the plastic material between Σ and the slip-lines OS is held rigidly to the non-plastic ends. There is evidently a discontinuity across OS in the tangential component of velocity; if the ends of the bar are drawn outwards with unit speed the discontinuity is of amount $\sqrt{2}$. The distribution of velocity within SOS can be found, if required, from the known velocity components normal to OS by the approximate integration of Geiringer's equations, or analytically by the application of Riemann's method (Chap. VI, equation (30)).

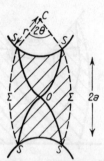

Fig. 68. Qualitative representation of the plastic region at the yield-point of a deeply-notched bar in tension.

We see, then, that at the yield-point the state of stress within SOS, and in particular across the minimum section, is uniquely determined by the contour of the root (the position of Σ depends, of course, on the notch depth). We can therefore calculate the yield-point load without needing to know the distribution of stress throughout the remainder of the bar. It should be carefully observed how this possibility depends on the assumption that the plastic region spreads directly across the neck;‡ we cannot similarly sidetrack detailed calculations of the plastic boundary in cases where it is impossible, or unsafe, to surmise the direction of spread.§ The distribution of longitudinal stress over the

† It will be appreciated, after the discussion of Sect. 2 (iv), that the equations (7), re-written with u' and v', should strictly have been used, rather than Geiringer's equations, in the proof of the theorem. The argument is unaffected since the slip-lines are the characteristics in both cases.

‡ And also on the circumstance that the stress and velocity characteristics are coincident. The possibility would not necessarily exist if the material yielded according to Mohr's criterion (see Chap. XI, Sect. 3).

§ The method should be reliable whenever the notch is sufficiently deep, provided the minimum section is given by (9). By integration the yield-point load is

$$L = 4ka\left(1+\frac{r}{a}\right)\ln\left(1+\frac{a}{r}\right). \tag{10}$$

$L/4ka$ is, by definition, the *constraint factor*; this is the factor by which the mean axial stress in the minimum section exceeds the true yield stress $2k$. The constraint factor rises steadily from the value unity with increasing a/r. If C is the centre of the circular contour of the root, and if 2θ is the angular span of the arc SS, it follows from the polar equation of the slip-lines OS that

$$\frac{a}{r} = e^{\theta}-1.$$

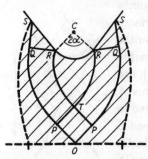

Fig. 69. Slip-line field round a deep wedge-shaped notch with a circular root.

If, then, the angular span of the circular root is 2α, the formula (10) is valid for $\theta \leqslant \alpha$, or for values of a/r such that

$$0 < \frac{a}{r} \leqslant e^{\alpha}-1.$$

Notice that values of α greater than $\tfrac{1}{2}\pi$ correspond to a keyhole notch.

To see how the solution can be extended to cover greater values of a/r, suppose, as an example, that the sides of the notch are straight and inclined at an angle $\pi-2\alpha$ ($\alpha \leqslant \tfrac{1}{2}\pi$), so that they join smoothly with the circular root (Fig. 69). The points S where Σ meets the surface now lie beyond the circular root RR. Since the surface is stress-free the slip-lines in QRS are straight lines meeting the surface at $45°$, while in RTR they are logarithmic spirals. The region $PQRT$ is uniquely defined by the slip-lines QR and RT; the slip-lines normal to RT are straight and equal in length (cf. Fig. 23). Hence the slip-lines in $OPTP$ are straight, and
$$RS = OT = a-r(e^{\alpha}-1).$$

The axial stress across the minimum section rises steadily according to (9) as far as T, but along TO it is constant since $OPTP$ is a region of uniform stress. By substituting $x = r(e^{\alpha}-1)$ in (9), or by a direct application of Hencky's theorem, the constant may be shown to be

curvature of the root does not change sign. The slip-line field near an elliptical notch has been computed for various eccentricities by W. W. Sokolovsky, *Theory of Plasticity*, p. 123 (Moscow, 1946), and also by P. S. Symonds, *Journ. App. Phys.* **20** (1949), 107. The method is probably less reliable when applied to a rectangular notch (R. Hill, Ministry of Supply, Armament Research Dept., Theoretical Research Rep. 9/46); the reader should have no difficulty in constructing the field qualitatively.

$2k(1+\alpha)$. Hence the yield-point load is

$$L = 4kr\alpha e^\alpha + 4k(1+\alpha)[a-r(e^\alpha-1)]$$

$$= 4ka\left[(1+\alpha)-\frac{r}{a}(e^\alpha-1-\alpha)\right] \quad \left(\frac{a}{r} \geqslant e^\alpha-1\right). \tag{11}$$

The constraint factor $L/4ka$ increases steadily with α, and is greatest when the notch has parallel sides ($\alpha = \frac{1}{2}\pi$). The maximum constraint factor obtainable with a parallel-sided notch corresponds to a vanishingly small root radius ($r/a \to 0$), and its value is $1+\frac{1}{2}\pi$, or about 2·571. This agrees, by chance, with the value found experimentally for *cylindrical* notched bars of a metal with a sharp yield-point; thus, Orowan, Nye, and Cairns† have observed a maximum constraint factor of about 2·6 for annealed mild steel.

The formulae (10) and (11) do not involve the width of bar. They are valid, as has been mentioned, only when the notch is sufficiently deep. How deep cannot be stated without accurate solutions, but a lower limit can be set at once. We must obviously have $L < 2kw$, for, if the load is greater than this, the implication is that the bar would already have yielded in uniform tension near its ends. Otherwise expressed, $w/2a$ must not be less than the constraint factor. Taking formula (10) as an example, we must have

$$\frac{w}{2a} > \left(1+\frac{r}{a}\right)\ln\left(1+\frac{a}{r}\right).$$

When $r/a = 1$, the least value of $w/2a$ is $2\ln 2$, or about 1·39. However, the range of validity of (10) certainly does not extend to such a shallow notch, since we know from Southwell and Allen's solution for $r/a = 1$ and $w/2a = 2$ that the plastic region spreads in a way very different from that contemplated in deriving the formula. Indeed, according to (10), the constraint factor for $r/a = 1$ is 1·39, whereas the value given by the numerical solution is 1·22. The same conclusion may also be reached in regard to a 90° wedge-shaped notch ($r = 0$, $\alpha = \frac{1}{4}\pi$) and $w/2a = 2$, where further calculations of Southwell and Allen suggest that plastic zones spread outwards from the roots in directions parallel to the longitudinal axis, rather like long fingers; the mean stress at the yield-point appears to be a little in excess of $0.64 \times 2k$, for the zones then bend sharply and rapidly in to the axis, which they intersect at distances from the centre somewhat greater than a. For this notch, (11) implies a constraint factor of 1·79, as against 1·28 from the numerical solution. The

† E. Orowan, J. F. Nye, and W. J. Cairns, Ministry of Supply, Armament Research Dept., Theoretical Research Rep. 16/45.

results are only qualitative since, for this notch, the problem is never statically determined at any stage. Using the same technique Jacobs[†] has found similar plastic zones for a slit notch ($r = 0$, $\alpha = \tfrac{1}{2}\pi$).

4. Plastic yielding round a cavity

If a body containing a cavity is loaded externally, the local concentration of stress on or near the perimeter of the cavity ultimately induces yielding. Suppose that the perimeter is subjected to given stress; in practice this will usually be zero or a uniform internal pressure. If, now, we are able to surmise the direction in which the plastic region spreads under a prescribed loading-path, we can at once calculate the distribution of stress in that part of the plastic region dependent only on the shape of the cavity. When most of some cross-section of the body is occupied by the cavity, so that the load is supported there only by two narrow strips, the conditions are most favourable for estimating the yield-point load by the method described for notched bars. On the other hand, when the cavity is far distant from the surface the problem is usually to find how local yielding relieves the stress concentration at loads well below the yield-point of the body as a whole; this problem is encountered, for example, in the theory of the rupture of a (normally) ductile metal where it is essential, if fracture is initiated at minute internal cracks, to estimate the depth and influence of any plastic zones. In such circumstances a detailed solution can hardly be avoided.

When the cavity is circular and is situated in an infinite medium, the surfaces of which are uniformly stressed by normal pressures, the distribution of stress is easily obtained by an obvious modification of the theory for an autofrettaged tube (Chap. V, Sect. 2). Galin[‡] has investigated the more difficult problem where the stress at infinity is not simply a uniform hydrostatic pressure but a uniform stress of any kind. The stresses in the part of the plastic region dependent only on the shape of the cavity are (equations (36) and (37), Chap. V)

$$\sigma_r = -p + 2k \ln \frac{r}{a}, \qquad \sigma_\theta = -p + 2k\left(1 + \ln \frac{r}{a}\right),$$

where p is the internal pressure and a is the radius of the cavity. The corresponding stress function, defined by

$$\frac{\sigma_r}{k} = \frac{1}{r}\frac{\partial \psi}{\partial r}, \qquad \frac{\sigma_\theta}{k} = \frac{\partial^2 \psi}{\partial r^2},$$

[†] J. A. Jacobs, *Phil. Mag.* **41** (1950), 349 and 458.
[‡] L. A. Galin, *Prikladnaia Matematika i Mekhanika*, **10** (1946), 365.

is
$$\psi = -\tfrac{1}{2}\Big(1+\frac{p}{k}\Big)r^2 + r^2 \ln\frac{r}{a}.$$

This happens to be bi-harmonic, that is, it satisfies the same equation $\nabla^4\psi = 0$ as the stress function in the elastic region; naturally, only a few plastic states have this property, another being the slip-line field consisting of radii and circular arcs. Galin took advantage of this circumstance to treat the stress in the plastic and elastic regions on a common footing, and replaced ψ by a combination of functions of the complex variable $z = re^{i\theta}$.[†] In this way he was able to construct an elastic-stress state satisfying the boundary conditions and such that the plastic-elastic interface is an ellipse surrounding the cavity. It is necessary that the ellipse should not have too great an eccentricity (ratio of axes $< \sqrt{2}$) for otherwise it is cut twice by certain slip-lines, and the problem is not statically determined. Furthermore, the solution can only apply when the internal and external loads are mutually varied in such a way that successive ellipses contain their predecessors, and provided also that the rate of plastic work is positive. Galin did not investigate the latter point, nor did he propose a solution corresponding to the early part of the loading-path, where the plastic region consists of two discrete zones. It is therefore not known how (or even whether) the loads can be applied to produce the first ellipse (touching the perimeter of the cavity).

A slight extension of this method of constructing elastic states of stress adjoining a known plastic state has been sketched by Parasyuk,[‡] for use when the plastic state is not bi-harmonic. He has briefly indicated how it may be applied when a circular cavity is subjected to a constant shear stress mk ($0 \leqslant m \leqslant 1$) in addition to a uniform pressure. The stresses in the plastic region bordering the cavity are

$$\frac{\tau_{r\theta}}{k} = \frac{ma^2}{r^2}, \qquad \sigma_\theta - \sigma_r = 2k\sqrt{[1-(m^2a^4/r^4)]},$$

$$\frac{\sigma_r}{2k} = -\frac{p}{2k} - \ln\frac{r}{a} + \tfrac{1}{2}\ln\left[\frac{1-\sqrt{(1-m^2)}}{1-\sqrt{\{1-(m^2a^4/r^4)\}}}\right] - \tfrac{1}{2}[\sqrt{\{1-(m^2a^4/r^4)\}} - \sqrt{(1-m^2)}].$$

These equations are due to Nadai.[§]

[†] A function ψ satisfying the bi-harmonic equation can always be expressed in the form $\mathrm{Re}\{\bar{z}\Omega(z)+\omega(z)\}$, where Ω and ω are certain functions of z; A. C. Stevenson, *Phil. Mag.* **34** (1943), 766, has shown how this method may be systematically employed in the solution of plane problems in elasticity. I. N. Sneddon, ibid. **36** (1945), 629, has briefly indicated how Stevenson's method can be adapted to generate plastic states of stress.

[‡] O. S. Parasyuk, *Prikladnaia Matematika i Mekhanika*, **13** (1948), 367.

[§] A. Nadai, *Zeits. Phys.* **30** (1924), 106.

5. Indentation and the theory of hardness tests

At loads below the yield-point, when the distortion of external surfaces is negligible, the distribution of stress and strain in a notched bar under *compression* is clearly identical, apart from a change of sign, with that in a similar bar under tension. Since, by symmetry, there is no displacement across the minimum section, the action of one half of the bar on the other is equivalent to that of a flat rigid die fixed at the minimum section and having the same width; since the shear stress there is zero the die must, in addition, be smooth. By superposing a uniform velocity

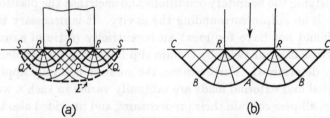

Fig. 70. Slip-line field and plastic region at the yield-point in the indenting of a semi-infinite medium by a flat die, showing (a) the author's solution, and (b) Prandtl's solution.

to bring one end of the bar to rest, we obtain the problem of the indenting of a medium by a smooth flat die. It follows that any solution for a notched bar under tension is immediately applicable, with only a change in sign, to the indenting of a medium similar in shape to one half of the bar. We may therefore take over the discussion and results of Section 3. It is to be observed that, although indentation by a die is generally expressible only through a velocity boundary condition, we have in this special case been able to formulate the problem entirely in terms of stress boundary conditions by considering the medium together with its mirror image; this is clearly only possible when the die is flat.

(i) *Semi-infinite medium with a plane surface.* By setting $\alpha = \tfrac{1}{2}\pi$ and $r = 0$ in (11) the estimated yield-point load for a plane semi-infinite medium indented by a smooth flat die is

$$L = 4ka(1+\tfrac{1}{2}\pi) \sim 2\cdot 571 \times 4ka, \tag{12}$$

where $2a$ is the width of the die.† The probable plastic boundary at the yield-point is indicated by the broken curve Σ in Fig. 70a; the slip-line

---

† The resistive pressure is uniformly distributed over the die. It should hardly be necessary to remark that the plastic region would be entirely different if a steadily increasing uniform pressure were applied over a section of width $2a$; it is known from elastic theory that plastic yielding first occurs along the semicircle having the section as diameter, at a pressure equal to $\tfrac{1}{2}\pi \times 2k$.

field is self-explanatory. The plastic material between $OPQS$ and Σ is rigid, while elements in $OPQSR$ move with speed $\sqrt{(2)}U$ along the slip-lines parallel to $OPQS$, where U is the downward speed of the die.

This problem was first investigated (for a plastic-rigid body) in 1920 by Prandtl,[†] who recognized the hyperbolic character of the stress equations and derived the two special fields needed in the solution (this work preceded the publication of Hencky's general equations by three years). Prandtl suggested the field of Fig. 70 b, for which the corresponding load is still given by (12). However, by the time the plastic region has spread so far, considerable indentation and distortion of the surface must have occurred; the condition determining the yield-point was apparently not known to Prandtl. The velocity distribution in the configuration of Fig. 70 b is indeterminate, and Prandtl assumed that the triangle RAR moved downward as a rigid body attached to the die. There is experimental evidence to show that this may, in fact, be true when the die is sufficiently *rough* (cf. the discussion of piercing in Chap. VII, Sect. 5); no theoretical solution has been suggested for intermediate amounts of friction.

(ii) *Truncated wedge.* Consider an infinite wedge of angle 2α truncated by a plane section of width $2a$ perpendicular to its axis of symmetry. The yield-point load, when the wedge is compressed by a smooth flat die of width not less than $2a$, is obtained from (11) by setting $r = 0$:

$$L = 4ka(1+\alpha). \tag{13}$$

The slip-line field is similar to that in Fig. 70 a, the angle PRQ being equal to α. The relation (13) and the assumed direction of development of the plastic region are in fair agreement with experiments on mild steel by Nadai.[‡] However, annealed mild steel is not a suitable material for comparison with theory when the overall strain is small, since the Lüders bands are well spaced. A Lüders band requires less stress for its propagation than for its formation, and in consequence there must be considerable variations from the stress state expected in a metal that deforms homogeneously.

The stress distribution in a severely compressed wedge has been described in Chapter VIII, Section 3.

(iii) *Finite medium with a plane surface.* A medium of depth h, resting on a plane foundation, is indented by a smooth flat die of width $2a$ ($a < h$). Provided h/a is not too great, experiment shows that the plastic

[†] L. Prandtl, *Nachr. Ges. Wiss. Göttingen* (1920), 74.
[‡] A. Nadai, *Zeits. ang. Math. Mech.* 1 (1921), 20. See also G. Sachs, *Zeits. tech. Phys.* 8 (1927), 132.

zones, originating at the corners of the die, fuse, and spread directly through the medium. Assuming this, the yield-point is determined by the condition that slip-lines should connect the corners R to the midpoint S of the supporting plane. The shape of these slip-lines must be chosen so that a velocity solution can be found such that (i) the component of velocity normal to RR is constant, and (ii) the normal component of velocity at any point on RS is compatible with the sideways displacement of the still-rigid material. These requirements are fulfilled by the field shown in Fig. 71,† where the slip-lines in PQR are radii and circular arcs, and where $PQSQ$ is defined by the equal arcs PQ. Since there is no change in volume the sideways speed of the rigid ends is equal to Ua/h, where U is the speed of the die. We can therefore construct the velocity solution in the order $PQSQ$, PQR, and RPR. Since the velocity component normal to the straight slip-line RQ is constant, the component normal to RP is also constant, by the now familiar theorem.

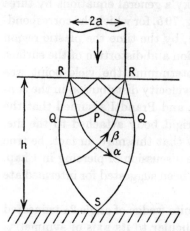

FIG. 71. Slip-line field at the yield-point in the indenting of a block of finite depth resting on a plane foundation.

Hence the velocity in RPR is uniform (and directed downward) and is therefore compatible with the movement of the rigid die; this velocity, when calculated, would turn out to be U since the speed Ua/h has been assigned to the rigid ends.

The relation between the angle PRQ and h/a is given by Table 1 (see end of book). The stress distribution is calculable by Hencky's theorem in terms of the mean compressive stress at one point of the field, say P. This is determined by the condition that the horizontal resultant of the stress acting across the vertical section through S should be equal to the resistance to sliding over the plane foundation. When the resistance is zero the relation‡ between the yield-point load and the ratio h/a is as

† This field was first suggested by L. Prandtl, *Zeits. ang. Math. Mech.* 3 (1923), 401, who did not, however, examine whether it could be associated with a velocity solution. A field closely resembling this has been observed in mild steel etched to show the Lüders bands, which coincide approximately with slip-lines; see F. Körber and E. Siebel, *Mitt. Kais. Wilh. Inst. Eisenf.* 10 (1928), 97; A. Nadai, *Plasticity*, p. 249 (McGraw-Hill Book Co., 1931).

‡ R. Hill, *Journ. Iron and Steel Inst.* 156 (1947), 513. A less accurate calculation is given by W. W. Sokolovsky, *Theory of Plasticity*, p. 143 (Moscow, 1946).

shown in Fig. 72; the theoretical relation when h/a is less than unity is not known. As h/a increases from unity the pressure necessary to begin indentation rises from $2k$, the compressive yield stress in plane strain, to $2k(1+\tfrac{1}{2}\pi)$ when $h/a \sim 8\cdot75$ (angle $PRQ \sim 77\cdot3°$). For values of h/a greater than this, the inference is that the plastic region develops in such a way that indentation begins at the load (12) and proceeds by displacing

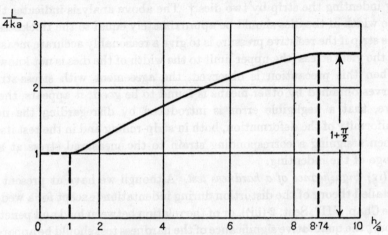

Fig. 72. Relation between the indentation pressure and h/a for a block of height h indented by a flat die of width $2a$.

material sideways to the free surface; the plastic zone beneath the die is still rigidly constrained by the non-plastic matrix (its depth of penetration naturally depends on h/a). Thus, for a true measure of hardness by a plane indentation test, the thickness of the material should not be less than about 4·4 times the width of the indenter. This is somewhat greater than the minimum thickness of a cold-worked metal recommended for the Brinell test,† where there is greater freedom of flow.

The effect of a frictional resistance F over the foundation plane is clearly to increase the mean compressive stress at every point by the amount F/h; the yield-point load is therefore

$$L' = L + \frac{2aF}{h},$$

where L is the load when the plane is smooth. If the resistance may be represented by a coefficient of friction μ, then $F = \tfrac{1}{2}\mu L'$ and so

$$L' = L\Big/\left(1-\frac{\mu a}{h}\right); \qquad L' \leqslant 4ka(1+\tfrac{1}{2}\pi). \tag{14}$$

† G. A. Hankins and C. W. Aldous, *Journ. Inst. Metals*, **54** (1934), 59.

The state of stress in a rectangular block of height $2h$, symmetrically indented on opposite sides by two dies, is evidently identical with that in a block of height h which is indented by one die and is resting on a smooth rigid foundation: the horizontal plane of symmetry in the former is replaced by the supporting plane in the latter. Now the plane stress-strain curve of a strip, before or after rolling, is conveniently obtained by indenting the strip by two dies.† The above analysis indicates that the width of the dies should be approximately equal to the thickness of the strip if the resistive pressure is to give a reasonably accurate measure of the yield stress (the upper limit to the width of the dies is not known). When this precaution is observed, the agreement with stress-strain curves obtained by other means is found to be good; it appears, therefore, that a negligible error is introduced by disregarding the non-uniformity of the deformation, both in strip-rolling and in the test itself, when assigning a corresponding strain to the measured stress at any stage of the indenting.

(iv) *Significance of a hardness test.* Although we have at present no detailed theory of the distortion during indentation (except for a wedge: see Chap. VIII, Sect. 2 (iii)), or of the relation between load and penetration,‡ the qualitative significance of the hardness test should be apparent from the foregoing analysis. For small loads the deformation is elastic and the impression disappears on removing the load; the diameter and depth of the impression are described by Hertz's theory of elastic contact.§ At a certain load, depending on the shape of the indenter, yielding occurs in some element, not necessarily on the surface. With an initially stress-free specimen and a ball indenter the first element to yield is distant about $0.25d$ directly below the centre of the impression, diameter d; the corresponding average resistive pressure on the ball is about $1.1\,Y$.‖ A plastic zone now spreads from this nucleus as the load is raised farther. If the load is subsequently removed the recovery is only partial; a permanent indentation remains and the specimen is left in a state of residual stress. Tabor†† has demonstrated experimentally that the recovery in many metals is elastic, so that a reapplication of the same

---

† H. Ford, *Proc. Inst. Mech. Eng.* **159** (1948), 115.

‡ Russian writers (see W. W. Sokolovsky, *Theory of Plasticity*, pp. 145–60) have computed slip-line fields and loads for indentations of various shapes and depths, but have assumed that the surface remains flat; the calculation of the surface distortion and its effect on the load is really the crux of the problem.

§ S. Timoshenko, *Theory of Elasticity*, pp. 339–50 (McGraw-Hill Book Co., 1934).

‖ Ibid. p. 344. For an experimental confirmation of this formula, see R. M. Davies, *Proc. Roy. Soc.* A, **197** (1949), 416.

†† D. Tabor, ibid. **192** (1948), 247.

load restores the original impression and brings every plastically strained element back to the point of yielding; he has shown further that the amount of elastic recovery is in accord with Hertz's theory. At first the plastic distortion is constrained to be of an elastic order of magnitude by the non-plastic matrix. In a pre-strained metal there is, as we have seen, a fairly well-defined load, the yield-point load, under which the plastic zone has spread so far that the constraint becomes ineffective near the indenter; the material there is displaced sideways towards the free surface, undergoing relatively severe distortion, to produce the raised coronet characteristic of indentation in a cold-worked metal. The yield-point load is normally much greater than the load at which the first plastic nucleus appears and corresponds, by definition, to the rapid bend in the load-penetration curve. The sharpness of the bend depends, in part, on the shape of the indenter, being more marked for a flat die than for a ball where a continually increasing contact-surface is presented to the medium; the yield-point load is vanishingly small for a cone where, after a penetration exceeding the radius of curvature of the tip, the configuration of stress and strain remains effectively similar and the load is proportional to the impression area. In a thoroughly annealed metal, on the other hand, the load-penetration curve is always well rounded, the transition from small to large plastic strains being gradual. Moreover, there is little or no displacement of material above the level of the original surface, and a 'sinking-in' impression is produced. We can understand the different behaviour of cold-worked and annealed metals by regarding the actual impression as a compromise between (i) flow out to the surface, necessitating severe and localized distortion, and (ii) an inward displacement accommodated by the resilience of the whole specimen, with relatively small strains spread through a much greater volume. When the work-hardening is slight (i) takes place before the elastic resistance of the bulk can be overcome; when the work-hardening is heavy the load which would be needed to effect (i) is so great that (ii) occurs first.

Consider, now, the dependence of the resistive pressure p (load/projected area of impression) on the shape of the indenter and on the properties of the material. If the indenter is a cone, pyramid, or wedge, geometric similarity implies that, in an initially stress-free specimen,

$$p = 4L/\pi d^2 = \text{constant},$$

where d is the diameter of the impression (measured in the plane of the original surface) when the load is L, and where the value of the constant depends on the angle 2α of the vertex, the properties of the material, and the coefficient of friction μ. If the work-hardening is negligible, we may write

$$p = cY, \qquad (15)$$

where Y is the tensile or compressive yield stress and c is a constant whose value depends on α, μ, and the shape of the indenter. When α is greater than about 60° and the lubrication is good, c is found to lie between about 2·5 and 3·0, the exact value depending slightly on α and on the shape of the indenter;[†] this is in agreement with the theory for a wedge (Chap. VIII, Sect. 2 (ii)) where, as $\alpha \to 90°$,

$$p \to 2k(1+\tfrac{1}{2}\pi) \sim 2\cdot 97Y,$$

if von Mises' criterion is applicable ($k = Y/\sqrt{3}$). When the indenter is a ball or flat die, dimensional considerations show that

$$p = 4L/\pi d^2 = f(d/D) \qquad (16)$$

for a ball of diameter D, and

$$p = 4L/\pi d^2 = g(t/D)$$

for a die of width D where t is the penetration. The functions f and g depend mainly on the properties of the material, and only slightly on μ. Now, when the material is pre-strained, the variation of p with depth can be neglected after the yield-point, so long as d and t are small compared with D (as is usual), and we may write $p = cY$ where c is again found to be of order 3.[‡] The general conclusion is, therefore, that the quantity measured in a hardness test, namely the resistance to penetration, is directly proportional to the tensile yield stress when the material is heavily pre-strained, and that the factor of proportionality depends on the shape of the indenter but does not differ greatly from 3 when a lubricant is used and the specimen is initially stress-free. If the specimen is initially in a state of residual stress (for example, when hardness micro-tests are performed across a section of a rolled sheet or a drawn wire) this affects the shape and size of the impression for a given load; the resistive pressure is therefore not necessarily equal to the hardness of the stress-free material.

We come now to annealed or lightly pre-strained metals. Since the plastic distortion around the indenter is non-uniform, different elements harden by different amounts and the hardness is therefore some complicated function of the whole stress-strain curve in the relevant range

[†] R. F. Bishop, R. Hill, and N. F. Mott, *Proc. Phys. Soc.*, **57** (1945), 147.
[‡] D. Tabor, op. cit., p. 258.

of strain. We can, if we wish, introduce a mean equivalent strain $\bar{\epsilon}$ and write
$$p = cY(\bar{\epsilon}),\qquad(17)$$
where $Y(\epsilon)$ is the tensile stress-strain relation. In effect, for any chosen value of c, this is a *definition* of $\bar{\epsilon}$ at each stage of the indentation; the justification of this procedure lies, of course, in the circumstance that when c is assigned a value of about 3, $\bar{\epsilon}$ is found to be reasonable. Consider, for example, the Brinell test. Meyer[†] has shown that for many metals the function f in (16) may be closely approximated by a power law
$$p = 4L/\pi d^2 = A(d/D)^m,\qquad(18)$$
where A is a constant with the dimensions of stress, depending only on the material, and m is a positive dimensionless exponent usually less than 0·5. m is smaller the greater the amount of pre-strain, and for heavily worked metals is almost zero; for the latter the power law (18) gives to p a very rapid rise for small d/D (corresponding to distortions of elastic order preceding the yield-point), followed by a sharp bend and an almost constant value A of p. Now the tensile stress-strain curve can be empirically represented by a power law
$$Y = B\epsilon^n,\qquad(19)$$
valid for moderate strains, (Chap. I, Sect. 4). Tabor has found that $m \sim n$, to a very rough approximation. Assuming this relation for broad descriptive purposes, and combining (17), (18), and (19), we obtain
$$cB\bar{\epsilon}^n \sim A(d/D)^n,$$
or
$$\bar{\epsilon} \sim (A/Bc)^{1/n}(d/D).\qquad(20)$$
Thus the mean strain in a Brinell test is proportional to the impression diameter. For annealed copper, where $A \sim 60$ kg./mm.², $B \sim 45$ kg./mm.², $n \sim 0.4$, $c \sim 2.8$, we have $(A/Bc)^{1/n} \sim 0.15$. Notice that for a heavily worked metal, where $n \to 0$, $A/Bc \sim 1 - \lambda n$, where λ is some positive constant; hence $(A/Bc)^{1/n} \to e^{-\lambda}$.

[†] O. E. Meyer, *Zeits. Ver. deutsch. Ing.* 52 (1908), 645, 740, and 835. See also R. H. Heyer, *Proc. Am. Soc. Test. Mat.* 37 (1937), 119.

X
AXIAL SYMMETRY

1. Fundamental equations

In an axially symmetric distribution of stress the non-vanishing components are σ_r, σ_θ, σ_z and τ_{rz}, referred to cylindrical coordinates (r, θ, z) with z as the axis of symmetry. The non-vanishing velocity components are u and w, respectively perpendicular and parallel to the axis; v, the component in the circumferential direction, is zero since the flow is confined to meridian planes. The stress and velocity are independent of θ, and are functions only of r, z, and the time. If there are no body forces, and inertial stresses are negligible, the equations of equilibrium are

$$\frac{\partial \sigma_r}{\partial r} + \frac{\partial \tau_{rz}}{\partial z} + \frac{\sigma_r - \sigma_\theta}{r} = 0,$$
$$\frac{\partial \tau_{rz}}{\partial r} + \frac{\partial \sigma_z}{\partial z} + \frac{\tau_{rz}}{r} = 0. \tag{1}$$

The yield criterion of von Mises (equation (9) of Chap. II) reduces to

$$(\sigma_r - \sigma_\theta)^2 + (\sigma_\theta - \sigma_z)^2 + (\sigma_z - \sigma_r)^2 + 6\tau_{rz}^2 = 6k^2. \tag{2}$$

In cylindrical coordinates the Lévy–Mises relations become (see Appendix II)

$$\dot{\epsilon}_r = \frac{\partial u}{\partial r} = \lambda(2\sigma_r - \sigma_\theta - \sigma_z),$$
$$\dot{\epsilon}_\theta = \frac{u}{r} = \lambda(2\sigma_\theta - \sigma_z - \sigma_r),$$
$$\dot{\epsilon}_z = \frac{\partial w}{\partial z} = \lambda(2\sigma_z - \sigma_r - \sigma_\theta),$$
$$2\dot{\gamma}_{rz} = \frac{\partial u}{\partial z} + \frac{\partial w}{\partial r} = 6\lambda\tau_{rz}. \tag{3}$$

If the plastic region includes a finite section of the axis of symmetry, it is necessary that $\sigma_\theta = \sigma_r$ when $r = 0$, if infinities in the stresses are to be avoided. Hence, since the axis is a principal stress direction, $\tau_{rz} = 0$ and $\sigma_r - \sigma_z = \pm\sqrt{3}k = \pm Y$ when $r = 0$; furthermore, $u = 0$ since the axis is a streamline. (1), (2), and (3) constitute a system of seven equations for the seven unknowns (4 stress components, 2 velocity components, and λ). There are only three equations involving the stresses alone, but a fourth can be derived from (3) by eliminating u, w, and λ

(this leads to a third-order differential equation). However, since the boundary values of the stresses, and not their derivatives, are specified in a physical problem, the velocity equations have to be resorted to, in any event, to secure uniqueness. In general, therefore, the problem is not statically determined.

It is easy to show also that the problem of axial symmetry is not hyperbolic.† Suppose the stress and velocity components are given on a curve C (in any meridian plane). If the distribution of stress is continuous, it is evident from (3) that in general the velocity gradient can only be discontinuous when λ is. But, since the velocity must be continuous (except possibly when C is a slip-line) it follows from the second equation in (3) that λ itself must be continuous (except, perhaps, when $u = 0$). In general, therefore, the velocity gradient is uniquely determined by the given stress and velocity on C. Thus the problem cannot be hyperbolic.‡ It should be noticed that if r is increased indefinitely the circumferential strain-rate $\dot\epsilon_\theta$ tends to zero, and we recover, in the limit, the plane strain equations which *are* hyperbolic and where λ may be discontinuous. This essential difference between axial symmetry and plane strain is directly due to the fact that the circumferential strain-rate is *finite* and continuous in the former, but is zero in the latter.

Whereas the theory of plane strain is well developed, and the method of solving specific problems is well understood, there is at present nothing similar for axial symmetry. It is, for example, not clear how to construct, in principle, the solution of mixed boundary-value problems such as cone indentation or wire-drawing. The axially symmetric stress distributions derived below are either approximate, or are obtained by inverse methods and afterwards related to a physical situation.

2. Extrusion from a contracting cylindrical container§

Consider a metal rod which is held within a closely fitting cylindrical sleeve over a part of its length. Suppose, now, that the sleeve contracts radially, extruding the rod from each end. This hypothetical process simulates the swaging of a rod between two diametrically opposed,

† R. Hill, *Dissertation*, p. 42 (Cambridge, 1948), issued by Ministry of Supply, Armament Research Establishment, as Survey 1/48.

‡ This conclusion has been restated by P. S. Symonds, *Quart. Journ. App. Math.* 6 (1949), 448. His method consists in showing that, if the stress and velocity components and their first derivatives are given on a curve C, the second derivatives of the stress and velocity components are uniquely defined. The conclusion appears not to be warranted by this theorem alone, since it is assumed without proof that the first derivatives are continuous; furthermore, in physical problems the boundary conditions generally specify only the stress or velocity, and not their gradients.

§ R. Hill, op. cit. (p. 184 of reference).

semi-cylindrical, dies. We assume that the frictional stress has the constant value mk ($0 \leqslant m \leqslant 1$), and seek a solution in which the inclinations of the principal axes are independent of z. This, by analogy with Prandtl's cycloid solution for a block compressed between rough plates (Chap. VIII, Sect. 5 (v)), should approximate the actual stress distribution in a long rod, except near the ends or the middle.

Let the plane $z = 0$ be taken through one end of the container, with the z-axis directed into the container. Let us try the expressions

$$\frac{\tau_{rz}}{k} = \frac{mr}{a}, \quad \frac{u}{U} = -\frac{r}{a}, \tag{4}$$

where U is the inward radial speed of the container and a is the radius of the rod. We then have $\dot{\epsilon}_r = \dot{\epsilon}_\theta$, and hence $\sigma_r = \sigma_\theta$. Substituting this in the yield criterion, we obtain

$$\sigma_z - \sigma_r = \sigma_z - \sigma_\theta = \sqrt{3}k\left(1 - \frac{m^2 r^2}{a^2}\right)^{\frac{1}{2}}.$$

Inserting these expressions, with (4), into the equations of equilibrium, we find that they are compatible, and that

$$\left.\begin{aligned}\frac{\sigma_r}{k} = \frac{\sigma_\theta}{k} &= -\frac{2mz}{a} - c, \\ \frac{\sigma_z}{k} &= -\frac{2mz}{a} + \sqrt{3}\left(1 - \frac{m^2 r^2}{a^2}\right)^{\frac{1}{2}} - c,\end{aligned}\right\} \tag{5}$$

where c is a constant. The ends of the rod being free from stress, the resultant axial force on a transverse section must balance the frictional resistance over the surface. Thus

$$\int_0^a 2\pi r \sigma_z \, dr = -2\pi mkaz,$$

and so

$$c = \frac{2}{\sqrt{3}m^2}[1 - (1-m^2)^{\frac{3}{2}}]. \tag{6}$$

The equation of incompressibility is

$$\frac{\partial w}{\partial z} = -\frac{\partial u}{\partial r} - \frac{u}{r} = \frac{2U}{a}, \quad \text{or} \quad \frac{w}{U} = \frac{2z}{a} + g(r).$$

The third independent relation remaining to be satisfied by u and w may be obtained by taking the ratio of the second and fourth equations of (3):
$$\frac{2\dot{\gamma}_{rz}}{\dot{\epsilon}_\theta} = -ag'(r) = -2\sqrt{3}\frac{mr}{a}\left(1 - \frac{m^2 r^2}{a^2}\right)^{-\frac{1}{2}}.$$

Integrating:
$$\frac{w}{U} = \frac{2z}{a} - \frac{2\sqrt{3}}{m}\left(1 - \frac{m^2 r^2}{a^2}\right)^{\frac{1}{2}} + C, \tag{7}$$

where C is a constant which must be determined so that the axial flow across a transverse section is equal to the rate at which material is displaced by the container. This requires

$$-\int_0^a 2\pi rw\, dr = 2\pi a(l-z)U,$$

where $2l$ is the length of the container. Therefore

$$C = -\frac{2l}{a} + \frac{4}{\sqrt{3}m^3}[1-(1-m^2)^{\frac{3}{2}}]. \tag{8}$$

The radicals in the above equations must all be positive in order that the rate of plastic work is positive. It may be shown, by the method used for Prandtl's solution, that an originally plane transverse section is distorted into an ellipsoid of revolution, with semi-major axis a/m and semi-minor axis $2(a_0-a)^3/\sqrt{3}ma^2$, where a_0 is the initial radius of the rod.

3. Compression of a cylinder under certain distributed loads†

In the present section an inverse method is applied to obtain a possible plastic state in a solid cylinder compressed by a certain distribution of stress over its plane ends. The incompressibility equation

$$\frac{\partial w}{\partial z} + \frac{\partial u}{\partial r} + \frac{u}{r} = 0$$

is satisfied by the velocity components

$$u = -AJ_0'\left(\frac{r}{b}\right)\cos\left(\frac{z}{b}\right), \qquad w = -AJ_0\left(\frac{r}{b}\right)\sin\left(\frac{z}{b}\right), \tag{9}$$

where A and b are positive constants, and $J_0(x)$ is the Bessel function of zero order defined by

$$J_0'' + \frac{J_0'}{x} + J_0 = 0; \qquad J_0(0) = 1, \qquad J_0'(0) = 0.$$

The expansion of J_0 in ascending powers of x is

$$J_0(x) = 1 - \frac{x^2}{2^2} + \frac{x^4}{2^2.4^2} - \frac{x^6}{2^2.4^2.6^2} + \dots.$$

Notice that $w = 0$ when $z = 0$, and that $u = 0$ when $r = 0$; the distribution of u over the surface represents the development of a bulge. The shear strain-rate $\dfrac{\partial u}{\partial z} + \dfrac{\partial w}{\partial r}$ is everywhere zero, and therefore

$$\tau_{rz} = 0. \tag{10}$$

† R. Hill, op. cit., p. 263 (p. 182 of reference).

This is consistent with the requirement that the cylindrical surface is stress-free. The remaining independent equation furnished by the velocity relations (3) is

$$\frac{\sigma'_z}{\sigma'_\theta} = \frac{\frac{r}{b}J_0\left(\frac{r}{b}\right)}{J'_0\left(\frac{r}{b}\right)} = f\left(\frac{r}{b}\right), \quad \text{say,} \tag{11}$$

where σ'_z and σ'_θ are, as usual, the reduced stresses. f takes values from -2 to infinity as r/b increases from zero to $3\cdot 83$, the approximate argument of the second zero of J'_0.

Now, when $\tau_{rz} = 0$, the yield condition (2) can be alternatively expressed as
$$\sigma'^2_r + \sigma'^2_\theta + \sigma'^2_z = 2k^2.$$
But $\sigma'_r = -\sigma'_\theta - \sigma'_z$, and so
$$\sigma'^2_\theta + \sigma'_\theta \sigma'_z + \sigma'^2_z = k^2.$$
Combining this with (11), we have
$$\sigma'_\theta = k/\sqrt{(1+f+f^2)}, \qquad \sigma'_z = kf/\sqrt{(1+f+f^2)}.$$
Hence
$$\left.\begin{aligned}\sigma_\theta - \sigma_r = \sigma'_\theta - \sigma'_r = 2\sigma'_\theta + \sigma'_z = k(f+2)/\sqrt{(1+f+f^2)},\\ \sigma_z - \sigma_r = \sigma'_z - \sigma'_r = 2\sigma'_z + \sigma'_\theta = k(2f+1)/\sqrt{(1+f+f^2)}.\end{aligned}\right\} \tag{12}$$

We have finally to satisfy the equilibrium equations (1). Since $\tau_{rz} = 0$, it follows from the second that σ_z is a function of r only, and so from (12) that σ_r and σ_θ are also functions of r only. This is compatible with the first equilibrium equation provided that

$$\frac{d\sigma_r}{dr} = \frac{\sigma_\theta - \sigma_r}{r} = \frac{k(f+2)}{r\sqrt{(1+f+f^2)}},$$

or
$$\frac{\sigma_r}{k} = -\int_r^a \frac{(f+2)}{\sqrt{(1+f+f^2)}} \frac{dr}{r}, \tag{13}$$

since σ_r is zero on the free surface $r = a$. The condition that λ should be positive requires

$$\lambda = u/3r\sigma'_\theta = -\sqrt{(1+f+f^2)}AJ'_0\left(\frac{r}{b}\right)\cos\left(\frac{z}{b}\right)\bigg/3kr > 0.$$

Thus, $J'_0(r/b)$ and $\cos(z/b)$ must not change sign, and the radical must be positive since J'_0 is negative between its first and second zeros. If $2l$ is the length of the cylinder, this demands that

$$\frac{l}{b} \leqslant \frac{\pi}{2}, \qquad \frac{a}{b} \leqslant 3\cdot 83 \quad \text{(approx.)}.$$

The distribution of stress over the ends of the cylinder, needed to produce this plastic state, is

$$\frac{\sigma_z}{k} = \frac{2f+1}{\sqrt{(1+f+f^2)}} - \int_r^a \frac{(f+2)}{\sqrt{(1+f+f^2)}} \frac{dr}{r}. \tag{14}$$

σ_z is compressive in the centre and decreases numerically towards the edge, becoming tensile there if $f(a/b) > -\frac{1}{2}$, that is, if $a/b > 2 \cdot 17$. The parameter b is still arbitrary and determines the total load.

The solution does not specify the manner in which this final external stress distribution is to be reached, and it is conceivable that for certain, or even all, loading paths the cylinder may not remain rigid (though partly plastic) until the final state is reached. Also, for certain loading paths the cylinder may not finally be completely plastic, even though the stresses (14) are applied. In short, it is impossible to say exactly what problem has been solved; this is a defect common to many inverse 'solutions' in plasticity.

4. Cylindrical tube under axial tension and internal pressure

In an experimental determination of the yield criterion, or of the (μ, ν) relation, a common method is to use a hollow tube stressed by combined axial tension σ and internal pressure p. Provided the ratio t/a of wall thickness to mean radius is sufficiently small, the variation of the axial and circumferential stress components through the wall may be neglected. The radial stress varies from $-p$ on the inside to zero on the outside, and is either entirely disregarded in comparison with the mean circumferential tension pa/t, or roughly allowed for by assigning to it the mean value $-p/2$. It is of interest to investigate the actual variation of the stresses through the thickness, on the assumption that the Lévy–Mises relations are valid.

Provided the tube is sufficiently long, the state of stress and strain should be effectively independent of z, except near the ends. Furthermore, transverse plane sections should remain plane. The combined radial and axial strain must therefore be expressible by the velocity components

$$u = -\frac{lr}{2l} + \frac{C}{r}, \qquad w = \frac{lz}{l}, \tag{15}$$

where l is the gauge length, and C is a parameter depending on the relative rates of extension and expansion. The incompressibility equation is satisfied, and (3) states that $\tau_{rz} = 0$, and that

$$\frac{\frac{\partial u}{\partial r}}{\frac{u}{r}} = \frac{-\frac{l}{2l} - \frac{C}{r^2}}{-\frac{l}{2l} + \frac{C}{r^2}} = \frac{\sigma'_r}{\sigma'_\theta}.$$

If this is substituted in the yield condition

$$\sigma_r'^2 + \sigma_r'\sigma_\theta' + \sigma_\theta'^2 = k^2,$$

there results

$$\frac{\sigma_r'}{k} = -\frac{\left(1+\dfrac{lr^2}{2lC}\right)}{\left(1+\dfrac{3l^2r^4}{4l^2C^2}\right)^{\frac{1}{2}}}, \quad \frac{\sigma_\theta'}{k} = \frac{\left(1-\dfrac{lr^2}{2lC}\right)}{\left(1+\dfrac{3l^2r^4}{4l^2C^2}\right)^{\frac{1}{2}}},$$

where the radical must be given the same sign as C, in order that the rate of work shall be positive. The second equilibrium equation in (1) is satisfied identically (since $\tau_{rz} = 0$ and the stresses are independent of z), and the first gives

$$\frac{d\sigma_r}{dr} = \frac{\sigma_\theta - \sigma_r}{r} = \frac{2k}{r\left(1+\dfrac{3l^2r^4}{4l^2C'^2}\right)^{\frac{1}{2}}}.$$

Hence

$$\frac{\sigma_r}{k} = -2\int_r^b \frac{dr}{r\left(1+\dfrac{3l^2r^4}{4l^2C^2}\right)^{\frac{1}{2}}} = \coth^{-1}\sqrt{\left(1+\frac{3l^2b^4}{4l^2C^2}\right)} - \coth^{-1}\sqrt{\left(1+\frac{3l^2r^4}{4l^2C^2}\right)}, \tag{16}$$

where b is the current external radius. The internal pressure is therefore

$$p = k\left[\coth^{-1}\sqrt{\left(1+\frac{3l^2a^4}{4l^2C^2}\right)} - \coth^{-1}\sqrt{\left(1+\frac{3l^2b^4}{4l^2C^2}\right)}\right]. \tag{17}$$

By allowing l to tend to zero, we regain the familiar expression

$$p = 2k\ln(b/a)$$

for the expansion of a closed tube, when elastic strains are neglected. When $C = 0$, p is zero, the tube being extended under uniaxial tension. The distribution of axial stress is

$$\frac{\sigma_z}{k} = \frac{\sigma_r - (2\sigma_r' + \sigma_\theta')}{k}$$

$$= \frac{\left(1+\dfrac{3lr^2}{2lC}\right)}{\left(1+\dfrac{3l^2r^4}{4l^2C^2}\right)^{\frac{1}{2}}} - \coth^{-1}\sqrt{\left(1+\frac{3l^2r^4}{4l^2C^2}\right)} + \coth^{-1}\sqrt{\left(1+\frac{3l^2b^4}{4l^2C^2}\right)}. \tag{18}$$

The total axial load and the pressure p are functions of l/lC; in other words, a definite combination of load and pressure must be applied to produce given relative rates of extension and expansion. The internal and external radii, a and b, vary during the deformation; according to (15) their rates of change are

$$\dot{a} = u(a) = -\frac{la}{2l} + \frac{C}{a}, \qquad \dot{b} = u(b) = -\frac{lb}{2l} + \frac{C}{b}.$$

This distribution of stress was derived, in essence, by Nadai,† employing the Hencky stress-strain relations. In Nadai's solution, therefore, u and w denote the (small) displacements and l/l is replaced by ϵ_z, the total axial strain. As shown in Chapter II (Sect. 6) the respective solutions agree only when the stress ratios are constant, that is, when l/lC is constant (the deformation being small).

5. Tube-sinking‡

Whereas the equations of Section 1 are satisfied exactly in the three previous problems, the stress distributions now to be derived only

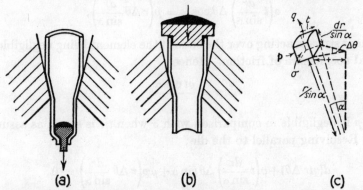

FIG. 73. Reduction of the wall-diameter of a tube by (a) drawing, and (b) pressing. The stresses on a small element of the wall are shown in (c).

satisfy them approximately. We consider first the process of tube-sinking, in which a thin-walled tube is drawn through a die, as in Fig. 73 a, or is pressed through, as in Fig. 73 b. A cup can be similarly drawn or pressed, the main difference being that in (a) the cup is pushed through the die by a loosely fitting internal mandrel, bearing on the base of the cup but not supporting the wall during its passage through the die. Apart from end conditions, the process is the same for both cup and tube. The object of the analysis is to calculate the load and the change in wall thickness. It will be supposed that the thickness is initially uniform, and that it is small in relation to the length of contact with the die. A steady state is then obtained (if the tube is sufficiently long), and it should be a good approximation to neglect the variation of stress through the

† A. Nadai, *Trans. Am. Soc. Mech. Eng.* **52** (1930), 193.
‡ H. W. Swift, *Phil. Mag.* **40** (1949), 883. See also G. Sachs and W. M. Baldwin, Jr., *Trans. Am. Soc. Mech. Eng.* **68** (1946), 655, who, however, do not allow for thickness changes.

thickness, by analogy with Sachs' theory of wire-drawing or von Karman's theory of sheet-rolling.

Let p be the pressure between the die and the wall at any point, q the mean tensile stress parallel to the die, and $\sigma \, (= -\sigma_\theta)$ the mean circumferential compressive stress. Let t be the thickness of the wall at radius r, μ the coefficient of friction, and α the semi-angle of the die (assumed conical). Consider the stresses acting on a small element of the wall, bounded by meridian planes inclined at an angle $\Delta\theta$, and by planes perpendicular to the local generator of the die and a distance $dr/\sin\alpha$ apart (Fig. 73c). Resolving perpendicular to the die:

$$\sigma\left(t\,\frac{dr}{\sin\alpha}\right)\Delta\theta\cos\alpha = p\left(r\,\Delta\theta\,\frac{dr}{\sin\alpha}\right),$$

the shear stresses acting over the ends of the element being negligible for normal coefficients of friction. Hence

$$p = \frac{\sigma t \cos\alpha}{r}. \tag{19}$$

Thus p is negligible in comparison with σ when t/r is small, as assumed here. Resolving parallel to the die:

$$d(qtr\,\Delta\theta) + \sigma\left(t\,\frac{dr}{\sin\alpha}\right)\Delta\theta\sin\alpha + \mu p\left(r\,\Delta\theta\,\frac{dr}{\sin\alpha}\right) = 0,$$

or
$$\frac{d}{dr}(rqt) + \sigma t + \frac{\mu p r}{\sin\alpha} = 0.$$

Using (19) to eliminate p, we have

$$\frac{d}{dr}(rqt) + \sigma t(1 + \mu\cot\alpha) = 0. \tag{20}$$

When the tube or cup is drawn, σ and q are both positive, and so the yield criterion of von Mises can be approximated by the equation

$$q + \sigma = mY \quad \text{(drawing)}, \tag{21}$$

where m is a disposable constant, to be selected to give closest overall agreement with the exact solution or with experiment. When the cup is pressed, σ is positive but q is negative. Now it may be verified *a posteriori* that σ is numerically greater than q in the practical range of reductions in diameter (< 30 per cent.), and so von Mises' criterion can be approximated by the equation

$$\sigma = mY \quad \text{(pressing)}, \tag{22}$$

where m again denotes a disposable constant.

In the steady state the ratio of the circumferential strain-rate to the strain-rate in the thickness direction is

$$\left(-\frac{dr}{r}\right)\bigg/\left(-\frac{dt}{t}\right) = \frac{-2\sigma-q}{\sigma-q};$$

i.e.
$$\frac{dt}{dr} = \frac{q-\sigma}{q+2\sigma}\frac{t}{r}. \tag{23}$$

The elimination of t between (20) and (23) leads to

$$r\frac{dq}{dr} + \frac{q(2q+\sigma)+(1+\mu\cot\alpha)\sigma(q+2\sigma)}{q+2\sigma} = 0. \tag{24}$$

If, now, σ is eliminated with the aid of (21) or (22), a differential equation for q in terms of r is obtained. The boundary condition is $q = 0$ when $r = a$ (drawing), and $q = 0$ when $r = b$ (pressing), where a and b are respectively the initial and final radii of the tube. When $r = b$ (drawing), $q = D$, the drawing stress; when $r = a$ (pressing), $-q = P$, the pressure needed for pressing. Similarly, by eliminating r between (20) and (23), we derive the relation between q and t:

$$t\frac{dq}{dt} + \frac{q(2q+\sigma)+(1+\mu\cot\alpha)\sigma(q+2\sigma)}{q-\sigma} = 0. \tag{25}$$

The boundary condition is $q = 0$ when $t = t_a$ (drawing), and $q = -P$ when $t = t_a$ (pressing), where t_a is the initial wall thickness. The integration of (24) gives the relation $q(r)$ between q and r, while the integration of (25) gives the relation $q(t)$ between q and t; combining these, we obtain the variation of t with r, and hence the final thickness. The formulae are too complicated to be worth giving; the calculated values of the stress and the relative increase in thickness are shown in Fig. 74 for $\mu\cot\alpha = 0, \frac{1}{2},$ and 1. The loads for drawing and pressing are respectively $2\pi Dbt_b\cos\alpha$ and $2\pi Pat_a\cos\alpha$ (it is a curious property of the present theory that $D = P$ when $\mu = 0$). Since the stress and relative thickening at any point are functions only of r/a, the load for a given value of b/a is directly proportional to the initial thickness. This is a consequence of the approximations in the analysis, and is only likely to be true when the tube is thin-walled; in fact, Baldwin and Howald[†] have observed that the thickness of a drawn tube *decreases* over a wide range of reductions when a/t_a is less than about 5. In the practical range the effect of strain-hardening has been shown by Swift to be comparatively slight in regard to thickness changes; the load can be estimated with sufficient accuracy by using a mean yield stress.

† W. M. Baldwin, Jr., and T. S. Howald, *Trans. Am. Soc. Metals*, 33 (1944), 88.

The changes in thickness shown in Fig. 74 may be understood qualitatively from equation (23). In drawing, when the reduction is small, σ is the predominant stress; thus $q-\sigma$ is negative and the tube thickens. For large reductions the axial tension q becomes predominant near the exit from the die; $q-\sigma$ is then positive and the tube thins in this region; however, there is still a resultant thickening except for very large reductions. (It should be noted that the value of t at a given position r/a is

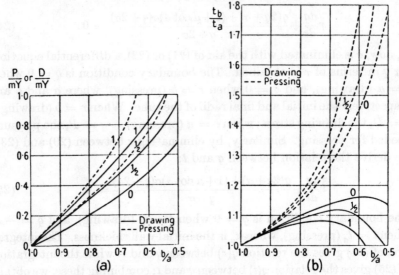

FIG. 74. (a) Relation between the load and the reduction in diameter, and (b) between the change in thickness and the reduction in diameter, for drawing and pressing (no work-hardening).

independent of the final reduction in drawing, but not in pressing.) In contrast, since the axial stress q is a compression in pressing, $q-\sigma$ is always negative and the tube thickens continuously. The resultant change is thus much smaller in drawing than in pressing; indeed, Swift has shown that it makes little difference to the calculated stresses and thickness changes in drawing if the variation of t is neglected in the equilibrium equation (20).

6. Stress distribution in the neck of a tension specimen

After a neck has formed in a cylindrical tensile specimen, the distribution of the stress across a transverse section is not necessarily uniform. The quantity (load)/(minimum sectional area), conventionally plotted in a stress-strain diagram, then measures, not the uniaxial yield stress, but the mean stress through the neck. The true yield stress beyond the point of maximum load can only be determined by this test if the state of stress and strain in the neck is known. Even if the material does not work-harden, the stress is likely to vary across the minimum section, by analogy with the notch effect described in Chapter IX (Sect. 3). The problem of calculating the stress distribution is much more complicated than for the notch, since the shape of the neck is unknown and can only be determined by tracing its gradual development. Also, elements in the neck harden by varying amounts because of the differential deformation.

Fortunately, a radical simplification is possible if we accept the experimental evidence of Bridgman,[†] indicating that elements *in the minimum section* are deformed uniformly (at any rate to a first approximation). These elements must therefore be in the same work-hardened state, with the same yield stress Y; moreover, the strain corresponding to this yield stress is known directly from the measured reduction in area at the neck. A further consequence of uniform radial strain is that the radial velocity is proportional to r; hence the circumferential strain-rate $\dot\epsilon_\theta \,(= u/r)$ is equal to the radial strain-rate $\dot\epsilon_r \,(= \partial u/\partial r)$ in the minimum section. It follows that $\sigma_r = \sigma_\theta$ there, whatever the (μ, ν) relation, provided that isotropy is maintained. Inserting this in the first of the equilibrium equations (1), we have

$$\frac{d\sigma_r}{dr} + \frac{\partial \tau_{rz}}{\partial z} = 0 \quad \text{when} \quad z = 0, \tag{26}$$

the origin being taken at the centre of the neck. Furthermore, since the state of stress at each point in the minimum section is just an axial tension, with a varying superposed hydrostatic stress, the yield condition is

$$\sigma_z - \sigma_r = Y \quad \text{when} \quad z = 0, \tag{27}$$

provided the influence of a hydrostatic stress on yielding is neglected. Combining (26) and (27):

$$\frac{d\sigma_z}{dr} + \frac{\partial \tau_{rz}}{\partial z} = 0 \quad \text{when} \quad z = 0. \tag{28}$$

The second equilibrium equation, applied in the plane $z = 0$, states that $\partial \sigma_z / \partial z = 0$ (τ_{rz} being zero from symmetry), and is not useful in the subsequent analysis.

We now introduce the change of variable which characterizes Bridgman's analysis[‡] of the stress distribution. Let ψ be the inclination of a

[†] P. W. Bridgman, *Rev. Mod. Phys.*, 17 (1945), 3. See also N. N. Davidenkov and N. I. Spiridonova, *Proc. Am. Soc. Test. Mat.*, 46 (1946), 1147.
[‡] P. W. Bridgman, *Trans. Am. Soc. Metals*, 32 (1943), 553.

principal stress direction in a meridian plane to the axis (Fig. 75a). If σ_3 and σ_1 denote the principal stresses (σ_3 being equal to σ_z when $z = 0$), the equations for the transformation of stress components in a plane give

$$\sigma_z \simeq \sigma_3, \qquad \sigma_r \simeq \sigma_1, \qquad \tau_{rz} \simeq (\sigma_3 - \sigma_1)\psi,$$

near the plane $z = 0$, to the first order of small quantities. Now the yield condition for an element near this plane can be written as

$$\sigma_3 - \sigma_1 = Y + \text{a term of order } \psi.$$

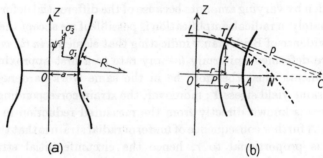

FIG. 75. (a) Coordinate axes of principal stress in the neck of a tension specimen, and (b) construction for the derivation of Bridgman's formula.

Thus $\tau_{rz} \simeq Y\psi$, and we obtain (*exactly*)

$$\left(\frac{\partial \tau_{rz}}{\partial z}\right)_{z=0} = Y\left(\frac{\partial \psi}{\partial z}\right)_{z=0} = \frac{Y}{\rho},$$

where ρ is the radius of curvature of a longitudinal line of principal stress where it crosses the plane $z = 0$. Substituting in (28):

$$\frac{d\sigma_z}{dr} + \frac{Y}{\rho} = 0 \quad \text{when} \quad z = 0.$$

This equation may alternatively be derived by considering the stresses on a small curvilinear element, bounded by principal planes.† Integrating, and introducing the boundary condition $\sigma_z = Y$ on the surface (since $\sigma_r = 0$ there), we find

$$\frac{\sigma_z}{Y} = 1 + \int_r^a \frac{dr}{\rho}, \tag{29}$$

$$\frac{\bar{\sigma}_z}{Y} = \frac{\int_0^a 2\pi r \sigma_z \, dr}{\pi a^2} = 1 + \int_0^a \frac{r^2 \, dr}{a^2 \rho}, \tag{30}$$

where a is the radius of the neck, and $\bar{\sigma}_z$ is the mean axial stress.

† N. N. Davidenkov and N. I. Spiridonova, op. cit., p. 273.

Failing an accurate solution for the stress distribution in the neck we may estimate $\bar{\sigma}_z$ by arbitrarily selecting a reasonable variation of ρ with r. On the surface, $\rho = R$, the radius of curvature (in an axial plane) of the neck at its root (Fig. 75a). Since the principal directions at the centre of the neck are along the axes of z and r, it appears improbable that ρ could do otherwise than increase steadily from the surface inwards. Hence σ_z steadily increases inwards from the value Y at the surface. The radial stress is then a tension,† and $\bar{\sigma}_z$ is greater than Y. It is probable that the neck forms in such a way that the lines of principal stress pass smoothly from coincidence with the surface to coincidence with the axis. If this is so, it can hardly be far wrong to assume, following Bridgman, that the transverse trajectories of principal stress in the neighbourhood of the minimum section are circular arcs meeting the surface and axis orthogonally (Fig. 75b). Let LM be such a trajectory, infinitesimally near the minimum section OA and intersecting it in N. Let C be the centre of curvature at P of a longitudinal principal stress trajectory PT. By simple geometry,

$$\rho^2 \simeq CT^2 = OC^2 - ON^2 \simeq (r+\rho)^2 - ON^2,$$

to the first order of small quantities. Hence, for any point P on OA,

$$r^2 + 2r\rho = \text{constant} = a^2 + 2aR,$$

on inserting the particular values of r and ρ corresponding to A. Thus, on passing to the limit,

$$\rho = \frac{a^2 + 2aR - r^2}{2r}. \tag{31}$$

Substituting this in (29) and (30), and integrating, we obtain formulae due to Bridgman:

$$\frac{\sigma_z}{Y} = 1 + \ln\left(\frac{a^2 + 2aR - r^2}{2aR}\right),$$

$$\frac{\bar{\sigma}_z}{Y} = \left(1 + \frac{2R}{a}\right)\ln\left(1 + \frac{a}{2R}\right). \tag{32}$$

This constraint factor is represented by the lower curve in Fig. 76.

The application of this formula in practice depends on the possibility of measuring R, the radius of curvature at the root of the neck. For widely differing steels Bridgman has found that, to a first approximation, a/R is a function only of the reduction of area (all the specimens having a length to diameter ratio of about three). His measurements, extending

---

† The initiation of fracture at the centre of the neck appears to be due to this state of triaxial tension. For an experimental investigation of the development of a cup-and-cone fracture, see S. L. Pumphrey, *Proc. Phys. Soc.* B, **62** (1949), 647.

up to reductions in area of 95 per cent., are scattered about a mean represented by the empirical formula

$$\frac{a}{R} = \sqrt{\left\{\ln\left(\frac{A_0}{A}\right) - 0 \cdot 1\right\}}, \qquad (33)$$

where A_0 is the initial sectional area and A is the current area of the minimum section. According to (33) R is infinite when the reduction of

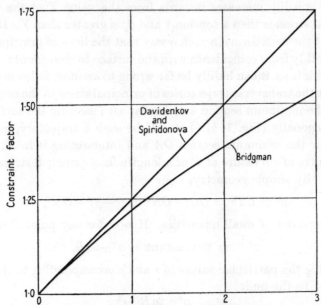

Fig. 76. Theoretical relation between the constraint factor and the ratio a/R, where a is the radius of the minimum section and R is the radius of curvature of the neck.

area is about 10 per cent.; this represents the uniform strain preceding necking. Since the latter varies with the amount of cold work previously given to the material, the formula is least accurate when the total reduction is small; however, the correction factor to be applied to the measured mean stress is then relatively unimportant.

Another formula for $\bar{\sigma}_z$ has been derived by Davidenkov and Spiridonova,† who proposed that

$$\rho = \frac{Ra}{r}.$$

† N. N. Davidenkov and N. I. Spiridonova, op. cit., p. 273. They sectioned a necked specimen and measured the curvatures (in the neck) of the distorted fibre lines originally parallel to the axis. They found that the curvature was very nearly proportional to $1/r$, and then assumed, without justification, that the longitudinal lines of principal stress coincided with the fibre lines.

This leads to
$$\frac{\bar{\sigma}_z}{Y} = 1 + \frac{a}{4R},$$
which is slightly greater than Bridgman's factor (see Fig. 76).

7. Compression of a cylinder between rough plates

Siebel† (1923) has derived a factor by which the mean stress measured in a cylinder compression test can be roughly corrected for the effect of a *small* amount of friction on the plates. The analysis is analogous to that described in Chapter VIII (Sect. 5 (vi)) for the plane compression test in that the variation of the stresses in the longitudinal direction is neglected. In addition σ_θ is set equal to σ_r; the error introduced by this assumption is not known. Integrating the first of the equilibrium equations (1) with respect to z, we then have

$$h\frac{d\sigma_r}{dr} - 2\mu p = 0,$$

where h is the height of the cylinder, and p is the pressure on the plate at radius r (τ_{rz} is equal to $-\mu p$ on the upper plate and to $+\mu p$ on the lower one). If μ is small, the yield condition may be written approximately as
$$p + \sigma_r = Y,$$
where Y is the yield stress corresponding to the current compressive strain. Eliminating σ_r between these equations:

$$\frac{dp}{dr} = -\frac{2\mu p}{h}.$$

Integrating:
$$p = Ye^{2\mu(a-r)/h}, \qquad (34)$$

on inserting the boundary condition $p = Y$ on $r = a$ (since $\sigma_r = 0$). The mean pressure on the plates is therefore

$$\bar{p} = \frac{\int_0^a 2\pi r p \, dr}{\pi a^2} \simeq \left(1 + \frac{2\mu a}{3h}\right) Y, \qquad (35)$$

to the first order in μ. The measured mean stress must be divided by the factor $1 + 2\mu a/3h$ to give the true yield stress; the correction is likely to be less good when there is appreciable bulging. The linear relation between $\bar{p}$ and a/h, which is predicted by (35) for specimens *in the same work-hardened state*, has been observed experimentally for copper by Cook and Larke‡

† E. Siebel, *Stahl und Eisen*, 43 (1923), 1295.

‡ M. Cook and E. C. Larke, *Journ. Inst. Metals*, 71 (1945), 377, particularly Fig. 11. See also W. Schroeder and D. A. Webster, *Journ. App. Mech.* 16 (1949), 289, for comparative data for thin disks (a/h extending up to 50); it is shown also how the theory must be modified when the frictional stress becomes equal to the yield stress in shear.

over a range of a/h from 0·5 to 1·5; their results correspond to values of μ between 0·2 and 0·3. Extrapolation of the experimental line back to a zero value of a/h appears to be reasonably safe, and should give the true yield stress. The complete stress-strain curve can be constructed in this way from three or four compression tests on specimens with different a/h ratios (preferably widely spread for greater accuracy).

8. Relations along slip-lines and flow-lines†

By taking the slip-lines in a meridian plane as curvilinear coordinates, the fundamental equations for the problem of axial symmetry can be

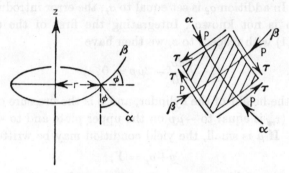

FIG. 77. Coordinate axes in a meridian plane for an axially-symmetric plastic state.

expressed in a more symmetrical form which is useful for certain purposes. The maximum shear stress in a meridian plane is denoted by τ, and the mean compressive stress $-\tfrac{1}{2}(\sigma_r+\sigma_z)$ in this plane by p. The principal stresses are then $-p \pm \tau$, σ, where σ is written for the circumferential stress σ_θ. The yield criterion (2) becomes

$$\tau^2 + (\sigma+p)^2/3 = k^2 = Y^2/3. \qquad (36)$$

The α and β families of slip-lines are distinguished by the sense of the shear stress τ (Fig. 77), the convention being the same as in the theory of plane strain. The angular coordinate of the β-lines is denoted by ϕ, measured away from the r direction towards the positive z direction. The equations of equilibrium are

$$\left. \begin{array}{l} \dfrac{\partial p}{\partial s_\alpha} + \dfrac{2\tau}{R} + \dfrac{(\sigma+p)}{r}\dfrac{\partial r}{\partial s_\alpha} - \dfrac{1}{r}\dfrac{\partial}{\partial s_\beta}(r\tau) = 0, \\[1.2ex] \dfrac{\partial p}{\partial s_\beta} + \dfrac{2\tau}{S} + \dfrac{(\sigma+p)}{r}\dfrac{\partial r}{\partial s_\beta} - \dfrac{1}{r}\dfrac{\partial}{\partial s_\alpha}(r\tau) = 0, \end{array} \right\} \qquad (37)$$

† R. Hill, *Dissertation*, p. 40 (Cambridge, 1948), issued by Ministry of Supply, Armament Research Establishment, as Survey 1/48.

where $\partial/\partial s_\alpha$ and $\partial/\partial s_\beta$ denote space derivatives along the slip-lines, their mutual sense being such that they form a right-handed pair. R and S are the (algebraic) radii of curvature of the slip-lines, defined by

$$\frac{1}{R} = \frac{\partial \phi}{\partial s_\alpha}, \qquad \frac{1}{S} = -\frac{\partial \phi}{\partial s_\beta}. \tag{38}$$

There are also the geometrical relations

$$\sin\phi = \frac{\partial r}{\partial s_\alpha}, \qquad \cos\phi = \frac{\partial r}{\partial s_\beta}. \tag{39}$$

The stress-velocity equations (3) transform into

$$\left. \begin{aligned} \frac{\partial u}{\partial s_\alpha} - \frac{v}{R} &= -\lambda(\sigma+p), \\ \frac{u\sin\phi + v\cos\phi}{r} &= 2\lambda(\sigma+p), \\ \frac{\partial v}{\partial s_\beta} - \frac{u}{S} &= -\lambda(\sigma+p), \\ \frac{\partial u}{\partial s_\beta} + \frac{\partial v}{\partial s_\alpha} + \frac{u}{R} + \frac{v}{S} &= 6\lambda\tau, \end{aligned} \right\} \tag{40}$$

where u and v denote the velocity components referred to the slip-lines (the meaning of u in the present section must not be confused with its previous meaning as the radial velocity).

The elimination of λ from the first two equations in (40) gives

$$\frac{\partial u}{\partial s_\alpha} - \frac{v}{R} + \frac{u\sin\phi + v\cos\phi}{2r} = 0.$$

With the help of (38), this becomes

$$\left. \begin{aligned} du - v\,d\phi + (u+v\cot\phi)\frac{dr}{2r} &= 0 \text{ along an } \alpha\text{-line.} \\ \text{Similarly:} \quad dv + u\,d\phi + (v+u\tan\phi)\frac{dr}{2r} &= 0 \text{ along a } \beta\text{-line.} \end{aligned} \right\} \tag{41}$$

These differ from Geiringer's equations in the theory of plane strain only in the extra terms in r (which vanish as $r \to \infty$). Equations (41) represent two of three independent relations which can be derived by eliminating λ from (40). To find a third, multiply the four equations of (40) by u^2, $\tfrac{1}{2}(u^2+v^2) - 3uv\tau/(\sigma+p)$, v^2, uv, respectively, and add. R, S, and λ vanish, and we find

$$\tfrac{1}{2}\left(u\frac{\partial}{\partial s_\alpha} + v\frac{\partial}{\partial s_\beta}\right)(u^2+v^2) + \left(\frac{u^2+v^2}{2} - \frac{3uv\tau}{\sigma+p}\right)\left(u\frac{\partial}{\partial s_\alpha} + v\frac{\partial}{\partial s_\beta}\right)\ln r = 0.$$

Now $u\dfrac{\partial}{\partial s_\alpha}+v\dfrac{\partial}{\partial s_\beta}$ is the space derivative along the momentary directions of flow. Hence, writing q for the total velocity $\sqrt{(u^2+v^2)}$, we have

$$d(rq^2) = \frac{6\tau uv}{\sigma+p}\,dr \qquad (42)$$

along the flow-lines at a given moment. On multiplying the equations (37) through by ds_α and ds_β, respectively, we obtain

$$\left.\begin{aligned}dp+2\tau\,d\phi+(\sigma+p-\tau\cot\phi)\frac{dr}{r} &= \frac{\partial\tau}{\partial s_\beta}ds_\alpha \text{ along an } \alpha\text{-line,}\\ dp-2\tau\,d\phi+(\sigma+p-\tau\tan\phi)\frac{dr}{r} &= \frac{\partial\tau}{\partial s_\alpha}ds_\beta \text{ along a } \beta\text{-line.}\end{aligned}\right\} \qquad (43)$$

These are not so simple as the relations for the velocities, owing to the derivatives of τ on the right-hand side.

If, however, we adopt Tresca's criterion instead of (36), then, *when σ is the intermediate principal stress*, $\tau = \tfrac{1}{2}Y$ and the equations become

$$\left.\begin{aligned}dp+2\tau\,d\phi+(\sigma+p-\tau\cot\phi)\frac{dr}{r} &= 0 \text{ along an } \alpha\text{-line,}\\ dp-2\tau\,d\phi+(\sigma+p-\tau\tan\phi)\frac{dr}{r} &= 0 \text{ along a } \beta\text{-line.}\end{aligned}\right\} \qquad (44)$$

These differ from the Hencky plane strain equations only in the terms in r. However, this does not mean that the slip-lines are characteristics here also, nor that axially-symmetric problems can be solved by the techniques of the theory of plane strain. The three velocity relations impose certain continuity restrictions on the stress derivatives, and must be resorted to in order to determine the circumferential stress σ.

Some writers have adopted an artificial and unreal yield condition, originally suggested by Haar and von Karman† (1909), in order to complete the analogy with the plane strain equations. This condition does not allow the stress point freedom to traverse a yield locus, but stipulates that it shall be fixed in the position corresponding either to uniaxial tension or to uniaxial compression. When applied to the problem of axial symmetry, it states that σ is equal to one of the other two principal stresses, so that $\sigma+p = \pm\tfrac{1}{2}Y$ and $\tau = \tfrac{1}{2}Y$. Thus σ can at once be eliminated from (44), and the slip-lines become characteristics for the stresses. However, if the stress distribution is determined in this manner, it will not generally be possible to find an associated

† A. Haar and Th. von Karman, *Nachr. Ges. Wiss. Göttingen, Math.-phys. Klasse*, (1909), 204.

velocity distribution satisfying the Lévy–Mises relations, since there are three independent equations and only two velocity components. The Haar–Karman criterion has been used in the problem of indenting a plane surface by a flat cylindrical punch. The indentation pressure was calculated approximately by Hencky,† who assumed that the slip-line field was identical with that for the analogous problem in plane strain, and then integrated (44) along the family of slip-lines passing from the free surface to the punch. Ishlinsky‡ improved Hencky's solution by determining the actual field by step-by-step integration of (44) along *both* families of slip-lines; he obtained the value $2 \cdot 85Y$ for the mean pressure of indentation. Ishlinsky also computed the field beneath a spherical indenter, assuming that the impression was there to begin with and so neglecting the displacement of the surface. Calculations such as these have little value since the Haar–Karman hypothesis is without physical reality for metals and introduces an error of unknown magnitude.

† H. Hencky, *Zeits. ang. Math. Mech.* 3 (1923), 241.
‡ A. Ishlinsky, *Prikladnaia Matematika i Mekhanika*, 8 (1944), 201.

XI
MISCELLANEOUS TOPICS

1. Deep-drawing

In the process of deep-drawing, a thin circular blank is formed into a cylindrical cup, open at the top and closed at the base. The blank is placed over a die with a circular aperture and rounded lip (Fig. 78a);

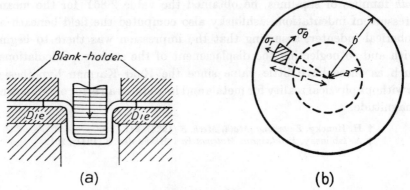

Fig. 78. Deep-drawing of a circular blank into a cylindrical cup, showing (a) side view, and (b) plan.

usually a blank-holder is employed to minimize crinkling.† A punch is forced down on to the blank, drawing the outer annulus over the die to make the wall of the cup. Among the variables are the radius of the lip, the shape of the punch, the lubrication, the clearance between punch and die and between blank-holder and die.‡ The following analysis is restricted to a calculation of the stress and strain in the outer annulus of the blank while it is being drawn radially towards the lip of the die.

(i) *Plane strain*. If there is a fixed clearance between the blank-holder and the die, equal to the thickness of the material, the blank is drawn under conditions of plane strain. (This is seldom realized in practice, even approximately, owing to the elasticity of the apparatus; there is generally a greater thickening than would be expected from the nominal clearance.) Friction between the blank and the die is dis-

† For a theoretical and experimental investigation of crinkling, see W. M. Baldwin, Jr., and T. S. Howald, *Trans. Am. Soc. Metals*, **38** (1946), 757.

‡ For an account of the effect of these variables, see H. W. Swift, *Proc. Inst. Auto. Eng.* **34** (1939), 361.

regarded. Let σ_r and σ_θ be the radial and circumferential stress components in the annulus at radius r (Fig. 78 b). The equation of equilibrium is

$$\frac{d\sigma_r}{dr} = \frac{\sigma_\theta - \sigma_r}{r},$$

as in the expansion of a thick cylinder. If there is no strain-hardening,

$$\sigma_r - \sigma_\theta = 2k, \qquad \sigma_r = 2k\ln\frac{b}{r},$$

where b is the external radius of the blank at the moment under consideration. The condition of plane strain can only be maintained by the blank-holder if σ_z is compressive. This requires that

$$\sigma_z = \tfrac{1}{2}(\sigma_r + \sigma_\theta) = 2k\left(\ln\frac{b}{r} - \frac{1}{2}\right) \leqslant 0 \quad \text{for} \quad a \leqslant r \leqslant b,$$

where a is the radius of the die aperture. This is true for all b provided that $\ln(b_0/a) \leqslant \tfrac{1}{2}$, or $b_0/a \leqslant \sqrt{e} \sim 1{\cdot}65$, where b_0 is the initial radius of the blank. If b_0/a is greater than this, the blank thins near the lip of the die, and the theory must be modified. The circumferential stress is numerically greatest on the rim, where it is a compression of amount $2k$. The radial tension where the blank is about to be drawn over the lip of the die is $2k\ln(b/a)$; this is greatest when drawing begins.

To investigate the influence of work-hardening we must find the strain in the annulus at any radius r. Let s be the initial radius of an element which has moved to the radius r when the external radius of the blank is b. Then

$$s^2 - r^2 = b_0^2 - b^2. \tag{1}$$

The strain-increment is a tangential compression of amount $(-dr)/r$; this is equivalent to a shear of amount $2(-dr)/r$ (engineering definition). The equivalent total shear is therefore $2\ln(s/r)$. Hence, if the relation between the shear stress τ and the engineering shear strain γ is $\tau = f(\gamma)$, the condition for yielding is

$$\sigma_r - \sigma_\theta = 2f\left(2\ln\frac{s}{r}\right) = 2f\left\{\ln\left(1 + \frac{b_0^2 - b^2}{r^2}\right)\right\},$$

with the help of (1). Substitution in the equation of equilibrium, followed by integration, leads to

$$\sigma_r = 2\int_r^b f\left\{\ln\left(1 + \frac{b_0^2 - b^2}{r^2}\right)\right\}\frac{dr}{r}.$$

The radial tension T at the lip $(r = a)$ is therefore

$$T = 2\int_a^b f\left\{\ln\left(1+\frac{b_0^2-b^2}{r^2}\right)\right\}\frac{dr}{r}.$$

If the material is annealed before drawing, the maximum value of T does not occur at the beginning of the draw (i.e., within a displacement of elastic order). The greatest initial slope of the stress-strain curve, for which T attains its maximum at the beginning, is given by the equation

$$0 = \left(\frac{\partial T}{\partial b}\right)_{b=b_0} = 2\left[\frac{f(0)}{b_0} - \int_a^{b_0} 2b_0 f'(0)\frac{dr}{r^3}\right].$$

Hence
$$f'(0) = f(0)\bigg/\left(\frac{b_0^2}{a^2}-1\right).$$

If the initial rate of hardening in shear is more than this, the tension at the lip is greatest at some later stage of the draw.

(ii) *Plane stress*.† If a fixed load is applied to the blank-holder, such that the mean pressure exerted on the blank is small compared with the current yield stress of the material, the blank is virtually drawn in a condition of plane stress (see Sect. 4 of this chapter for the general theory of plane stress). Since σ_r is tensile and σ_θ compressive, Tresca's criterion gives
$$\sigma_r - \sigma_\theta = 2k,$$

where $k = Y/2$. This expression is also a fair approximation to von Mises' criterion if for $2k$ we substitute mY, where m is an empirical constant slightly greater than unity (see p. 21); the best value for m increases with the drawing ratio. It is verifiable *a posteriori* that the equivalent strain in any element is never more than about 3 per cent. greater than the numerical value of the circumferential strain,

$$|\epsilon_\theta| = \ln(s/r).$$

Hence, if the tensile stress-strain relation is $\sigma = H(\epsilon)$, we may write

$$\sigma_r - \sigma_\theta = H\left(\ln\frac{s}{r}\right)$$

to a good approximation (comparable with the accuracy with which H would be known in industrial practice).

Inserting this in the equation of equilibrium

$$\frac{d}{dr}(h\sigma_r) = \frac{h(\sigma_\theta - \sigma_r)}{r},$$

† R. Hill, British Iron and Steel Research Association, Rep. MW/E/48/49, reproduced here by kind permission of the Director of B.I.S.R.A.

where h is the local thickness of the blank, we obtain

$$\frac{d}{dr}(h\sigma_r) = -\frac{h}{r}H\left(\ln\frac{s}{r}\right). \tag{2}$$

If v denotes the radial velocity of a particle in the sheet, measured with respect to b as the time-scale, the Lévy–Mises relations give

$$\frac{\dot\epsilon_r}{\dot\epsilon_\theta} = \frac{\partial v/\partial r}{v/r} = \frac{2\sigma_r - \sigma_\theta}{2\sigma_\theta - \sigma_r} = \frac{\sigma_r + H}{\sigma_r - 2H}. \tag{3}$$

The thickness changes are found from the equation of incompressibility

$$\frac{1}{h}\frac{Dh}{Db} + \frac{\partial v}{\partial r} + \frac{v}{r} = 0, \quad \text{where } \frac{D}{Db} \equiv \frac{\partial}{\partial b} + v\frac{\partial}{\partial r}.$$

With the use of (3) the differential form of this becomes

$$\frac{dh}{h} = -\left(\frac{H - 2\sigma_r}{2H - \sigma_r}\right)\frac{dr}{r} \tag{4}$$

taken along the paths of the particles, $dr - v\, db = 0$. Equations (2), (3), and (4) are hyperbolic and are solved by numerical integration along the characteristics $db = 0$ and $dr - v\, db = 0$ in the (r, b) plane. If the initial radius of the blank is taken to be unity, without loss of generality, the boundary conditions are

$$\sigma_r = 0, \quad \frac{h}{h_0} = \frac{1}{\sqrt{b}}, \quad v = 1, \quad \text{along the edge } r = b;$$

$$\sigma_r = H(0)\ln\frac{1}{s}, \quad \frac{h}{h_0} = 1, \quad v = s\left(1 - \frac{1}{2}\ln\frac{1}{s}\right)^{-3}, \quad \text{initially.}$$

In order that the blank does not neck at the lip before it is completely plastic, it is necessary that the radial tensile stress should not exceed $H(0)$; this is ensured if the drawing ratio is less than e (2·718...). It is found that the variation of thickness across the blank at any instant can be neglected in the equation of equilibrium, with at most a 5 per cent. error in σ_r for drawing ratios less than 2. The maximum error in the position of any particle occurs locally near the lip and is of order 1 in 3,000, while that in the thickness is about 1 in 500. For most purposes, then, it is sufficient to write

$$\frac{d\sigma_r}{dr} = -\frac{1}{r}H\left(\ln\frac{s}{r}\right). \tag{2'}$$

When H is constant we can obtain an analytic expression for the paths of the particles. Substitution of $\sigma_r/H = \ln(b/r)$ in (3), followed by an easy integration, leads to

$$v = \left(\frac{dr}{db}\right)_s = \frac{r}{b}\left(1 - \frac{1}{2}\ln\frac{b}{r}\right)^{-3}; \quad r = s \text{ when } b = 1.$$

The exact integral of this equation is unwieldy and for practical purposes it is more convenient to use the approximate solution

$$r\sqrt{b}\ln\frac{b}{r} = s\ln\frac{1}{s},$$

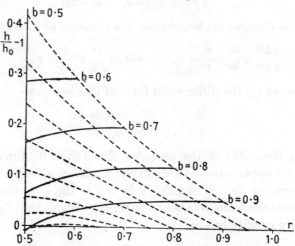

Fig. 79. Fractional increase in thickness of the annulus at various stages during deep-drawing (no work-hardening).

which is correct to 1 in 1,000 over the relevant range. Now the incompressibility equation can be written as

$$hr\,dr = h_0 s\,ds \quad (b \text{ constant}).$$

Combined with the previous formula this gives

$$\frac{h}{h_0} = \frac{s\sqrt{b}}{r}\left(\frac{1-\ln b/r}{1-\ln 1/s}\right).$$

However, this expression underestimates the total thickness at certain points by some 3 per cent., owing to magnification of errors in the empirical formula for r. This is avoided by integrating (4) numerically to give the *change* in thickness. Accurate thickness contours for H = constant, computed from equations (2), (3), and (4), are shown in Fig. 79 for various stages of the drawing. The thickness of the drawn cup is somewhat less than this owing to bending over the die† (see Sect. 2).

† H. W. Swift, *Engineering*, 166 (1948), 333 and 357; the same paper can also be found in *Proc. 7th Int. Cong. App. Mech.* London (1948).

It is found that the positions of the particles depend little on the stress-strain properties of the material; for example, the positions for $H =$ constant and $H \propto \sqrt{\epsilon}$ differ by less than 1 per cent. over most of the range. There are greater differences in the thickness, the general tendency of strain-hardening being to reduce the variation in thickness across the annulus at any moment. Thus, when $H \propto \sqrt{\epsilon}$, the thickness is very slightly greater at the die than at the edge when $b = 0.9$; the opposite is true by the time b has decreased to 0.7. The thickness of the outside edge for a given reduction in the external radius is, of course, independent of strain-hardening since it is always stressed in pure compression. When the stress-strain curve is well rounded, the paths of the particles are closely approximated by

$$b^2 - r^2 = \sqrt{b}(1-s^2),$$

which is obtained by neglecting the variation in thickness across the blank at any moment.

In an earlier investigation by Sachs the particles were assumed to move along the paths (1).[†] This is a rather poor approximation involving positional errors of up to 10 per cent. Sachs then used (4) to calculate the thickness, assuming the work-hardening to be zero; his results overestimate the true theoretical thickness *at a given point in the drawn blank* by several per cent., but agree closely with the final thickness of a *given element*.

2. General theory of sheet-bending[‡]

In the elementary theory of sheet bending (Chap. IV, Sect. 6), it is assumed that the strains are so small that the transverse stresses induced by the curvature can be neglected. It is supposed, too, that the neutral surface coincides with the central plane of the sheet throughout the distortion. This theory is a good approximation provided the final radius of curvature is not less than four or five times the thickness of the sheet. We now examine the state of stress, and the movement of the neutral surface and individual elements, when the strains are of any magnitude. The analysis is restricted to the bending of a wide sheet or bar, where there is negligible strain in the width direction. It is supposed,

[†] G. Sachs, *Mitt. d. deutschen Materialsprüfungsanstalten*, **16** (1931), 11. The approximation $b-r = 1-s$ has been used by K. L. Jackson, *Journ. Inst. Prod. Eng.* **27** (1948) 709; *Sheet Metal Industries*, **26** (1949), 723 and 1447.

[‡] R. Hill, British Iron and Steel Research Association, Rep. MW/B/6/49. J. D. Lubahn and G. Sachs, *Trans. Am. Soc. Mech. Eng.* **72** (1950), 201, obtained similar formulae for the stresses but evaluated the strains incorrectly. For experimental data see G. S. Sangdahl, Jr., E. L. Aul, and G. Sachs, *Proc. Soc. Exp. Stress Analysis*, **6** (1948), 1.

to begin with, that the bending is enforced by couples applied along opposite edges and that there are no other external forces. For greater clarity in describing the main features of the process it is assumed that elastic strains can be neglected and that the material does not work-harden.

(i) *The distribution of stress*. It is to be expected that, in the part of the sheet well away from the region of application of the couples, the stress and strain are the same at all points along each longitudinal fibre. This part of the sheet deforms in such a way that its thickness remains

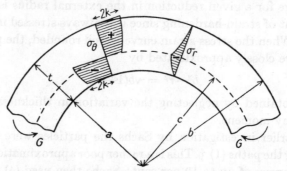

Fig. 80. Distribution of radial and circumferential stresses in a sheet bent in plane strain (no work-hardening).

uniform and its plane surfaces become cylindrical; we shall verify this *a posteriori*. Let a and b be the internal and external radii of curvature at any moment, and let t be the thickness (Fig. 80). It follows from the symmetry of the deformation that the principal stresses in the planes of bending act in the radial and circumferential directions. If σ_θ is the circumferential stress and σ_r is the radial stress, the equation of equilibrium is
$$\frac{d\sigma_r}{dr} = \frac{\sigma_\theta - \sigma_r}{r}.$$

Let c be the radius of fibres which, in a further infinitesimally small strain, undergo no change of length; the cylindrical surface containing these fibres is called the *neutral surface*. Fibres outside this surface are *momentarily* extended and those inside compressed. Since the bending is assumed to take place under conditions of plane strain the yield criterion is
$$\left.\begin{array}{l}\sigma_\theta - \sigma_r = 2k \quad (c \leqslant r \leqslant b),\\ \sigma_r - \sigma_\theta = 2k \quad (a \leqslant r \leqslant c).\end{array}\right\} \quad (5)$$

Substituting (5) in the equilibrium equation, and integrating:
$$\left.\begin{array}{l}\sigma_r = -2k \ln b/r \quad (c \leqslant r \leqslant b),\\ \sigma_r = -2k \ln r/a \quad (a \leqslant r \leqslant c),\end{array}\right\} \quad (6)$$

after using the fact that σ_r is zero on both surfaces. Since, for equilibrium, σ_r must be continuous across the neutral surface, we must have

$$2k\ln\frac{b}{c} = 2k\ln\frac{c}{a}.$$

Thus the radius of the neutral surface is

$$c = \sqrt{(ab)}. \tag{7}$$

From (5) and (6) the circumferential stress is

$$\left.\begin{array}{l}\sigma_\theta = 2k\{1-\ln(b/r)\} \quad (c \leqslant r \leqslant b), \\ \sigma_\theta = -2k\{1+\ln(r/a)\} \quad (a \leqslant r \leqslant c).\end{array}\right\} \tag{8}$$

The distribution of σ_r and σ_θ is shown diagrammatically in Fig. 80. The radial stress has a maximum numerical value $k\ln(b/a)$ on the neutral surface. It is obvious that the distribution (8), over a cross-section of the sheet, must automatically be equivalent to a couple since we have satisfied the condition that no external forces act over the surfaces of the sheet. In fact, the resultant force over a section is

$$\int_a^b \sigma_\theta\, dr = \int_a^b \frac{d}{dr}(r\sigma_r)\, dr = [r\sigma_r]_a^b = 0.$$

The couple per unit width is

$$G = \int_a^b \sigma_\theta r\, dr,$$

where, for convenience in performing the integration, moments have been taken about the centre of curvature. Substituting from (8):

$$\frac{G}{2k} = \tfrac{1}{4}(a^2+b^2-2c^2)+\tfrac{1}{2}c^2\ln\frac{ab}{c^2}$$

after an easy integration. Inserting the value of c from (7)

$$\frac{G}{2k} = \tfrac{1}{4}(b-a)^2 = \tfrac{1}{4}t^2. \tag{9}$$

(ii) *The deformation in bending.* Let $u\,d\alpha$ be the *inward* radial component, and $v\,d\alpha$ the circumferential component, of the displacement of an element during a further small strain in which the angle of bending α (*per unit original length*) increases by $d\alpha$. Since we are neglecting elastic compressibility, the components of displacement must be such that there is no change in volume. Furthermore, they must be consistent with the stress distribution (6) and (8); this demands that the associated strain is a circumferential extension of elements outside the neutral surface, and a compression of those inside. These conditions are satisfied if

$$u = \frac{1}{2\alpha}\left(r + \frac{c^2}{r}\right), \qquad v = \frac{r\theta}{\alpha}, \qquad (10)$$

where θ is the angle between a radius and the plane of symmetry. The corresponding strain-increment has components

$$d\epsilon_\theta = -d\epsilon_r = \left(1 - \frac{c^2}{r^2}\right)\frac{d\alpha}{2\alpha}, \qquad d\gamma_{r\theta} = 0. \qquad (11)$$

Since u is constant where r is constant, the surfaces of the sheet remain cylindrical, and, since v is proportional to r, radial sections remain plane. An arbitrary multiple of $\cos\theta$ may be added to u, and the same multiple of $\sin\theta$ to v, to represent a rigid-body displacement of the sheet as a whole parallel to the plane of symmetry. In this way we can satisfy a further condition: for example, that some point of the sheet is fixed in space. For present purposes it is simplest to make the centre of curvature the fixed point, as in (10). The incremental distortion can be visualized as the sum of two strains:

$$\text{(i)} \quad u = \frac{c^2}{2r\alpha}, \quad v = 0; \qquad \text{(ii)} \quad u = \frac{r}{2\alpha}, \quad v = \frac{r\theta}{\alpha}.$$

The first represents an inward radial displacement as in the uniform shrinkage of a hollow tube; the second represents a uniform circumferential strain of amount $d\alpha/2\alpha$.

An immediate deduction from (10) is that the thickness does not change. Thus, the decrements in the internal and external radii are

$$-da = \left(a + \frac{c^2}{a}\right)\frac{d\alpha}{2\alpha} = (a+b)\frac{d\alpha}{2\alpha},$$

$$-db = \left(b + \frac{c^2}{b}\right)\frac{d\alpha}{2\alpha} = (b+a)\frac{d\alpha}{2\alpha},$$

and so $\qquad db - da = dt = 0.$

t can therefore be treated as a constant henceforward, and it follows from (9) that the couple G is independent of the amount of plastic bending, a result which is true only for non-hardening material. The relation between the internal radius and the angle of bending is found by equating the initial and final areas; if L is the original length of the sheet then

$$Lt = \tfrac{1}{2}(b^2 - a^2)L\alpha, \qquad \alpha = \frac{2}{a+b}. \qquad (12)$$

The work done (per unit width) is therefore

$$GL\alpha = 2k\frac{t^2}{4}\frac{2L}{(a+b)}.$$

This is
$$k\left(\frac{b-a}{b+a}\right) = k\bigg/\left(1+\frac{2a}{t}\right) \tag{13}$$
per unit volume.

(iii) *The movement of individual elements.* Although the thickness of the sheet remains constant, the distances of internal elements from the surfaces vary. In fact, according to (11), elements are radially compressed outside the neutral surface, and radially extended inside. Consider the movement of a fibre distant $mt/2$ from the central plane in the unbent sheet ($-1 \leqslant m \leqslant 1$; m is regarded as positive on the convex side). Let r be the radius of curvature of the fibre when the internal radius is a. Since the material is assumed incompressible, the fibre must continue to divide the section into the same areas. Hence

$$\frac{1+m}{1-m} = \frac{r^2-a^2}{b^2-r^2},$$

or
$$r = \sqrt{[\tfrac{1}{2}(a^2+b^2)+\tfrac{1}{2}(b^2-a^2)m]}. \tag{14}$$

Thus the final radius r_0 of the original centre fibre ($m = 0$) is

$$r_0 = \sqrt{\left(\frac{a^2+b^2}{2}\right)}.$$

This fibre is, of course, nearer to the convex surface in its final position (Fig. 81). By setting $r = c$ in (14) it is found that the initial position of the fibre finally coinciding with the neutral surface is such that

$$m = -\left(\frac{b-a}{b+a}\right).$$

Now the neutral surface initially coincides with the central plane, and approaches the inner surface during the bending. Hence all fibres *initially* situated to the tension side of the central plane ($m \geqslant 0$) are progressively extended, while all fibres *finally* situated inside the radius c have been progressively compressed. Fibres for which

$$-\left(\frac{b-a}{b+a}\right) < m < 0 \quad \text{or} \quad c < r < \sqrt{\left(\frac{a^2+b^2}{2}\right)}$$

are overtaken by the neutral surface at some intermediate stage, and are therefore first compressed and afterwards extended (the Bauschinger effect would thus be operative in practice). A fraction $t/4a$ of the width (approximately) is traversed by the neutral surface.

At each stage of the bending there is one fibre which has undergone equal amounts of compression and extension, and whose *resultant* change in length is zero. The radius of this fibre is, from (12),

$$r = \frac{1}{\alpha} = \tfrac{1}{2}(a+b).$$

Thus, the fibre which has undergone no resultant change in length is

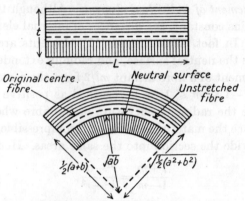

FIG. 81. Relative movement of the longitudinal fibres during sheet-bending.

that which coincides, at the moment under consideration, with the central surface of the sheet. The initial position, according to (14), is

$$m = -\frac{1}{2}\left(\frac{b-a}{b+a}\right).$$

It may be shown that the effect of work-hardening is to displace the neutral surface nearer to the concave side and so lead to a thinning of the sheet.

(iv) *Bending under tension.* Suppose, now, that the bending is carried out while the sheet is stressed by tensions T (per unit width) applied to the ends. The line of action of each tension is assumed to be normal to the terminal radial section (that is, in the circumferential direction). The tensions have an inward resultant which is assumed to be balanced by a uniform pressure p applied over the inner surface of the sheet. For equilibrium it is evident that

$$T = ap. \tag{15}$$

The stress components are

$$\sigma_r = -2k\ln\frac{b}{r}, \qquad \sigma_\theta = 2k\left(1-\ln\frac{b}{r}\right) \qquad (c \leqslant r \leqslant b);$$

$$\sigma_r = -p-2k\ln\frac{r}{a}, \qquad \sigma_\theta = -p-2k\left(1+\ln\frac{r}{a}\right) \qquad (a \leqslant r \leqslant c).$$

Since σ_r must be continuous across the neutral surface, c satisfies

$$c^2 = abe^{-p/2k}. \tag{16}$$

The neutral surface is therefore displaced inwards by the application of tension. It may be shown that the bending moment G per unit width, referred to the central fibre, is given by

$$\frac{G}{2k} = \tfrac{1}{4}(a^2+b^2-2abe^{-T/2ka})-\frac{Tb}{4k}. \tag{17}$$

The surfaces remain cylindrical and the displacements are still given by (10). However, the thickness of the sheet does not remain constant. In fact,

$$-da = \left(a+\frac{c^2}{a}\right)\frac{d\alpha}{2\alpha} = (a+be^{-p/2k})\frac{d\alpha}{2\alpha},$$

$$-db = \left(b+\frac{c^2}{b}\right)\frac{d\alpha}{2\alpha} = (b+ae^{-p/2k})\frac{d\alpha}{2\alpha},$$

and so

$$\frac{dt}{da} = \frac{db}{da}-1 = \frac{t(e^{p/2k}-1)}{t+a(e^{p/2k}+1)}. \tag{18}$$

Thus the application of tension causes the thickness to decrease. If the variation of T, and hence of p, with the internal radius a is prescribed, equation (18) determines the relation between t and a.

When t/a is small (less than $\tfrac{1}{5}$ say) a first approximation to the thinning may be obtained by neglecting the transverse stresses and the compression and subsequent extension of the fibres overtaken by the moving neutral surface; the influence of these factors only enters to the second order of approximation. Let the applied mean tensile stress be a fraction x of the yield stress, so that

$$\frac{p}{2k} = \frac{T}{2ka} = \frac{xt}{a} \quad (0 \leqslant x \leqslant 1).$$

The neutral surface is a distance $\tfrac{1}{2}xt$ from the central plane. From (18):

$$\frac{dt}{da} \sim \frac{xt^2}{2a^2}.$$

Hence, x being maintained constant during the bending, the proportional thinning is

$$-\frac{\Delta t}{t} = \frac{xt}{2a}. \tag{19}$$

To the present order of approximation the change in length of each fibre is regarded as achieved by a *continuous* extension or compression; thus, the deformation work per unit volume of an element distant y

from the neutral surface is $2ky/a$ since y/a is the resultant strain. The mean work per unit volume of the whole bar is therefore

$$\frac{2k}{at}\left[\int_0^{\frac{1}{2}(1+x)t} y\,dy + \int_0^{\frac{1}{2}(1-x)t} y\,dy\right] = (1+x^2)\frac{kt}{2a}, \quad \text{to order } \frac{t}{a}. \quad (20)$$

If the sheet is re-straightened under the same tension there is a further thinning of equal amount. Thus the total work done per unit volume in producing a thinning of amount xt/a is $k(1+x^2)t/a$. Since the work needed to produce the same thinning by uniform extension is $2kxt/a$, the efficiency of bending and re-straightening under tension as a means of reducing thickness is $2x/(1+x^2)$. The efficiency is higher the greater the tension, and becomes unity (to the present approximation only) when the applied tension is equal to the yield stress ($x = 1$).

Swift[†] has shown how these results can be applied with suitable modifications to the calculation of thinning produced in a strip bent under tension by pulling it round a roller (like a rope over a pulley). A part of the thinning takes place during the bending of the strip immediately before passing over the roller, and the remainder as the strip is re-straightened on leaving the roller. Since this is a continuous steady-motion process, in which there is a variation of thickness along the strip where it is being bent, the deformation and state of stress are strictly not identical with that contemplated in the above analysis. However, the difference is probably negligible when t/a is small, and in fact Swift has found good agreement with the observed thinning when strain-hardening is suitably allowed for (strain-hardening acts to reduce the amount of thinning when the applied tension is large).

3. Plane strain of a general plastic material

(i) *Significance of the envelope of stress circles in Mohr's diagram.* The plastic yielding of some non-metallic materials, such as clay and ice, may be broadly described (if all secondary phenomena are disregarded) by a yield criterion in which the hydrostatic component of stress has a significant effect. The study of the plane deformation of such a material is included in this book because it illuminates certain features of the plane deformation of a metal which are apt to pass unnoticed in the less general theory.

Elastic strains are neglected, and it is supposed that the plastic strain-increment ratios depend only on the applied stress. The condition

[†] H. W. Swift, op. cit., p. 286.

that there is no strain in the z direction then implies the functional dependence of σ_z on the principal stresses σ_1 and σ_2 in the planes of flow; in particular, if the reversal of the stress reverses the strain-increment (so that $\mu = 0$ when $\nu = 0$), σ_z is equal to $\tfrac{1}{2}(\sigma_1+\sigma_2)$. σ_z may thereby be eliminated from the yield criterion, which is reduced to

$$F(p,\tau_m) = 0, \qquad (21)$$

where
$$p = -\tfrac{1}{2}(\sigma_1+\sigma_2), \qquad \tau_m = \tfrac{1}{2}(\sigma_2-\sigma_1).$$

p is the mean compressive stress in the plane, and τ_m is the (algebraic)

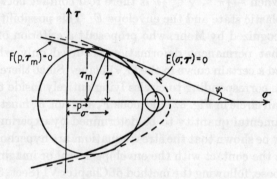

Fig. 82. Yield envelope in the plane deformation of a material whose yielding is influenced by hydrostatic pressure.

maximum shear stress. The yield criterion may be represented by a locus referred to coordinate axes $\tfrac{1}{2}(\sigma_1+\sigma_2)$, $\tfrac{1}{2}(\sigma_2-\sigma_1)$ (Fig. 82). If a pure hydrostatic tension of sufficient amount produces yielding, the locus is closed on one side. The yield condition $\tau_m = \pm k$, for an ideal metal, appears as a special case in which F is independent of p, and the F locus degenerates into a pair of parallel lines.

Following Mohr's well-known representation we may also plot in the same diagram the shear stress τ and normal stress σ, over any plane parallel to the z-axis, the circle of stress for a plastic state being

$$(\sigma+p)^2+\tau^2 = \tau_m^2. \qquad (22)$$

For a point (σ,τ) on the envelope E of these circles we have

$$(\sigma+p+dp)^2+\tau^2 = (\tau_m+d\tau_m)^2,$$

or
$$(\sigma+p)\,dp = \tau_m\,d\tau_m.$$

Also:
$$\frac{\partial F}{\partial p}dp + \frac{\partial F}{\partial \tau_m}d\tau_m = 0.$$

Hence
$$\frac{\sigma+p}{\tau_m} = -\frac{\partial F}{\partial p}\bigg/\frac{\partial F}{\partial \tau_m} = \frac{d\tau_m}{dp}. \qquad (23)$$

The elimination of p and τ_m between (21), (22), and (23) furnishes the (σ, τ) equation of the envelope. If the inclination of the tangent to E is denoted by ψ (regarded as positive where $|\tau|$ decreases with increasing tension σ), and the inclination of the tangent to F by χ, equation (23) expresses the fact that at corresponding points on E and F

$$\sin\psi = \tan\chi. \qquad (24)$$

Thus, only when $-\tfrac{1}{4}\pi \leqslant \chi \leqslant \tfrac{1}{4}\pi$ is there real contact between a stress circle for a plastic state and the envelope E. This possibility was seemingly not recognized by Mohr, who proposed† a criterion of yielding to the effect that permanent deformation occurred only when a stress circle touched a certain curve in the (σ, τ) plane. Since there are possible plastic states corresponding to circles lying entirely inside the envelope (e.g., the small circle in Fig. 82) the locus F, and not E, must be regarded as the fundamental quantity to be determined by experiment.‡

It will now be shown that the stress equations are hyperbolic or elliptic according as the contact with the envelope is real or imaginary, respectively. Suppose, following the method of Chapter VI (Sect. 3), that some curve C is given along which the stresses are known, and that it is required to find whether the stress derivatives normal to this curve are uniquely determined. Consider any point P on C, and take Cartesian axes (x, y) parallel respectively to the normal and tangent at P. The equilibrium equations limit the only possible discontinuity to $\partial\sigma_y/\partial x$. However, this derivative can be found from

$$0 = \frac{\partial F}{\partial x} = \frac{\partial}{\partial x} F[-\tfrac{1}{2}(\sigma_x+\sigma_y), \{\tfrac{1}{4}(\sigma_x-\sigma_y)^2+\tau_{xy}^2\}^{\frac{1}{2}}]$$

$$= -\frac{1}{2}\frac{\partial F}{\partial p}\left(\frac{\partial\sigma_x}{\partial x}+\frac{\partial\sigma_y}{\partial x}\right) + \frac{1}{\tau_m}\frac{\partial F}{\partial \tau_m}\left\{\tfrac{1}{4}(\sigma_x-\sigma_y)\left(\frac{\partial\sigma_x}{\partial x}-\frac{\partial\sigma_y}{\partial x}\right)+\tau_{xy}\frac{\partial\tau_{xy}}{\partial x}\right\}, \qquad (25)$$

unless the coefficient of $\partial\sigma_y/\partial x$ vanishes. This happens when

$$\frac{\sigma_x+p}{\tau_m} = -\frac{\partial F}{\partial p}\bigg/\frac{\partial F}{\partial \tau_m}. \qquad (26)$$

---

† O. Mohr, *Zeits. Ver. deutsch. Ing.* **44** (1900).

‡ In soil mechanics E is often approximated by two oblique straight lines, the contact then being always real. The present modification of Mohr's statement of the yield criterion does not seem to be widely known to workers in that field, who generally regard the shear-box test as giving a direct measure of points on the envelope. Since the deformation is so constrained as to be effectively a simple shear in a narrow zone, it seems that the test really gives a direct measure, not of E, but of F, for if the material is isotropic the directions of maximum shear stress and maximum shear strain-rate must coincide.

Comparing (26) with (23), we observe that C is a characteristic when, at every point, the normal stress component acting *across* C corresponds to the point of contact with the envelope E. Since there are two points of contact for a given plastic state, there are two characteristic directions; these are inclined at an angle $\frac{1}{4}\pi + \frac{1}{2}\psi$ to the direction of the algebraically greater principal stress (Fig. 83). Thus the stress equations are hyperbolic if and only if the contact is real. When $\chi = \pm \frac{1}{4}\pi$ and $\psi = \pm \frac{1}{2}\pi$ the characteristics are coincident. These points on the locus F correspond to the maximum and minimum values of σ_1 and σ_2; the characteristics coincide with the axis of the numerically lesser principal stress. For the ideal metal ψ is always zero and the characteristics are the slip-lines. The relation between characteristics and the envelope was known to Prandtl† (1920) for the special states of stress encountered in his investigation of the problem of indentation by a flat punch. The first general demonstration is apparently due to Mandel‡ (1942), who reached the result by geometrical reasoning and also by the standard method of the characteristic determinant (Appendix III). The present method is due to the author.§

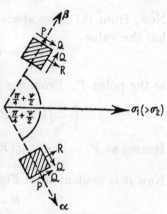

Fig. 83. Stress components referred to the characteristic directions.

(ii) *Relations along the characteristics.* When the equations are hyperbolic there exist relations, analogous to those of Hencky, for the variation of the stresses along the characteristics. Let the stress components, referred to the directions tangential and normal to a characteristic at *any* point, be denoted by P, Q, R (Fig. 83). The sense of Q, and the distinction between the α and β families of characteristics, follow the convention in the theory for an ideal metal. Now from (25) and (26) we have

$$\frac{\partial \tau_{xy}}{\partial x} = \pm \tan\psi \frac{\partial \sigma_x}{\partial x}$$

at the point P, when C is a characteristic (the upper and lower signs correspond respectively to α- and β-curves). With the first equilibrium

---

† L. Prandtl, *Nachr. Ges. Wiss. Göttingen* (1920), 74.
‡ J. Mandel, *Équilibres par tranches planes des solides* (Louis-Jean, Paris, 1942); *Proc. 6th Int. Cong. App. Mech.*, Paris (1946).
§ R. Hill, *Phil. Mag.* **40** (1949), 971.

equation (the weight of the material being neglected) this becomes

$$\frac{\partial \tau_{xy}}{\partial x} \pm \tan\psi \frac{\partial \tau_{xy}}{\partial y} = 0. \tag{27}$$

If a β-characteristic be regarded as the curve C, and if ϕ_β is its anti-clockwise orientation at any point with respect to the x-axis, then (Appendix IV)

$$\tau_{xy} = \tfrac{1}{2}(P-R)\sin 2\phi_\beta - Q\cos 2\phi_\beta.$$

Now, from (27), the space derivative of τ_{xy} along an α-characteristic has the value

$$\frac{\partial \tau_{xy}}{\partial s_\alpha} = \cos\psi \frac{\partial \tau_{xy}}{\partial x} - \sin\psi \frac{\partial \tau_{xy}}{\partial y} = 0$$

at the point P. From the last two equations:

$$\left\{\frac{\partial}{\partial s_\alpha}[\tfrac{1}{2}(P-R)\sin 2\phi_\beta - Q\cos 2\phi_\beta]\right\}_{\phi_\beta = \frac{1}{4}\pi} = 0.$$

Hence, at P,
$$(R-P)\frac{\partial \phi_\beta}{\partial s_\alpha} + \frac{\partial Q}{\partial s_\alpha} = 0. \tag{28}$$

Now it is evident from Fig. 82 that

$$\frac{R-P}{2Q} = \tan\psi = -\frac{dQ}{dR}. \tag{29}$$

Combining (28) and (29):

$$dR - 2Q\,d\phi_\beta = 0 \text{ along an } \alpha\text{-curve}, \tag{30}$$

where ϕ_β may be measured from any fixed direction. Similarly, by regarding an α-characteristic as the curve C,

$$dR + 2Q\,d\phi_\alpha = 0 \text{ along a } \beta\text{-curve}, \tag{30'}$$

where ϕ_α is the anti-clockwise orientation of an α-curve from the fixed direction. For an ideal metal, where

$$\psi = 0, \quad R = -p, \quad Q = k, \quad d\phi_\alpha = d\phi_\beta = d\phi,$$

equations (30) and (30') reduce to the familiar Hencky relations.

If ω is the anti-clockwise orientation of a principal stress direction,

$$d\omega = d(\phi_\alpha + \tfrac{1}{2}\psi) = d(\phi_\beta - \tfrac{1}{2}\psi).$$

Hence, by defining a quantity λ such that

$$\lambda = \frac{1}{2}\left(\int \frac{dR}{Q} - \psi\right), \tag{31}$$

the equations (30) and (30') can be written more symmetrically as

$$\left.\begin{array}{l} \lambda - \omega = \text{constant along an } \alpha\text{-curve,} \\ \lambda + \omega = \text{constant along a } \beta\text{-curve.} \end{array}\right\} \tag{32}$$

XI. 3] PLANE STRAIN OF A GENERAL PLASTIC MATERIAL 299

λ is a known function of ψ, calculable from the envelope E. Equations (32) were derived by Mandel.†

In soil mechanics the envelope E is often approximated by two oblique straight lines:
$$c - Q = R \tan\psi,$$
where c is the so-called cohesion, and ψ is constant. By a slight extension of the preceding theory to allow for body forces, it may be shown that

$$\left.\begin{aligned} dR - 2Q\, d\omega &= \frac{\cos\psi \sin(\omega + \tfrac{1}{2}\psi - \tfrac{1}{4}\pi)}{\sin(\omega - \tfrac{1}{2}\psi - \tfrac{1}{4}\pi)} g\rho\, dh \text{ on an } \alpha\text{-curve,} \\ dR + 2Q\, d\omega &= \frac{\cos\psi \cos(\omega - \tfrac{1}{2}\psi - \tfrac{1}{4}\pi)}{\cos(\omega + \tfrac{1}{2}\psi - \tfrac{1}{4}\pi)} g\rho\, dh \text{ on a } \beta\text{-curve,} \end{aligned}\right\} \quad (33)$$

where ρ is the density, h is the height relative to some fixed horizontal, and ω here refers to the direction of the *algebraically greater* principal stress. These equations are due to Kötter‡ (1903). When $\psi = 0$ and $Q = c$, the characteristics are the slip-lines, and it is easy to show that the geometrical theorems of Hencky apply (Chap. VI, Sect. 4).

If it is assumed that the material is incompressible, and that the principal axes of stress and strain-rate coincide, the Cartesian components of velocity are governed by Saint-Venant's equations ((4) and (7) of Chap. VI). The slip-lines are therefore always characteristics for the velocities, whose variations along slip-lines conform to Geiringer's equations ((14) of Chap. VI). Thus, the solution of a problem defined by both stress and velocity boundary conditions may either require the consideration of four real and distinct characteristics, or of only two, namely those corresponding to the velocity equations. All investigations of special problems known to the author are defective in that the velocity equations are not discussed. Furthermore, in initial-motion problems (for example, stability of earth slopes§ or indentation by a flat die‖) the complete loading path has been neither specified nor followed in the solution, so that, as shown in Chapter IX, the extent of the plastic region at the single moment under consideration could only be a matter for conjecture. Finally, it is to be remarked that a *unique* mathematical solution may not necessarily exist, since the yield criterion and plastic potential are not identical here. Uniqueness has so far only been proved

---

† J. Mandel, op. cit., p. 297.

‡ F. Kötter, *Berlin Akad. Bericht* (1903), 229. For an alternative but less elegant derivation see W. W. Sokolovsky, *Comptes Rendus (Doklady)*, **32** (1939), 153, and **33** (1939), 4.

§ W. W. Sokolovsky, *Trans. Am. Soc. Mech. Eng.* **68** (1946), A-1; *Isvestia Akad. Nauk SSSR*, **2** (1939), 107.

‖ L. Prandtl, op. cit., p. 297.

when they are identical (Chap. III, Sect. 2 (ii), and again for the plane problem in Chap. IX, Sect. 2 (iii), where the coincidence of stress and velocity characteristics seems to be a necessary property for uniqueness).

4. The theory of plane plastic stress, with applications

A state of stress is said to be plane with respect to Cartesian axes (x, y) when the stress components σ_z, τ_{xz}, τ_{yz}, in the z direction, are zero. If a plate of uniform thickness is loaded along its edge, by forces acting in its plane, the distribution of stress should be very nearly plane sufficiently far from the edge. If, during plastic deformation under these applied loads, the thickness does not remain uniform, the state of stress may still be treated as approximately plane so long as dh/ds is small compared with unity, where h is the local thickness and s is the distance in any direction parallel to the surface. For greater generality the theory of plane plastic stress will be developed for non-uniform plates, subject to this condition being satisfied.

(i) *The stress equations.* If σ_x, σ_y, and τ_{xy} denote values of the stress components averaged through the thickness, the equations of equilibrium are (in the absence of body forces)

$$\frac{\partial}{\partial x}(h\sigma_x)+\frac{\partial}{\partial y}(h\tau_{xy}) = 0, \qquad \frac{\partial}{\partial x}(h\tau_{xy})+\frac{\partial}{\partial y}(h\sigma_y) = 0. \tag{34}$$

The averaged stresses may be inserted in the yield criterion, with only a small error. Von Mises' criterion reduces to

$$\sigma_x^2-\sigma_x\sigma_y+\sigma_y^2+3\tau_{xy}^2 = \sigma_1^2-\sigma_1\sigma_2+\sigma_2^2 = 3k^2, \tag{35}$$

where σ_1 and σ_2 are the principal stresses in the plane of the plate, and $k = Y/\sqrt{3}$. Equation (35), considered as a locus in the (σ_1, σ_2) plane, represents an ellipse (Fig. 84); it is in fact the section of the Mises cylinder in $(\sigma_1, \sigma_2, \sigma_3)$ space by the plane $\sigma_3 = 0$. Referred to its principal axes, the equation of the ellipse is

$$F(p, \tau_m) \equiv \tfrac{1}{3}p^2+\tau_m^2 = k^2, \tag{36}$$

where $\qquad p = -\tfrac{1}{2}(\sigma_1+\sigma_2), \qquad \tau_m = \tfrac{1}{2}(\sigma_2-\sigma_1).$

Now (36) is, *formally*, a special case of the yield criterion (21). It is evident, therefore, that the investigation of characteristics for plane stress is mathematically analogous to that for plane strain of a general material. The additional terms in h do not fundamentally alter the investigation since h is to be regarded as a known function of x and y at the moment under consideration, while $\partial h/\partial x$ and $\partial h/\partial y$ must in general

XI. 4] THEORY OF PLANE PLASTIC STRESS

be continuous. The envelope E, corresponding to (36), may be shown (from (23)) to have the equation

$$E(\sigma, \tau) \equiv \tfrac{1}{4}\sigma^2 + \tau^2 = k^2. \tag{37}$$

Now it was proved in Section 3 that the stress equations are hyperbolic when the inclination χ of the locus F to the σ-axis is numerically less than $\tfrac{1}{4}\pi$; this corresponds to the thickened arcs of the ellipse in Fig. 84. Over the remainder of the ellipse, where both $|\sigma_1|$ and $|\sigma_2|$ are greater than k (or, equivalently, where the maximum shear stress in the plane

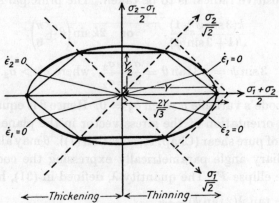

FIG. 84. Yield criteria of von Mises (ellipse) and Tresca (hexagon) in plane stress.

is numerically less than $\tfrac{1}{2}k$), the equations are elliptic. The characteristic directions are coincident with the axis of the numerically lesser principal stress when the principal stresses have the values $\pm(k, 2k)$.

The equations analogous to (30) and (30'), obtained by transforming (34), are

$$\left. \begin{aligned} d(Rh) - 2Qh\, d\phi_\beta &= -(R\sin\psi + Q\cos\psi)\frac{\partial h}{\partial s_\beta} ds_\alpha \text{ along an } \alpha\text{-curve,} \\ d(Rh) + 2Qh\, d\phi_\alpha &= -(R\sin\psi + Q\cos\psi)\frac{\partial h}{\partial s_\alpha} ds_\beta \text{ along a } \beta\text{-curve.} \end{aligned} \right\} \tag{38}$$

In the present problem it is convenient to express P, Q, and R in terms of ψ, or a related angle θ such that

$$\tan\theta = \sqrt{3}\sin\psi \quad (-\tfrac{1}{2}\pi \leqslant \psi \leqslant \tfrac{1}{2}\pi; \ -\tfrac{1}{3}\pi \leqslant \theta \leqslant \tfrac{1}{3}\pi). \tag{39}$$

Now, from (37), $\quad \tfrac{1}{4}R^2 + Q^2 = k^2,$

and so, from (29),

$$\tan\psi = \frac{R-P}{2Q} = -\frac{dQ}{dR} = \frac{R}{4Q}.$$

Hence
$$R = 2P = \pm 2\sqrt{(k^2-Q^2)}, \qquad \tan\psi = \pm\frac{\sqrt{(k^2-Q^2)}}{2Q}.$$

From (39):
$$\left.\begin{aligned}\frac{R}{k} &= \frac{2P}{k} = \frac{4\sin\psi}{\sqrt{(1+3\sin^2\psi)}} = \frac{4}{\sqrt{3}}\sin\theta, \\ \frac{Q}{k} &= \frac{\cos\psi}{\sqrt{(1+3\sin^2\psi)}} = \sqrt{(1-\tfrac{4}{3}\sin^2\theta)},\end{aligned}\right\} \quad (40)$$

where the positive radical is to be taken. The principal stresses σ_1 and σ_2 are
$$\frac{k(3\sin\psi\pm 1)}{\sqrt{(1+3\sin^2\psi)}}, \qquad \text{or} \quad 2k\sin\left(\theta\pm\frac{\pi}{6}\right). \tag{41}$$

Thus
$$3\sin\psi = \sqrt{3}\tan\theta = \frac{\sigma_1+\sigma_2}{\sigma_1-\sigma_2}, \text{ where } \sigma_1 > \sigma_2. \tag{42}$$

But this is Lode's variable μ when $\sigma_3 = 0$. Hence θ is equal to the angle defining the orientation of the stress vector in the plane Π relative to the position of pure shear (Chap. II, equation (7)). θ may also be regarded as the auxiliary angle parametrically expressing the coordinates of a point on the ellipse F. The quantity λ, defined in (31), has the value
$$\begin{aligned}\lambda &= \tan^{-1}(2\tan\psi) - \tfrac{1}{2}\psi \\ &= \sin^{-1}\left(\frac{2}{\sqrt{3}}\sin\theta\right) - \tfrac{1}{2}\sin^{-1}\left(\frac{1}{\sqrt{3}}\tan\theta\right) \quad (-\tfrac{1}{4}\pi \leqslant \lambda \leqslant \tfrac{1}{4}\pi),\end{aligned} \tag{43}$$

where the inverse sine and tangent denote angles in the range $(-\tfrac{1}{2}\pi, \tfrac{1}{2}\pi)$. λ is tabulated as a function of ψ or θ in Table 2 at the end of the book. Inserting in (38) the values of P, Q, and R from (40) we obtain, finally,

$$\left.\begin{aligned}d(\lambda-\omega)+2\tan\psi\frac{dh}{h} &= -\frac{(1+3\sin^2\psi)}{2\cos\psi}\frac{\partial h}{\partial s_\beta}\frac{ds_\alpha}{h} \text{ on an } \alpha\text{-curve}, \\ d(\lambda+\omega)+2\tan\psi\frac{dh}{h} &= -\frac{(1+3\sin^2\psi)}{2\cos\psi}\frac{\partial h}{\partial s_\alpha}\frac{ds_\beta}{h} \text{ on a } \beta\text{-curve}.\end{aligned}\right\} \quad (44)$$

Alternatively, written in terms of θ, the equations are

$$\left.\begin{aligned}d(\lambda-\omega)+\frac{(2/\sqrt{3})\sin\theta}{\sqrt{(1-\tfrac{4}{3}\sin^2\theta)}}\frac{dh}{h} &= -\frac{\tfrac{1}{2}\sec\theta}{\sqrt{(1-\tfrac{4}{3}\sin^2\theta)}}\frac{\partial h}{\partial s_\beta}\frac{ds_\alpha}{h}\text{on an }\alpha\text{-curve}, \\ d(\lambda+\omega)+\frac{(2/\sqrt{3})\sin\theta}{\sqrt{(1-\tfrac{4}{3}\sin^2\theta)}}\frac{dh}{h} &= -\frac{\tfrac{1}{2}\sec\theta}{\sqrt{(1-\tfrac{4}{3}\sin^2\theta)}}\frac{\partial h}{\partial s_\alpha}\frac{ds_\beta}{h}\text{on a }\beta\text{-curve}.\end{aligned}\right\} \quad (45)$$

The thickness h being given, these equations express the variation of the stress components (specified by the parameter θ) along the characteristics in terms of the angle through which the principal axes rotate.† The field of characteristics corresponding to various boundary conditions is constructed by methods closely resembling those described in Chapter VI (Sect. 5). When the thickness is uniform an analogy to Hencky's first theorem may be stated: there is a constant difference in the values of both λ and ω where two given characteristics of one family are cut by any member of the other. It follows that if a section of one characteristic is straight so also are the corresponding sections of other members of the same family.

The F locus corresponding to Tresca's criterion is a hexagon inscribed in the Mises ellipse (Fig. 84). When σ_1 and σ_2 have the same sign, the maximum shear stress lies outside the plane and the greater of $|\sigma_1|$ and $|\sigma_2|$ is equal to $2k$, where k is now written for $\tfrac{1}{2}Y$. When σ_1 and σ_2 have opposite signs, the maximum shear stress lies in the plane and

$$|\sigma_1-\sigma_2| = 2k.$$

The envelope E coincides with F for stress states such that $|\sigma_1-\sigma_2| = 2k$, but for all other states it degenerates into the pair of points $(\pm 2k, 0)$. In the first case the equations are hyperbolic, with the slip-lines as characteristics. Setting $Q = k$, $P = R = -p$, $d\phi_\alpha = d\phi_\beta = d\phi$, $\psi = 0$, we have from (38):

$$\left.\begin{aligned} d(ph)+2kh\,d\phi &= k\frac{\partial h}{\partial s_\beta}ds_\alpha \text{ along an } \alpha\text{-line,}\\ d(ph)-2kh\,d\phi &= k\frac{\partial h}{\partial s_\alpha}ds_\beta \text{ along a } \beta\text{-line}\end{aligned}\right\} \quad (0 \leqslant |p| \leqslant k). \quad (46)$$

In the second case there is only a single characteristic, which coincides with the direction of the numerically lesser principal stress. Returning to equation (25) we find that

$$\frac{\partial \sigma_x}{\partial x} = 0,$$

since $\tau_{xy} = 0$ at the point P. Thus, at the point P, the equations of equilibrium (34) reduce to

$$\frac{\partial \tau_{xy}}{\partial y} = -\frac{R}{h}\frac{\partial h}{\partial x}, \qquad \frac{\partial \tau_{xy}}{\partial x}+\frac{\partial \sigma_y}{\partial y} = -\frac{P}{h}\frac{\partial h}{\partial y}.$$

Now,

$$\tau_{xy} = \tfrac{1}{2}(P-R)\sin 2\omega, \qquad \sigma_y = \tfrac{1}{2}(P+R)-\tfrac{1}{2}(P-R)\cos 2\omega,$$

where ω denotes the anti-clockwise orientation of that principal stress

† Unpublished work of the writer (1949). When the thickness is uniform the equations reduce to those derived (in different fashion) by W. W. Sokolovsky, *Comptes Rendus (Doklady)*, **51** (1946), 175; *Theory of Plasticity*, p. 201 (Moscow, 1946).

direction which is also the characteristic. Substituting these expressions in the equilibrium equations, and inserting values at P after differentiation, we find

$$(P-R)\frac{\partial \omega}{\partial y} = \frac{R}{h}\frac{\partial h}{\partial x}, \qquad \frac{\partial P}{\partial y}+\frac{P}{h}\frac{\partial h}{\partial y} = (P-R)\frac{\partial \omega}{\partial x}.$$

Since only the derivative of ω occurs, we may measure ω relative to any fixed axis, and write

$$\left.\begin{aligned}\left(1-\frac{|P|}{2k}\right)d\omega &= -\frac{1}{h}\frac{\partial h}{\partial n}ds, \\ \frac{\partial}{\partial s}\left(\frac{h|P|}{2k}\right) &= -\left(1-\frac{|P|}{2k}\right)h\frac{\partial \omega}{\partial n},\end{aligned}\right\} \quad (k \leqslant |p| \leqslant 2k), \qquad (47)$$

where use has been made of the facts that R is a constant and that $P/R = |P|/2k$ since P and R have the same sign. ds and dn are arc-lengths measured along the trajectories of principal stress, with ds along the characteristic; their relative sense is such that the direction of s is inclined at an anti-clockwise angle of $90°$ to that of n. When the thickness is uniform ω is constant along each characteristic; that is, the characteristics are straight lines, and their parametric equation referred to Cartesian axes (X, Y) may be expressed in the form

$$Y = X\tan\omega + F(\omega),$$

where $F(\omega)$ is a function determined by the stress boundary conditions prescribed along some curve.† It is not difficult to show that the curvature of the other principal stress trajectories is

$$\frac{\partial \omega}{\partial n} = -\frac{1}{X\sec\omega + \cos\omega F'(\omega)}.$$

Therefore, from the second equation of (47):

$$\frac{d}{dX}\left(\frac{|P|}{2k}\right) = \sec\omega\frac{\partial}{\partial s}\left(\frac{|P|}{2k}\right) = \frac{1-\dfrac{|P|}{2k}}{X+\cos^2\omega F'(\omega)}$$

along a characteristic. On integrating, we obtain an equation due to Sokolovsky:‡

$$1-\frac{|P|}{2k} = \frac{G(\omega)}{X+\cos^2\omega F'(\omega)}, \qquad (48)$$

where $G(\omega)$ is a function to be determined from the boundary values of P.

(ii) *The velocity equations.* Let the (x, y) components of velocity

---

† It is implicitly assumed that the slopes vary. If the lines are parallel the stress is uniform; this case is not included in the following general integral.

‡ W. W. Sokolovsky, *Comptes Rendus (Doklady)*, 51 (1946), 421; *Theory of Plasticity*, p. 211 (Moscow, 1946).

averaged throughout the thickness of the sheet be denoted by (u_x, v_y). To the usual order of approximation, these, together with the averaged stresses, may be inserted in the Lévy–Mises relations:

$$\frac{\partial u_x/\partial x}{2\sigma_x-\sigma_y} = \frac{\partial v_y/\partial y}{2\sigma_y-\sigma_x} = \frac{\partial u_x/\partial y + \partial v_y/\partial x}{6\tau_{xy}}. \tag{49}$$

We begin by investigating whether these equations possess characteristics. Let C be a curve along which u_x, v_y, and the components of stress are given, and let the (x, y) axes be taken to coincide respectively with the normal and tangent to C at some point P. Then, if the velocity is continuous across C, $\partial u_x/\partial y$ and $\partial v_y/\partial y$ are known at P. Equations (49) suffice to determine $\partial u_x/\partial x$ and $\partial v_y/\partial x$ uniquely at P unless $2\sigma_y - \sigma_x = 0$, so that

$$\dot{\epsilon}_y = \frac{\partial v_y}{\partial y} = 0 \quad \text{at } P. \tag{50}$$

Thus C is a characteristic for the velocity components if it coincides at every point with a direction of zero rate of extension. There are two such directions through any point, but, because of the z component of strain, they are not generally orthogonal, or necessarily even real. The characteristics are inclined at an angle $\tfrac{1}{4}\pi + \tfrac{1}{2}\psi$ to the σ_1 principal stress axis ($\sigma_1 > \sigma_2$), where

$$0 = \dot{\epsilon}_y = \tfrac{1}{2}(\dot{\epsilon}_1 + \dot{\epsilon}_2) - \tfrac{1}{2}(\dot{\epsilon}_1 - \dot{\epsilon}_2)\sin\psi.$$

Now

$$\frac{\dot{\epsilon}_1}{\dot{\epsilon}_2} = \frac{2\sigma_1 - \sigma_2}{2\sigma_2 - \sigma_1},$$

and so

$$\sin\psi = \frac{\dot{\epsilon}_1 + \dot{\epsilon}_2}{\dot{\epsilon}_1 - \dot{\epsilon}_2} = \frac{1}{3}\left(\frac{\sigma_1 + \sigma_2}{\sigma_1 - \sigma_2}\right). \tag{51}$$

ψ is therefore real, and the equations hyperbolic, when $|p| \leqslant 3|\tau_m|$.

When von Mises' yield criterion is used, the characteristics for the stresses coincide with those for the velocities. This is clear on comparing (42) and (51), when it will be seen that they define the same value of ψ; alternatively, it is evident from (40) that the stress characteristics are directions of zero extension since P is equal to $\tfrac{1}{2}R$, and is therefore the mean of the other two normal stress components. The statement may also be demonstrated directly by recalling that the function $f(\sigma_x, \sigma_y, \tau_{xy})$ in von Mises' yield criterion (35) is identical with the plastic potential when $\sigma_z = 0$. Thus (50) implies that

$$\frac{\partial f}{\partial \sigma_y} = 0.$$

Now if the stress components are given on C, their derivatives in the y

direction are known. The equilibrium equations then give $\partial\sigma_x/\partial x$ and $\partial\tau_{xy}/\partial x$, while $\partial\sigma_y/\partial x$ is uniquely determined from

$$\frac{\partial f}{\partial x} = \frac{\partial f}{\partial \sigma_x}\frac{\partial \sigma_x}{\partial x} + \frac{\partial f}{\partial \sigma_y}\frac{\partial \sigma_y}{\partial x} + 2\frac{\partial f}{\partial \tau_{xy}}\frac{\partial \tau_{xy}}{\partial x}$$

unless $\partial f/\partial\sigma_y = 0$; this, however, is also the condition that C should be a velocity characteristic. If the Lévy–Mises relations are used in conjunction with Tresca's criterion, the stress and velocity characteristics are in general distinct.

Let (u, v) be the velocity components along the (α, β) velocity characteristics. Then, if an α-characteristic is regarded as the curve C, simple considerations of the parallelogram of velocities show that

$$v_y = [v\cos\phi_\alpha + u\sin(\phi_\alpha+\psi)]/\cos\psi.$$

On substituting in (50), and setting $\phi_\alpha = \tfrac{1}{2}\pi$ after differentiation, we derive

$$du - \left(\frac{v+u\sin\psi}{\cos\psi}\right)d\phi_\alpha = 0 \quad \text{on an } \alpha\text{-curve.} \tag{52}$$

Similarly, if a β-characteristic is regarded as the curve C,

$$v_y = [v\sin(\phi_\beta-\psi) - u\cos\phi_\beta]/\cos\psi.$$

Hence
$$dv + \left(\frac{u+v\sin\psi}{\cos\psi}\right)d\phi_\beta = 0 \quad \text{on a } \beta\text{-curve.} \tag{52'}$$

The stress distribution having been found, equations (52) and (52') are the basis for the calculation of the velocity in terms of prescribed boundary conditions.† ψ must be calculated beforehand at each point from (51), while $d\phi_\alpha$ and $d\phi_\beta$ are equal to $d(\omega \mp \tfrac{1}{2}\psi)$, where ω is the anti-clockwise orientation of a principal stress axis.

The sign of the rate of strain in the z direction is that of $-(\sigma_x+\sigma_y)$, or of p (this follows from the Lévy–Mises relation for the z component of strain-rate). The amount of thickening or thinning is determined from the condition for zero volume change. The rate of strain in the z direction, averaged through the thickness, is

$$\frac{1}{h}\frac{Dh}{Dt} = \frac{1}{h}\left(\frac{\partial h}{\partial t} + u_x\frac{\partial h}{\partial x} + v_y\frac{\partial h}{\partial y}\right),$$

where D/Dt is the operator denoting rate of change following an element. The equation of incompressibility is therefore

$$\left.\begin{aligned}\frac{\partial u_x}{\partial x} + \frac{\partial v_y}{\partial y} + \frac{1}{h}\frac{Dh}{Dt} &= 0, \\ \frac{\partial h}{\partial t} + \frac{\partial}{\partial x}(hu_x) + \frac{\partial}{\partial y}(hv_y) &= 0.\end{aligned}\right\} \tag{53}$$

or

† Unpublished work of the writer (1949).

It follows from (53) that, wherever the discontinuity in velocity gradient allowed by (49) occurs, there must also be one in $\partial h/\partial t$ (assuming that the slope of the surface is initially continuous). This implies that the ensuing strain produces a discontinuity in the surface slope.

As in the theory of plane strain, where the velocity equations are also hyperbolic, plastic material may either be undergoing deformation or be rigid. It can be shown, by the reasoning of Section 6 (ii) in Chapter VI, that the plastic material is rigid everywhere within the triangular area bounded by the plastic boundary and the intersecting velocity characteristics through its terminal points. A change in thickness of the sheet is only possible outside such an area; in a plastic-elastic material, on the other hand, thickening strains could occur within the corresponding area, but would only be of an elastic order of magnitude.

(iii) *Expansion of a circular hole in a plate.* The only work on special problems of plane stress in which the progressive changes in thickness have been adequately incorporated in the solution, other than where the state of stress is uniform, appears to be that on deep-drawing (Sect. 1 (ii)) and that of Taylor,[†] described below. In certain Russian work[‡] on the drawing of thin strips through a die, under conditions of plane stress, the appreciable thickening that must occur is ignored and it is not shown that the proposed plastic region satisfies the velocity boundary conditions. Other investigations on the plastic zones around notches[§] or holes[||] of arbitrary shape in a tensioned plate are subject to the same limitations as the similar work on plane strain, discussed in Chapter IX, and to the additional defect that the circumstances under which local thinning is initiated are not examined.

Consider a circular hole of radius a in a uniform infinite plate of plastic-rigid material, and let a gradually increasing pressure P be applied uniformly over the edge of the hole. While the plate is stressed below the yield limit the radial and circumferential components of stress are known from elastic theory to be

$$\sigma_r = -\frac{Pa^2}{r^2}, \qquad \sigma_\theta = \frac{Pa^2}{r^2}.$$

The state of stress in every element is a pure shear. Yielding begins first on the edge of the hole, at a pressure k, where k is equal to $Y/\sqrt{3}$

[†] G. I. Taylor, *Quart. Journ. Mech. App. Math.* **1** (1948), 103.
[‡] W. W. Sokolovsky, *Trans. Am. Soc. Mech. Eng.* **68** (1946), A–1; *Theory of Plasticity*, pp. 229–35 (Moscow, 1946).
[§] R. V. Southwell and D. N. de G. Allen, *Phil. Trans. Roy. Soc.* A, **242** (1950), 379.
[||] W. W. Sokolovsky, *Theory of Plasticity*, chap. x (Moscow, 1946).

for von Mises' criterion and to $Y/2$ for Tresca's. If the pressure is now raised further the material round the hole becomes plastic within some radius c. The stresses in the non-plastic material are

$$\sigma_r = -\frac{kc^2}{r^2}, \qquad \sigma_\theta = \frac{kc^2}{r^2} \qquad (c \leqslant r).$$

We have to decide whether or not the plate immediately begins to thicken; as shown in the preceding section, the plate does not thicken in a certain area near the plastic boundary if the velocity equations are hyperbolic. This is so if $|p| \leqslant 3|\tau_m|$, and this condition is certainly satisfied on the plastic boundary where $p = -\tfrac{1}{2}(\sigma_r + \sigma_\theta) = 0$. Furthermore, according to (51), the velocity characteristics at points on the plastic boundary are inclined to the radial direction at a finite angle, namely 45°. Thus the rigid part of the plastic region must extend over a finite annulus, its inner boundary being the circle where the velocity characteristics are coincident. In the rigid plastic annulus, then, the thickness is unaltered and the equation of equilibrium is simply

$$\frac{d\sigma_r}{dr} = \frac{\sigma_\theta - \sigma_r}{r}.$$

To facilitate the integration if von Mises' criterion is adopted, we introduce the auxiliary angle θ defined in (39). From (41):

$$\sigma_\theta = 2k\sin(\tfrac{1}{6}\pi + \theta), \qquad \sigma_r = -2k\sin(\tfrac{1}{6}\pi - \theta), \qquad \sigma_\theta - \sigma_r = 2k\cos\theta,$$

where θ is zero on the plastic boundary. Substituting in the equilibrium equation:

$$\cos(\tfrac{1}{6}\pi - \theta)\frac{d\theta}{dr} = \frac{\cos\theta}{r}.$$

Integrating:
$$\frac{c^2}{r^2} = e^{-\sqrt{3}\theta}\cos\theta. \tag{54}$$

The internal pressure producing a plastic region of radius c is given† parametrically by

$$P = 2k\sin(\tfrac{1}{6}\pi - \theta_a), \qquad \frac{c^2}{a^2} = e^{-\sqrt{3}\theta_a}\cos\theta_a. \tag{55}$$

θ_a becomes increasingly negative with increasing P, and the characteristics coincide when $\theta_a = -\tfrac{1}{3}\pi$, or $c/a = \rho$ where

$$\rho = (\tfrac{1}{2}e^{\pi/\sqrt{3}})^{\frac{1}{2}} \sim 1\cdot751. \tag{56}$$

The corresponding value of the angle ψ is $-\tfrac{1}{2}\pi$, and the angle at which the characteristics are inclined to the direction of the algebraically greater principal stress (viz. σ_θ) is zero; thus the characteristics envelop

† A. Nadai, *Plasticity*, p. 191 (McGraw-Hill Book Co., 1934).

the edge of the hole. The pressure required just to advance the plastic region to the radius ρa is equal to $2k$; the circumferential stress at the edge is then a compression of amount k. If a still greater internal *load* is applied, the plastic boundary continues to expand and the inner radius of the rigid annulus of plastic material is still the constant fraction $1/\rho$ ($\sim 0{\cdot}571$) of the radius of the whole plastic region (Fig. 85). The material inside the radius $0{\cdot}571c$ is not constrained to remain rigid by the material farther out, and a thickening becomes possible. That such a

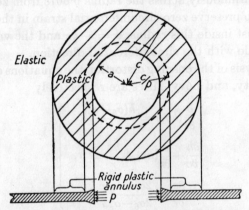

Fig. 85. The plastic region round an expanded hole in a thin plate, showing the rigid plastic annulus.

thickening *does* occur is due to the fact that the plastic material cannot sustain a greater stress than $2k$, the pressure already applied; hence, if the load is increased, the plate must thicken to support it, and the pressure can be expected to fall.

Turning, now, to the corresponding analysis for Tresca's criterion, we have
$$\sigma_\theta - \sigma_r = 2k$$
in a certain annulus near the plastic boundary where the principal stresses have opposite signs. Thus, from the equation of equilibrium,
$$\sigma_r = -k\left(1 + 2\ln\frac{c}{r}\right), \qquad \sigma_\theta = k\left(1 - 2\ln\frac{c}{r}\right).$$

When the internal pressure is raised to its maximum possible value $2k$ the plastic boundary is advanced to a radius ρa where
$$\rho = \sqrt{e} \sim 1{\cdot}649. \tag{57}$$

The application of a greater load causes a thickening, and the rigid annulus of plastic material is confined between the radii $0{\cdot}607c$ and c.

That the velocity equations are hyperbolic in this annulus may be seen from (51), since the angle
$$\psi = -\sin^{-1}\left(\frac{2}{3}\ln\frac{c}{r}\right)$$
is real ($-\sin^{-1}\frac{1}{3} \leqslant \psi \leqslant 0$) when $c/\sqrt{e} \leqslant r \leqslant c$. When a greater load is applied σ_θ becomes negative on the edge of the hole, and the yield criterion changes to
$$\sigma_r = -2k.$$
σ_θ varies discontinuously across the radius $0\cdot607c$ from zero to $-k$, the value needed to preserve zero circumferential strain in the presence of a thickening. Just inside this radius, $\psi = -\frac{1}{2}\pi$ and the velocity characteristics coincide with the circumferential direction.

For the analysis of the ensuing distortion the equations of equilibrium, incompressibility, and Lévy–Mises are respectively

$$\left. \begin{aligned} \frac{\partial}{\partial r}(h\sigma_r) &= \frac{h(\sigma_\theta - \sigma_r)}{r}, \\ \frac{\partial h}{\partial c} + \frac{\partial}{\partial r}(hv) + \frac{hv}{r} &= 0, \\ \frac{\partial v}{\partial r} &= \frac{2\sigma_r - \sigma_\theta}{2\sigma_\theta - \sigma_r}\frac{v}{r}, \end{aligned} \right\} \qquad (58)$$

where v denotes the radial velocity with the parameter c as the time-scale.† For both Tresca's and von Mises' criteria $\partial\sigma_r/\partial r$ is zero ($\partial\sigma_\theta/\partial r$ being finite), and $\sigma_\theta - \sigma_r$ is equal to k, just within the inner boundary of the rigid annulus. It follows from the equation of equilibrium that $\partial h/\partial r$ is equal to $-h_0\rho/2c$ just within the radius c/ρ; thus, the slope of the surface changes discontinuously. It may be shown from the Lévy–Mises equation that $\partial\sigma_\theta/\partial r$ has the value $-3k\rho/2c$ just within the radius c/ρ, and from the incompressibility equation that $\partial v/\partial r$ has the value $-1/2c$ ($\partial h/\partial c = (-1/\rho)\partial h/\partial r$ at this point).

Since it is clearly immaterial whether the pressure at any radius is applied by an external agency or through the displacement of an inner annulus of the plate, the stress and velocity in an element depend only on what happens beyond this radius. Since the plate is infinite they are functions only of the relative distance from the plastic boundary; that

---

† This system of equations is hyperbolic, with $dc = 0$ and $dr - v\,dc = 0$ as characteristic directions in the (r, c) plane. If the system is considered by itself, and not as a specialization of the general (x, y) equations, the fundamental reason for the existence of a rigid annulus of plastic material is largely obscured. Similarly, the inner boundary of the rigid annulus no longer appears as the locus of points where the velocity characteristics coincide, but merely as the circle where $2\sigma_\theta - \sigma_r = 0$.

is, they are functions of the single parameter r/c. In particular, the stress distribution around a hole expanded from some finite radius is identical with that in the corresponding annulus in a plate in which a hole has been enlarged from zero radius and in which, therefore, the distribution of stress remains geometrically similar.† The latter problem has been investigated by Taylor,‡ using Tresca's criterion. Let

$$\theta = r/c, \qquad \sigma = -\sigma_\theta/2k \qquad (0 \leqslant \sigma \leqslant \tfrac{1}{2})$$

and put $\sigma_r = -2k$ in (58). Then

$$\frac{\partial}{\partial r} = \frac{1}{c}\frac{d}{d\theta}, \qquad \frac{\partial}{\partial c} = -\frac{\theta}{c}\frac{d}{d\theta};$$

$$\left.\begin{aligned} \frac{\theta h'}{h} &= \sigma-1, \\ (v-\theta)\frac{h'}{h}+\frac{v}{\theta}+v' &= 0, \\ v' &= \frac{2-\sigma}{2\sigma-1}\frac{v}{\theta}, \end{aligned}\right\} \qquad (59)$$

where a dash denotes differentiation with respect to θ. Eliminating h'/h:

$$v' = \frac{(2-\sigma)(\sigma-1)}{2(\sigma^2-\sigma+1)}, \qquad \frac{v}{\theta} = \frac{(2\sigma-1)(\sigma-1)}{2(\sigma^2-\sigma+1)}. \qquad (60)$$

Eliminating v:

$$\theta\frac{d}{d\theta}\left[\frac{(2\sigma-1)(\sigma-1)}{(\sigma^2-\sigma+1)}\right] = -\frac{3(\sigma-1)^2}{(\sigma^2-\sigma+1)}.$$

Integrating, and using the boundary condition $\sigma = \tfrac{1}{2}$ when $\theta = 1/\rho$:

$$\ln\!\left(\frac{1}{\rho\theta}\right) = -\frac{1}{3}\!\left(\frac{1-2\sigma}{1-\sigma}\right)+\tfrac{1}{2}\ln\!\left[\frac{3(1-\sigma)^2}{(\sigma^2-\sigma+1)}\right]+\frac{1}{\sqrt{3}}\tan^{-1}\!\left(\frac{1-2\sigma}{\sqrt{3}}\right). \qquad (61)$$

Now the yield criterion $\sigma_r = -2k$ is valid only when σ_θ is negative. As θ decreases, σ decreases steadily from $\tfrac{1}{2}$ and becomes zero when $\theta = 1/\rho'$, where

$$\ln\frac{\rho'}{\rho} = -\tfrac{1}{3}+\tfrac{1}{2}\ln 3+\frac{\pi}{6\sqrt{3}} \sim \cdot 5183; \qquad \rho' \sim 2\cdot 769. \qquad (62)$$

The value for ρ' obtained by Taylor, by approximate numerical integration of equations equivalent to (59), agrees very closely with (62). It is not difficult to show that

$$\frac{h}{h_0} = \left[\frac{1}{2(1-\sigma)}\right]^{\tfrac{1}{3}}\exp\!\left[\frac{2}{\sqrt{3}}\tan^{-1}\!\left(\frac{1-2\sigma}{\sqrt{3}}\right)\right], \qquad (63)$$

---

† If the plate is *finite*, the stress distribution is the same in the plastically deforming material provided the entire plate is not plastic; the breadth of the rigid plastic annulus is, of course, different.   ‡ G. I. Taylor, op. cit., p. 307.

where h_0 is the initial thickness of the plate. The thickness at the radius where σ_θ is zero is thus

$$h = h_0(\tfrac{1}{2})^{\frac{1}{2}} \exp\left(\frac{\pi}{3\sqrt{3}}\right) \sim 1{\cdot}453 h_0. \tag{64}$$

Equations (60), (61) and (63) give v and h/h_0 as functions of $\theta = r/c$, parametrically in terms of σ. The total displacement $u(r,c)$ satisfies the equation

$$\frac{Du}{Dc} = \left(\frac{\partial}{\partial c} + v\frac{\partial}{\partial r}\right)u = v.$$

Now the displacement must be of the form

$$u(r,c) = cF(\theta).$$

Hence $\qquad\dfrac{\partial u}{\partial r} = F', \qquad \dfrac{\partial u}{\partial c} = F - \theta F',$

and $\qquad (v-\theta)F' + F = v; \qquad F = 0$ when $\theta = 1/\rho$. (65)

$v(\theta)$ having been determined, this equation serves to calculate $F(\theta)$. Alternatively, we can proceed directly from (63), by writing the equation of incompressibility in the form

$$h_0 s\, ds = hr\, dr \quad (c \text{ constant}),$$

where $s = r - u$ is the initial radius to an element. Integrating:

$$\frac{c^2}{\rho^2} - s^2 = \frac{c^2}{\rho^2} - (r-u)^2 = 2c^2 \int_\theta^{1/\rho} \frac{h\theta}{h_0} d\theta. \tag{66}$$

When θ is less than $1/\rho'$, that is when r is less than $0{\cdot}361c$, σ_θ becomes positive again and the yield criterion reverts to $\sigma_\theta - \sigma_r = 2k$. The corresponding modification of (59) is such that an analytic solution appears to be impossible and the equations must be integrated numerically. The theory can only be regarded as approximate near the edge of the hole, where the slope of the surface becomes infinite and the coronet is knife-edged. The results are shown in Fig. 86; they have been recalculated† since Taylor's method of integration was insufficiently accurate. At the edge of the hole ($\theta = 0{\cdot}2805$) h/h_0 has the value $3{\cdot}84$ (approx.); the almost exact agreement with the value for h/h_0 found experimentally by Taylor is certainly fortuitous. The hole was expanded by piercing a lead sheet with a lubricated cone of very small taper. The experiments indicate that the mode of deformation contemplated in this analysis may be unstable; once the hole reached a critical size in

† R. Hill, *Phil. Mag.* **40** (1949), 971.

relation to the plate thickness, the configuration became unsymmetrical and the plate bent out of its plane.

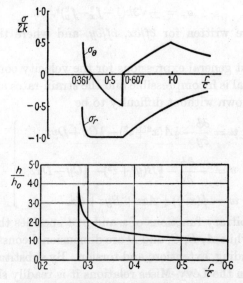

Fig. 86. Distribution of the stresses and the thickness round a hole expanded in a thin plate (Tresca criterion).

5. Completely plastic states of stress in a prismatic bar

A prismatic bar of plastic-rigid material, with an arbitrary uniform cross-section, is deformed by forces applied at its ends. It is supposed that the forces are applied in such a way, and in such combination, that the *entire* bar is in a plastic state; surfaces across which certain components of stress are discontinuous are permitted (for example, the neutral surface in pure bending). Let Cartesian axes of reference be taken so that the z-axis is parallel to the generators, and the axes of x and y lie in the plane of a cross-section. Only those states will be considered in which the stress and strain-rate are independent of z. This rules out the possibility of flexural forces but allows certain combinations of bending couple, twisting couple, and longitudinal force. By analogy with the corresponding problem in elasticity let it be assumed tentatively that σ_x, σ_y, and τ_{xy}, are identically zero. The equilibrium equations then require the existence of a stress function $f(x,y)$ such that

$$\tau_{xz} = -k\frac{\partial f}{\partial y}, \qquad \tau_{yz} = k\frac{\partial f}{\partial x}. \tag{67}$$

f is zero on the contour of the section since no external forces are applied to the cylindrical surface. According to the yield criterion of von Mises, the longitudinal stress has the value

$$\sigma_z = \pm\sqrt{3}k(1-f_x^2-f_y^2)^{\frac{1}{2}}, \qquad (68)$$

where f_x, f_y are written for $\partial f/\partial x$, $\partial f/\partial y$, and where the radicals are positive.

Now the most general expressions for the velocity components, such that the material is incompressible and the strain-rates are independent of z, may be shown without difficulty to be

$$\left. \begin{array}{l} u = \dfrac{\partial \phi}{\partial y} - \tfrac{1}{2}A(x^2+z^2) - \tfrac{1}{2}Cx + Dyz, \\[4pt] v = -\dfrac{\partial \phi}{\partial x} - \tfrac{1}{2}B(y^2+z^2) - \tfrac{1}{2}Cy - Dxz, \\[4pt] w = \psi(x,y) + Axz + Byz + Cz. \end{array} \right\} \qquad (69)$$

ϕ and ψ are arbitrary functions of x and y; ψ specifies the warping of a cross-section, while A, B, C, and D are dimensional constants related to the rates of bending, extension, and torsion. By substitution from (67), (68), and (69) in the Lévy–Mises relations it is readily shown that

$$\phi(x,y) = \tfrac{1}{12}(Ay^3 - Bx^3) + \tfrac{1}{4}xy(Ax-By); \qquad (70)$$

$$\left. \begin{array}{l} \dfrac{\partial \psi}{\partial x} = -Dy \mp \dfrac{\sqrt{3}(Ax+By+C)f_y}{\sqrt{(1-f_x^2-f_y^2)}}, \\[8pt] \dfrac{\partial \psi}{\partial y} = Dx \pm \dfrac{\sqrt{3}(Ax+By+C)f_x}{\sqrt{(1-f_x^2-f_y^2)}}. \end{array} \right\} \qquad (71)$$

Hence, combining (69) and (70), possible velocities for a completely plastic state are

$$\left. \begin{array}{l} u = \tfrac{1}{4}A(y^2-x^2-2z^2) - \tfrac{1}{2}Bxy - \tfrac{1}{2}Cx + Dyz, \\[2pt] v = -\tfrac{1}{2}Axy + \tfrac{1}{4}B(x^2-y^2-2z^2) - \tfrac{1}{2}Cy - Dxz, \\[2pt] w = \psi(x,y) + Axz + Byz + Cz. \end{array} \right\} \qquad (72)$$

Equations (71) are compatible if

$$\frac{\partial}{\partial x}\left[\frac{(Ax+By+C)f_x}{\sqrt{(1-f_x^2-f_y^2)}}\right] + \frac{\partial}{\partial y}\left[\frac{(Ax+By+C)f_y}{\sqrt{(1-f_x^2-f_y^2)}}\right] \pm \frac{2D}{\sqrt{3}} = 0, \qquad (73)$$

or
$$(Ax+By+C)[(1-f_y^2)f_{xx} + 2f_xf_yf_{xy} + (1-f_x^2)f_{yy}] +$$
$$+ (Af_x+Bf_y)(1-f_x^2-f_y^2) \pm \frac{2D}{\sqrt{3}}(1-f_x^2-f_y^2)^{\frac{3}{2}} = 0. \qquad (73')$$

The solution of this (elliptic) equation for given ratios A/D, B/D, C/D, and subject to the boundary condition $f = 0$, determines the stress function f and thereby the stress distribution. When $A = B = C = 0$,

$D \neq 0$, we obtain the familiar solution $f_x^2+f_y^2 = 1$ corresponding to pure torsion.

The external loads applied at the ends of the bar are statically equivalent to a longitudinal force
$$Z = \sqrt{3}k \iint \pm(1-f_x^2-f_y^2)^{\frac{1}{2}} \, dx\,dy,$$
a twisting couple
$$G_z = \iint (x\tau_{yz}-y\tau_{xz}) \, dx\,dy = -2k \iint f \, dx\,dy,$$
and a bending couple with components
$$G_x = \sqrt{3}k \iint \pm y(1-f_x^2-f_y^2)^{\frac{1}{2}} \, dx\,dy, \quad G_y = \sqrt{3}k \iint \mp x(1-f_x^2-f_y^2)^{\frac{1}{2}} \, dx\,dy.$$
In the above equations either the upper or the lower sign must be taken throughout in any one element. The correct sign is decided by the consideration that the rate of work must be positive. Thus
$$0 \leqslant \tau_{xz}\left(\frac{\partial u}{\partial z}+\frac{\partial w}{\partial x}\right)+\tau_{yz}\left(\frac{\partial v}{\partial z}+\frac{\partial w}{\partial y}\right)+\sigma_z\frac{\partial w}{\partial z}$$
$$= -kf_y(\psi_x+Dy)+kf_x(\psi_y-Dx)\pm\sqrt{3}k(Ax+By+C)(1-f_x^2-f_y^2)^{\frac{1}{2}}$$
$$= \pm\sqrt{3}k\frac{Ax+By+C}{\sqrt{(1-f_x^2-f_y^2)}} \quad \text{from (71)}.$$
Hence the upper sign must be taken where $Ax+By+C$ is positive, and the lower sign where it is negative. If the longitudinal plane
$$Ax+By+C = 0$$
divides the bar in two parts, σ_z is tensile in one and compressive in the other; the rate of extension of elements lying in the plane is zero.

The equation (73) may alternatively be reached† by a simple application of the maximum work principle (Chap. III, Sect. 3 (iii)). The rate of work of the external forces is
$$\dot{W} = \iint (\tau_{xz}u+\tau_{yz}v+\sigma_z w) \, dx\,dy,$$
where the integral is taken over the ends of the bar. This expression must be made a maximum for varying stresses satisfying (67) and (68), and for given surface values of u, v, w, tentatively assumed to satisfy (69). In the integral, terms not involving z cancel out at opposite ends of the bar, while the remaining terms lead to
$$\frac{\dot{W}}{kL} = \sqrt{3}\iint \pm(Ax+By+C)(1-f_x^2-f_y^2)^{\frac{1}{2}} \, dx\,dy -$$
$$-D\iint (xf_x+yf_y) \, dx\,dy + \tfrac{1}{2}L\iint (Af_y-Bf_x) \, dx\,dy,$$

† R. Hill, *Quart. Journ. Mech. App. Math.* **1** (1948), 18.

where L is the length of the bar. On the right-hand side the third term is zero since $f = 0$ on the contour, while the second term can be transformed to give

$$\frac{\dot{W}}{kL} = \sqrt{3} \iint \pm (Ax+By+C)(1-f_x^2-f_y^2)^{\frac{1}{2}}\,dxdy + 2D \iint f\,dxdy.$$

By the well-known Euler–Lagrange formula of the calculus of variations, it may be verified that stationary values of $\dot{W}$ are obtained from functions f satisfying (73). If this method is adopted, it is necessary to verify afterwards that corresponding velocities exist such that the Lévy–Mises equations are satisfied and such that the rate of work in every element is positive. This justifies the assumption that a terminal distribution of velocities (69) produces a completely plastic state, in which, moreover, the stress components σ_x, σ_y, and τ_{xy} are zero.

XII
PLASTIC ANISOTROPY

THERE are certain important and striking phenomena which cannot be described by the theory constructed in Chapter II. The assumption that every material element remains isotropic is an approximation that becomes less good as the deformation continues. Individual crystal grains are elongated in the direction of the greatest tensile strain and the texture of the specimen appears fibrous. Now it is a consequence of the glide process that a single crystal rotates during the straining so that it approaches an orientation characteristic of the particular strain-path. For example, when hexagonal single crystals are stretched in tension, the basal planes gradually turn towards positions parallel to the direction of the applied load. Similarly, the grains of a polycrystal tend to rotate towards some limiting orientation (not necessarily equivalent to that in a single crystal, owing to the mutual constraints between the grains); thus, in face-centred cubic metals compressed between lubricated plates, face diagonals tend to align parallel to the direction of compression. By this mechanism a metal in which the grains are initially oriented at random, and which is therefore isotropic, is rendered anisotropic during plastic deformation. The distribution of orientations between the grains (measured, for example, on a percentage basis) then has one or more maxima. If such a maximum is well defined it is referred to as a preferred orientation. If the orientations of the individual crystals are not randomly distributed, the yield stress and the macroscopic stress-strain relations vary with direction. For example, in heavily cold-rolled brass, the tensile yield stress transverse to the direction of rolling may be as much as 10 per cent. greater than that parallel to the rolling direction.† Greater variations may be obtained by a critical sequence of mechanical and heat treatments which produces a final recrystallization texture approaching that of a single crystal (for example, rolled copper sheet can be prepared so that varying proportions of the grains have their cubic axes parallel to the edges of the sheet‡).

We now consider how the theoretical framework may be broadened to include anisotropy. As usual, we neglect the effects of those internal

† M. Cook, *Journ. Inst. Metals*, **60** (1937), 159.
‡ W. Köster, *Zeits. Metallkunde*, **18** (1926), 112. See also W. M. Baldwin, Jr., *Trans. Am. Inst. Min. Met. Eng.* **166** (1946), 591.

stresses which result directly from the differential orientation of the grains. These effects can be largely removed by a mild annealing which does not alter the preferred orientation. To change the latter the heat treatment has generally to be carried out above the recrystallization temperature.

1. The yield criterion

For simplicity we shall only consider states of anisotropy that possess three mutually orthogonal planes of symmetry at every point; the intersections of these planes are known as the principal axes of anisotropy. These axes may vary in direction throughout the specimen; for example, if anisotropy is developed in a hollow tube uniformly expanded by internal pressure, the principal axes must lie in the radial, circumferential, and axial directions. A strip cut from the centre of a cold-rolled sheet provides an example of uniformly directed anisotropy; it is observed, in accordance with expectation, that the principal axes lie in the direction of rolling, transversely in the plane of the sheet, and normal to this plane.† The principal axes in a given element can also vary relatively to the element itself during continued deformation, as in simple shear.

Let us fix our attention on a particular element in a certain state of anisotropy, and choose the principal axes as Cartesian axes of reference. The criterion approximately describing the yielding of isotropic material is that of von Mises. The simplest yield criterion for anisotropic material is therefore one which reduces to von Mises' law when the anisotropy is vanishingly small; the method adopted will be to follow the implications of this hypothesis, and generalize it later if necessary. If, then, the yield criterion is assumed‡ to be a quadratic in the stress components, it must be of the form

$$2f(\sigma_{ij}) \equiv F(\sigma_y - \sigma_z)^2 + G(\sigma_z - \sigma_x)^2 + H(\sigma_x - \sigma_y)^2 + \\ + 2L\tau_{yz}^2 + 2M\tau_{zx}^2 + 2N\tau_{xy}^2 = 1, \quad (1)$$

where F, G, H, L, M, N are parameters characteristic of the current state of anisotropy. Linear terms are not included since it is assumed, as already mentioned, that there is no Bauschinger effect. Quadratic terms in which any one shear stress occurs linearly are rejected in view of the symmetry restriction. Finally, only the differences of the normal components can appear if it is assumed that the superposition of a hydro-

---

† L. J. Klingler and G. Sachs, *Journ. Aero. Sci.* **15** (1948), 599.
‡ R. Hill, *Proc. Roy. Soc.* A, **193** (1948), 281.

static stress does not influence yielding. It must be remembered that the yield criterion only has this form when the principal axes of anisotropy are the axes of reference; otherwise the form changes in a way that can be found by transforming the stress components.

If X, Y, Z are the tensile yield stresses in the principal directions of anisotropy, it is easily shown that

$$\left.\begin{aligned} \frac{1}{X^2} &= G+H, & 2F &= \frac{1}{Y^2}+\frac{1}{Z^2}-\frac{1}{X^2}, \\ \frac{1}{Y^2} &= H+F, & 2G &= \frac{1}{Z^2}+\frac{1}{X^2}-\frac{1}{Y^2}, \\ \frac{1}{Z^2} &= F+G, & 2H &= \frac{1}{X^2}+\frac{1}{Y^2}-\frac{1}{Z^2}. \end{aligned}\right\} \quad (2)$$

It is clear that only one of F, G, H can be negative, and that this is possible only when the yield stresses differ considerably. Also, $F \geqslant G$ if and only if $X \geqslant Y$, together with two similar inequalities. If R, S, T are the yield stresses in shear with respect to the principal axes of anisotropy, then

$$2L = \frac{1}{R^2}, \qquad 2M = \frac{1}{S^2}, \qquad 2N = \frac{1}{T^2}. \quad (3)$$

L, M, N are thus essentially positive. If there is rotational symmetry of the anisotropy in an element about the z-axis, the form of the expression (1) remains invariant for arbitrary (x, y) axes of reference. Now (1) can be written alternatively as

$$[(G+H)\sigma_x^2 - 2H\sigma_x\sigma_y + (F+H)\sigma_y^2 + 2N\tau_{xy}^2] - 2(G\sigma_x + F\sigma_y)\sigma_z + \\ + 2(L\tau_{yz}^2 + M\tau_{zx}^2) + (F+G)\sigma_z^2 = 1.$$

Let other axes (x', y', z') be taken so that the z'-axis coincides with the z-axis, while the x'-axis is inclined at a clockwise angle θ to the x-axis. The equations of transformation (Appendix IV) are

$$\sigma_x = \sigma_{x'}\cos^2\theta + \sigma_{y'}\sin^2\theta + 2\tau_{x'y'}\sin\theta\cos\theta, \quad \sigma_z = \sigma_{z'},$$
$$\sigma_y = \sigma_{x'}\sin^2\theta + \sigma_{y'}\cos^2\theta - 2\tau_{x'y'}\sin\theta\cos\theta, \quad \tau_{yz} = \tau_{y'z'}\cos\theta - \tau_{z'x'}\sin\theta,$$
$$\tau_{xy} = (\sigma_{y'}-\sigma_{x'})\sin\theta\cos\theta + \tau_{x'y'}(\cos^2\theta - \sin^2\theta), \quad \tau_{zx} = \tau_{y'z'}\sin\theta + \tau_{z'x'}\cos\theta.$$

In order that the coefficient of $\sigma_{z'}$ should be equal to $-2(G\sigma_{x'}+F\sigma_{y'})$ after the transformation, it is obviously necessary and sufficient that $F = G$. The coefficients of the terms in $\sigma_{x'}\tau_{x'y'}$ and $-\sigma_{y'}\tau_{x'y'}$ are then both equal to

$$4(F+2H-N)\sin\theta\cos\theta(\cos^2\theta-\sin^2\theta).$$

Since this is to vanish for all θ, $F+2H-N$ must be zero. It may be verified that the coefficients of $\sigma_{x'}^2$, $\sigma_{y'}^2$, and $\sigma_{x'}\sigma_{y'}$ are then invariant. Finally, the term in $\tau_{z'x'}\tau_{y'z'}$ vanishes if and only if $L = M$, in which case the coefficients of $\tau_{y'z'}^2$ and $\tau_{z'x'}^2$ are also invariant. To sum up, the necessary and sufficient conditions for the anisotropy to be rotationally symmetric about the z-axis are

$$N = F+2H = G+2H, \qquad L = M. \tag{4}$$

If there is complete spherical symmetry, or isotropy,

$$L = M = N = 3F = 3G = 3H,$$

and the expression (1) reduces to von Mises' criterion when $2F$ is equated to $1/Y^2$ (Chap. II, equation (9)).

To describe fully the state of anisotropy in an element, we need to know the orientations of the principal axes and the values of the six independent yield stresses X, Y, Z, R, S, T. They must be considered as functions of the mechanical and heat treatments since the element was isotropic; in general, they will also vary during a further deformation. It is not yet known how to relate the yield stresses quantitatively with the microstructure, for example with the degree of preferred orientation, and it must be supposed, therefore, that they have been determined by experiment.

2. Relations between stress and strain-increment

By analogy with the Lévy–Mises equations for isotropic material it is supposed that $f(\sigma_{ij})$, in equation (1), is the plastic potential (Chap. III, Sect. 1). The strain-increment relations, referred to the principal axes of anisotropy, are then[†]

$$\begin{aligned}
d\epsilon_x &= d\lambda[H(\sigma_x-\sigma_y)+G(\sigma_x-\sigma_z)], & d\gamma_{yz} &= d\lambda L\tau_{yz}, \\
d\epsilon_y &= d\lambda[F(\sigma_y-\sigma_z)+H(\sigma_y-\sigma_x)], & d\gamma_{zx} &= d\lambda M\tau_{zx}, \\
d\epsilon_z &= d\lambda[G(\sigma_z-\sigma_x)+F(\sigma_z-\sigma_y)], & d\gamma_{xy} &= d\lambda N\tau_{xy}.
\end{aligned} \tag{5}$$

It will be noticed that $d\epsilon_x+d\epsilon_y+d\epsilon_z = 0$ identically, and that if the stress is reversed, so also is the strain-increment. Furthermore, if the principal axes of stress coincide with the axes of anisotropy, so do the principal axes of strain-increment. Otherwise, the principal axes of stress and strain-increment are not generally coincident.

For an experimental determination of the state of anisotropy it is

[†] R. Hill, op. cit., p. 318. In special cases, similar relations (derived from different assumptions) have been stated by L. R. Jackson, K. F. Smith, and W. T. Lankford, *Metals Technology*, Tech. Pub. 2440, (1948), and by J. E. Dorn, *Journ. App. Phys.*, **20** (1949), 15.

necessary that the anisotropy should be distributed *uniformly* through a volume sufficient to allow the cutting of tensile test pieces of arbitrary orientation. Then, if a pure tension X is applied to a strip or cylinder cut parallel to the x-axis of anisotropy, the incremental strains are in the ratios

$$d\epsilon_x : d\epsilon_y : d\epsilon_z = G+H : -H : -G.$$

The strain in each transverse direction is a contraction unless the yield stresses differ so much that one of G or H is negative. The contraction in the y direction is the larger if $H > G$, that is, if $Z > Y$; the strain is therefore less in the direction of the greater yield stress. Similarly, tension tests in the y and z directions furnish the ratios F/H and G/F. In principle, this allows an immediate test of the theory in view of the identity $(H/G) \times (G/F) \times (F/H) = 1$. Klingler and Sachs,† using rolled aluminium sheet $1\frac{1}{2}$ inches thick, measured the strains in tensile specimens cut at various orientations in, and obliquely to, the plane of the sheet. When the specimen was normal to the plane of rolling, the two transverse strain components were equal within experimental error; if x is the rolling direction and y the transverse direction in the plane of rolling, this implies that $F \sim G$. It was also observed that one principal strain was always in a direction parallel to the rolling plane. Where the theory is applicable, measurements of the strain ratios in tension specimens cut in the x and y directions provide an indirect method for determining the ratios of the three tensile yield stresses, with the use of equation (2); this is preferable to the direct method if the yielding is not sufficiently sharp and well defined. It is a particularly convenient means of determining the through-thickness yield stress when the material is in the form of a thin sheet. On the other hand, independent measurements of the strain ratios and the yield stresses provide further tests of the validity of the theory.

3. Plastic anisotropy of rolled sheet

(i) *Variation of yield stress with orientation.* We consider now what information can be obtained about the anisotropy of a rolled sheet from the behaviour of tensile specimens cut in its plane. Let axes of reference be chosen so that x is the rolling direction, y the transverse direction in the plane, and z normal to the plane. If any element of the sheet is now subjected to stresses applied in the plane of the sheet, the criterion of yielding is

$$(G+H)\sigma_x^2 - 2H\sigma_x\sigma_y + (H+F)\sigma_y^2 + 2N\tau_{xy}^2 = 1. \qquad (6)$$

† L. J. Klingler and G. Sachs, op. cit., p. 318.

In particular, for a tensile specimen cut at an angle α to the rolling direction,
$$\sigma_x = \sigma\cos^2\alpha, \quad \sigma_y = \sigma\sin^2\alpha, \quad \tau_{xy} = \sigma\sin\alpha\cos\alpha,$$
where σ is the tensile yield stress. Substitution in (6) gives
$$\sigma = [F\sin^2\alpha + G\cos^2\alpha + H + (2N - F - G - 4H)\sin^2\alpha\cos^2\alpha]^{-\frac{1}{2}}. \quad (7)$$
Values of F, G, H, and N (but not L or M) can be deduced from the observed dependence of the yield stress on the orientation. It may be shown that the maxima and minima of σ occur along the anisotropic axes, and also in directions $\bar{\alpha}$ such that
$$\tan^2\bar{\alpha} = \frac{N - G - 2H}{N - F - 2H}. \quad (8)$$
If $N > F+2H$ and $G+2H$ the yield stress has maximum (unequal) values in the x and y directions and minimum (equal) values in the $\bar{\alpha}$ directions. Cook,† and Palmer and Smith,‡ have observed variations of this kind in brass, after various rolling and annealing treatments. Klingler and Sachs§ found that for aluminium alloy sheet the yield stress had a minimum near the 45° positions, and that $F \sim G$. If $N < F+2H$ and $G+2H$, the yield stress has minimum (unequal) values in the x and y directions and maximum (equal) values in the $\bar{\alpha}$ directions. If N is intermediate to $F+2H$ and $G+2H$, there is no real value of $\bar{\alpha}$, and the yield stress has a maximum in the x direction and a minimum in the y direction when $F > G$,∥ and vice versa when $F < G$. It is evident from this that the relation of N to the quantities $F+2H$ and $G+2H$ has a definite physical significance (cf. also equation (4)); this will be further exemplified later.

(ii) *Strain ratios in a tensile specimen*. The equations for the components of the strain-increment corresponding to a uniaxial tension σ at an angle α to the x-axis are

$$\left. \begin{aligned} d\epsilon_x &= [(G+H)\cos^2\alpha - H\sin^2\alpha]\sigma\,d\lambda, \\ d\epsilon_y &= [(F+H)\sin^2\alpha - H\cos^2\alpha]\sigma\,d\lambda, \\ d\epsilon_z &= -(F\sin^2\alpha + G\cos^2\alpha)\sigma\,d\lambda, \\ d\gamma_{xy} &= (N\sin\alpha\cos\alpha)\sigma\,d\lambda. \end{aligned} \right\} \quad (9)$$

† M. Cook, *Journ. Inst. Metals*, **60** (1937), 159.
‡ E. W. Palmer and C. S. Smith, *Trans. Am. Inst. Min. Met. Eng.* **147** (1942), 164.
§ L. J. Klingler and G. Sachs, op. cit., p. 318.
∥ This kind of variation has been observed in aluminium sheet pre-strained 12 per cent. in tension (the direction of this pre-strain is taken to be the x-axis). See L. J. Klingler and G. Sachs, *Journ. Aero. Sci.* **15** (1948), 151.

The principal axes of the strain-increment coincide with the principal stress axes (i.e., along and perpendicular to the direction of pull) when

$$(d\epsilon_x - d\epsilon_y)/d\gamma_{xy} = (\sigma_x - \sigma_y)/\tau_{xy},$$

or

$$[(G+2H)\cos^2\alpha - (F+2H)\sin^2\alpha]/N \sin\alpha\cos\alpha = (\cos^2\alpha - \sin^2\alpha)/\sin\alpha\cos\alpha.$$

This is satisfied by $\alpha = 0$, $\bar{\alpha}$, or $\tfrac{1}{2}\pi$, where $\bar{\alpha}$ has the value (8). Thus the principal axes of stress and strain-increment coincide for the orientations where the yield stress is a maximum or minimum.

The ratio of the transverse to the through-thickness strain is

$$r = (d\epsilon_x \sin^2\alpha + d\epsilon_y \cos^2\alpha - 2d\gamma_{xy}\sin\alpha\cos\alpha)/d\epsilon_z$$

$$= \frac{H + (2N - F - G - 4H)\sin^2\alpha\cos^2\alpha}{F\sin^2\alpha + G\cos^2\alpha}. \tag{10}$$

This expression is in good agreement with the experimental data of Baldwin, Howald, and Ross† for copper with a cubic recrystallization texture of 50 per cent. or less; in this material $F = G$, by symmetry, since the cubic axes in the preferred orientation are parallel to the edges of the sheet. Values of N/F and H/F can be deduced by fitting (10) to their results; it is found, for example, that $N/F = 2 \cdot 0$ and $H/F = 0 \cdot 8$ for a 50 per cent. cubic texture, while for copper which had been rolled to a final reduction of 80 per cent., and was then annealed at 815° C., $N/F = 4 \cdot 8$ and $F = G = H$. In general, when $F = G$, the $r(\alpha)$ relation is concave upward when $N < F+2H$, and concave downward when $N > F+2H$. The latter type of relation was obtained in aluminium by Klingler and Sachs;‡ this is consistent with their measured yield stresses (see (i) above).

(iii) *Necking in tension*.§ When thin strips are pulled in tension (their width being at least five times their thickness), it is observed that the neck does not form directly across the specimen, but at an oblique angle which depends on the state of anisotropy. Necking begins, after some preliminary uniform extension, at a point where there is a slight non-uniformity, either geometrical or structural. Theoretically, the line of the neck should coincide with a characteristic, in view of the property of characteristics as curves along which small disturbances propagate. We must therefore begin an analysis of necking by determining the characteristics in a state of plane stress. Proceeding as in Chapter VI

† W. M. Baldwin, Jr., T. S. Howald, and A. W. Ross, *Metals Technology*, Tech. Pub. 1808 (1945). See also L. Bourne and R. Hill, *Phil. Mag.* 41 (1950), 671.
‡ L. J. Klingler and G. Sachs, *Journ. Aero. Sci.* 15 (1948), 599.
§ R. Hill, op. cit., p. 318.

(Sect. 3), with $\sigma_n, \sigma_s, \tau_{ns}$ as the stress components referred to the tangent and normal to the curve C, we find that all stress derivatives are uniquely determined unless $\partial f/\partial \sigma_s = 0$, where $f(\sigma_n, \sigma_s, \tau_{ns}) = 0$ is the yield criterion. Since f is also the plastic potential, $d\epsilon_s = (\partial f/\partial \sigma_s)\, d\lambda = 0$ when $\partial f/\partial \sigma_s = 0$. Thus C is a characteristic for the stresses if it coincides at every point with a direction of zero extension. There are thus two characteristics through a point, namely the directions of zero extension. It may be shown similarly that these are also the characteristics for the velocities. The characteristics are not generally orthogonal because of

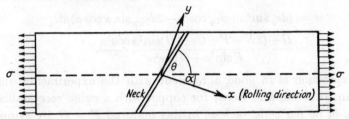

FIG. 87. Coordinate axes for analysis of the oblique necking of a strip in tension.

the strain normal to the sheet, nor are they always real (cf. Chap. XI, Sect. 4). The slopes dy/dx of the characteristics satisfy the equation

$$d\epsilon_x\, dx^2 + 2\, d\gamma_{xy}\, dx\, dy + d\epsilon_y\, dy^2 = 0,$$

or

$$[(G+H)\sigma_x - H\sigma_y]\, dx^2 + 2N\tau_{xy}\, dx\, dy + [(F+H)\sigma_y - H\sigma_x]\, dy^2 = 0. \quad (11)$$

In the present problem the characteristics are straight since the stress is uniform. Let θ be the inclination of a possible neck, measured *away* from the rolling direction (Fig. 87), so that $dy/dx = \tan(\theta+\alpha)$. Inserting this in (11), with $\sigma_x = \sigma\cos^2\alpha$, etc., we obtain

$$a\tan^2\theta + 2b\tan\theta - c = 0, \quad (12)$$

where
$$a = H + (2N - F - G - 4H)\sin^2\alpha\cos^2\alpha,$$
$$b = [(N - F - 2H)\sin^2\alpha - (N - G - 2H)\cos^2\alpha]\sin\alpha\cos\alpha,$$
$$c = a + F\sin^2\alpha + G\cos^2\alpha = 1/\sigma^2 \text{ from (7)}.$$

The anisotropic parameters refer to the state of anisotropy immediately preceding necking; this is effectively the same as in the rolled sheet since the additional anisotropy introduced by the preliminary uniform extension is usually negligible.

In an isotropic sheet $F = G = H = N/3$, $b = 0$, and $c = 2a$, so that $\tan\theta = \sqrt{2}$ or $\theta \sim \pm 54\cdot 7°$. There are thus two, equally possible, necking directions equally inclined to the specimen axis; if the origin of the disturbance which initiates necking lies, not on the edge, but in the middle of the specimen, a V-shaped neck is sometimes observed, with its branches coinciding with parts of both characteristics.

When the sheet is anisotropic there are still two, equally possible, necking directions corresponding to the roots of the quadratic (12) in $\tan\theta$, but generally with different inclinations. The roots are numerically equal, but opposite in sign, if $b = 0$, which happens when $\alpha = 0$, $\bar{\alpha}$, or $\pi/2$, where $\bar{\alpha}$ is given by (8). For these values of α the two possible necks are symmetrically situated with respect to the specimen axis; this was to be expected since the α directions are, as we have seen, those for which the principal axes of stress and strain-increment coincide (and for which the yield stress has stationary values). If N is greater than both $F+2H$ and $G+2H$, b is negative when $\alpha < \bar{\alpha}$ and positive when $\alpha > \bar{\alpha}$. Now a is generally positive (it can only be negative in the unlikely event of N being less than $\frac{1}{2}(F+G)$). Hence the sum $(-2b/a)$ of the roots is positive when $\alpha < \bar{\alpha}$ and negative when $\alpha > \bar{\alpha}$. This means that, when $\alpha < \bar{\alpha}$, θ is numerically larger for the neck tending to lie across the rolling direction, and vice versa when $\alpha > \bar{\alpha}$. These inequalities are reversed when N is less than both $F+2H$ and $G+2H$. The measurements by Körber and Hoff[†] of the necking angles in aluminium, copper, brass, and nickel, are in qualitative agreement with this analysis; it may be deduced that the state of anisotropy in their materials after cold-rolling 98 per cent. must have been such that $N > G+2H > F+2H$. This is consistent with their direct measurements of the yield stress which, for this reduction, was less in the rolling direction than in the transverse direction (i.e., $X < Y$). A closer quantitative comparison is prevented by the scatter in the data. Bitter,[‡] in work on iron-silicon sheets concluded that the two possible necks were approximately symmetrical to the directions of maximum elongation; this is in accord with the theory since the neck is a direction of zero elongation. It may be deduced that the state of anisotropy (which was very pronounced) was such that $N > F+2H$ and $F \sim G$.

4. Length changes in a twisted tube

A hollow isotropic tube twisted about its axis does not change in length so long as the internal stresses directly resulting from the differential grain orientations are negligible (p. 36). In the following discussion of the torsion of an anisotropic tube it is assumed that the

[†] F. Körber and H. Hoff, *Mitt. Kais. Wilh. Inst. Eisenf.* **10** (1928), 175.
[‡] F. Bitter, *Proc. Roy. Soc.* A, **145** (1934), 668.

tube is gripped at the ends in such a way that it is free to change its length (if this is prevented, an axial stress is induced).

Consider first the deformation of a small element of the tube, assuming that the material remains isotropic. Let TT' be the direction of shear, and ON be *unit* length of the generator through the centre O of the element (Fig. 88 a). During a shear strain γ (engineering definition) a direction OP, fixed in the element, rotates into a new direction OP' such that $PP' = \gamma$, and simultaneously undergoes a resultant extension

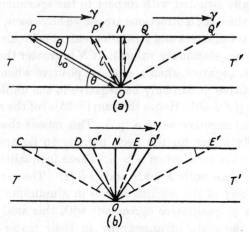

Fig. 88. Deformation of a small element in shear.

or contraction. If Q and Q' are the reflections of P and P' in ON, OQ rotates into the position OQ', being lengthened in the same ratio that OP is shortened; this is merely a consequence of the fact that the same amount of shear, applied in the reverse sense, restores the original configuration. There are two directions that undergo no resultant change in length: one of these is obviously parallel to TT', while the initial and final positions of the other are OD and OD' (Fig. 88b) where D and D' are reflections in ON, and $DD' = \gamma$. It is evident that there must be some direction between OT and OD which is contracted the most by a given shear, and one direction between OT' and OD which is extended the most. Let OP be inclined at θ to OT, and let $OP = l_0$ and $OP' = l$. Then

$$l_0 \sin\theta = 1, \qquad l^2 = l_0^2 + \gamma^2 - 2l_0\gamma\cos\theta.$$

Hence
$$\frac{l}{l_0} = (1 - 2\gamma\sin\theta\cos\theta + \gamma^2\sin^2\theta)^{\frac{1}{2}}.$$

By differentiation with respect to θ it is found that the minimum value of l/l_0 is such that
$$\frac{l}{l_0} = \tan\theta, \quad \text{where} \quad 2\cot 2\theta = \gamma.$$

OQ is then perpendicular to OP, and
$$\frac{l_0}{l} - \frac{l}{l_0} = \gamma. \tag{13}$$

The direction which undergoes the greatest contraction will be denoted

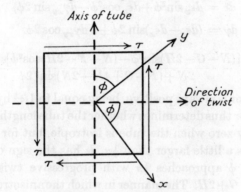

FIG. 89. Coordinate axes for the analysis of length changes in a twisted tube.

by OC (Fig. 88 b); the direction undergoing the greatest extension is, of course, OE where E is the reflection of C' in ON. Since
$$DN/ON = \tfrac{1}{2}\gamma = \cot 2\theta,$$
it follows that angle ODN is equal to 2θ, and hence that OC and OE are the internal and external bisectors of the angle between OD and OT; similarly, OC' and OE' are the bisectors of the angle between OD' and OT''.

It is now assumed† that the anisotropic axes coincide, at each moment, with the directions of greatest extension and contraction. Since the axial strain of the tube is very small compared with the shear strain, these directions are effectively the same as those in an isotropic tube. Suppose that at some stage the x, y axes of anisotropy (corresponding to $C'O$ and OE') make an angle ϕ with the direction of twist and the axis of the tube (Fig. 89). If the tube is isotropic to begin with, ϕ increases steadily from $\pi/4$ to $\pi/2$. If τ is the shear stress,
$$\sigma_x = -\sigma_y = -\tau\sin 2\phi, \qquad \tau_{xy} = \tau\cos 2\phi,$$
and
$$\tau = [2N + (F+G+4H-2N)\sin^2 2\phi]^{-\frac{1}{2}}. \tag{14}$$

† R. Hill, op. cit., p. 318.

The components of the increment of strain are

$$d\epsilon_x = -\tau \, d\lambda \, (G+2H)\sin 2\phi,$$
$$d\epsilon_y = \tau \, d\lambda \, (F+2H)\sin 2\phi,$$
$$d\epsilon_z = \tau \, d\lambda \, (G-F)\sin 2\phi,$$
$$d\gamma_{xy} = \tau \, d\lambda \, N \cos 2\phi.$$

The axial strain-increment $d\epsilon$ and the (engineering) shear strain-increment $d\gamma$ are given by

$$d\epsilon = d\epsilon_x \sin^2\phi + d\epsilon_y \cos^2\phi - d\gamma_{xy}\sin 2\phi,$$
$$d\gamma = (d\epsilon_y - d\epsilon_x)\sin 2\phi + 2\, d\gamma_{xy}\cos 2\phi.$$

Hence
$$\frac{d\epsilon}{d\gamma} = \frac{[(N-G-2H)\sin^2\phi - (N-F-2H)\cos^2\phi]\sin 2\phi}{2N + (F+G+4H-2N)\sin^2 2\phi}. \qquad (15)$$

The denominator is always positive, being equal to $1/\tau^2$ by (14). The sign of the numerator thus determines whether the tube lengthens or shortens. $d\epsilon/d\gamma$ is initially zero when the tube is isotropic, but for small angles of twist, when ϕ is a little larger than $\tfrac{1}{4}\pi$, $d\epsilon$ has the sign of $(F-G)$ or of $(X-Y)$. Since ϕ approaches $\tfrac{1}{2}\pi$ with progressive twist, $d\epsilon$ is finally positive if $N > G+2H$. The manner in which the anisotropic parameters vary during torsion is not yet known, but Swift[†] has found that initially isotropic specimens of aluminium, copper, brass, mild steel, 0.5 per cent. carbon steel, stainless steel, and cupro-nickel *lengthen* continuously by varying amounts, the extension being of the order of 5–10 per cent. for a shear of 3. Exceptionally, a lead specimen shortens. The axial strain cannot be attributed solely to the development of the internal stresses due to the differential orientations of the grains since on reversing the couple the ensuing axial strain-increment was also reversed (see p. 37).

5. The earing of deep-drawn cups

When a cup is deep-drawn from a circular blank cut from rolled sheet, it is often found that the height of the rim above the base is not uniform, as would be expected in a symmetrical operation on an isotropic blank. Instead it is observed that 'ears' form in positions symmetrically situated with respect to the direction of rolling in the original sheet. Generally four ears are found, either at the ends of the two diameters making 45° with the rolling direction, or at the ends of the diameters making 0° and 90° with the rolling direction. The positions and height of the ears

† H. W. Swift, *Engineering*, **163** (1947), 253.

depend among other things on the particular metal and on the prior mechanical and heat treatment. Both types of earing can be produced in the same metal by suitably varying the treatment before drawing, for example, in copper and steel.† With brass, six ears have also been observed in the 0° and 60° positions.‡ It is recognized that earing is due to anisotropy in the rolled sheet, and many attempts have been made to correlate the observed behaviour with the crystal texture and the mechanical properties of the material. In the following analysis§ it is supposed that the anisotropy is specified by the six parameters F, G, H, L, M, N, whose values are related in some complicated way to the previous treatment of the material.

Earing begins while the blank is being drawn towards the shoulder of the die (Fig. 78), and it is observed that the final positions of the ears coincide approximately with their initial positions.‖ It should be sufficient, therefore, to analyse the stress and strain distribution immediately after drawing begins, when the rim has just started to move towards the die aperture. Thickening of the sheet is controlled by the blank-holder. Two extreme possibilities will be considered: (i) the blank-holder is fixed in a position such that the space between the holder and the die is equal to the original sheet thickness (this is rarely attained in practice owing to the elasticity of the apparatus), and (ii) the space between the holder and the die is so much wider than the sheet thickness that negligible normal force is exerted in the early stages. If friction can be neglected, (i) corresponds to plane strain, and (ii) to plane stress.

The ears and hollows must begin to form at those points on the rim where the radial direction is one of the principal axes of the strain-increment. Now on the rim the circumferential stress is the only non-zero stress component in the plane of the sheet. Hence the ears and hollows form where the principal axes of stress and strain-increment coincide. Let axes of reference be chosen to coincide with the axes of anisotropy, the z-axis being normal to the blank and the x-axis along the direction of rolling. If the blank is drawn under conditions of plane strain, the ears and hollows can only be in the 0°, 45°, and 90° positions (see Sect. 7 below). It is clear that either four ears form at 0° and 90° with hollows at 45°, or that the reverse configuration occurs. If the blank is drawn under conditions of plane stress, the ears and hollows can

† C. S. Barrett, *The Structure of Metals*, p. 443 (McGraw-Hill Book Co., 1943).
‡ M. Cook, *Journ. Inst. Metals*, **60** (1937), 159.
§ R. Hill, op. cit., p. 318.
‖ F. H. Wilson and R. M. Brick, *Trans. Am. Inst. Min. Met. Eng.* **161** (1945), 173.

only be at $0°$, $90°$, and $(90°-\bar{\alpha})$ to the rolling direction, where $\bar{\alpha}$ is given by (8). There are thus four ears in the $0°$ and $90°$ positions, or in the $(90°-\bar{\alpha})$ positions. Notice that $\bar{\alpha} = 45°$ when $F = G$, irrespective of the values of N and H. If N is intermediate to $F+2H$ and $G+2H$, $\bar{\alpha}$ is not real and there are only two ears, either in the $0°$ or the $90°$ positions.

In both plane stress and plane strain the tangents at the points where the ears and hollows develop are in the directions for which the yield stress has stationary values (since these are the directions in which the principal axes of stress and strain-increment coincide). In order to distinguish between the possible arrangements of ears in terms of the relative magnitudes of the anisotropic parameters we must know the distribution of displacements on the rim. Failing this, it may reasonably be surmised that the ears and hollows develop respectively from points where the tangents to the rim are in the directions of the minimum and maximum values of the uniaxial yield stress. This hypothesis is supported by the results of a detailed investigation of earing by Baldwin, Howald, and Ross.† No blank-holder was used, and the material (copper) was such that $F = G$. The relative values of N and $F+2H$, as indicated by the strain ratios in tensile strips cut from the rolled sheet (see Sect. 3 (ii) above), were such that $N < F+2H$ for the copper producing $0°$ and $90°$ ears, and $N > F+2H$ for that producing $45°$ ears. It was proved in Section 3 (i) that when $N < F+2H$ the yield stress has minimum values in the $0°$ and $90°$ positions, and that when $N > F+2H$ the minimum values are in the $45°$ positions (since $\bar{\alpha} = 45°$ here).

The present theory predicts four ears at most, and must be generalized if it is to represent the anisotropy in a material such as rolled cartridge brass which, after a final annealing near $700°$ C., produces six ears.‡ Let us assume for the yield criterion and plastic potential a polynomial of degree n in the reduced stress components (the polynomial is assumed not to be a mere power of some polynomial of lower degree). For a state of plane stress or strain, the polynomial takes the form

$$\sum A_{ijk} \sigma_x^i \sigma_y^j \tau_{xy}^k,$$

where the powers i, j, k are positive integers or zero $(i+j+k \leqslant n)$, and k must be even when the x, y directions are the principal axes of aniso-

---

† W. M. Baldwin, Jr., T. S. Howald, and A. W. Ross, *Metals Technology*, Tech. Pub. 1808 (1945). See also L. Bourne and R. Hill, *Phil. Mag.* **41** (1950), 671.

‡ H. L. Burghoff and E. C. Bohlen, *Trans. Am. Inst. Min. Met. Eng.* **147** (1942), 144.

tropy. The strain-increments are

$$d\epsilon_x = d\lambda \frac{\partial f}{\partial \sigma_x} = d\lambda \sum i A_{ijk} \sigma_x^{i-1} \sigma_y^j \tau_{xy}^k,$$

$$d\epsilon_y = d\lambda \frac{\partial f}{\partial \sigma_y} = d\lambda \sum j A_{ijk} \sigma_x^i \sigma_y^{j-1} \tau_{xy}^k,$$

$$d\gamma_{xy} = d\lambda \frac{\partial f}{\partial \tau_{xy}} = \tfrac{1}{2} d\lambda \sum k A_{ijk} \sigma_x^i \sigma_y^j \tau_{xy}^{k-1},$$

where the factor $\tfrac{1}{2}$ is introduced in the last equation since τ_{xy} and τ_{yx} must be treated as distinct in the differentiation. In uniaxial tension σ at an angle α to the x-axis,

$$\sigma_x = \sigma \cos^2\alpha, \quad \sigma_y = \sigma \sin^2\alpha, \quad \tau_{xy} = \sigma \sin\alpha \cos\alpha.$$

The principal axes of stress and strain-increment coincide in directions $\bar{\alpha}$ where

$$\frac{\sigma_x - \sigma_y}{2\tau_{xy}} = \frac{d\epsilon_x - d\epsilon_y}{2 d\gamma_{xy}},$$

or

$$\frac{\cos^2\bar{\alpha} - \sin^2\bar{\alpha}}{2\sin\bar{\alpha}\cos\bar{\alpha}} = \frac{\sum A_{ijk}[i(\cos\bar{\alpha})^{2i+k-2}(\sin\bar{\alpha})^{2j+k} - j(\cos\bar{\alpha})^{2i+k}(\sin\bar{\alpha})^{2j+k-2}]}{\sum k A_{ijk}(\cos\bar{\alpha})^{2i+k-1}(\sin\bar{\alpha})^{2j+k-1}}.$$

This reduces to

$$\sum A_{ijk}(\cos\bar{\alpha})^{2i+k-1}(\sin\bar{\alpha})^{2j+k-1}[(2j+k)\cos^2\bar{\alpha} - (2i+k)\sin^2\bar{\alpha}] = 0. \quad (16)$$

Since k is even there must be a factor of $\sin\bar{\alpha}$ and $\cos\bar{\alpha}$ in every term, so that possible solutions are $0°$ and $90°$. Including these, there are altogether $4n$ roots in the range $0 \leqslant \bar{\alpha} < 2\pi$, symmetrically disposed in relation to the axes of anisotropy. The positions of the ears and hollows are $0°$, $90°$, and $90° - \bar{\alpha}$; if all the roots are real the number of ears is $2n$. For certain values of the parameters A_{ijk} some of the roots will not be real (cf. plane stress with $n = 2$ and N intermediate to $F + 2H$ and $G + 2H$). Thus the observed number of ears only enables us to set a lower limit to the degree of the plastic potential.

It is interesting to notice that, whatever the yield criterion, the ears and hollows form where the tangents to the rim are in the directions of stationary yield stress.† These directions are given by $d\sigma/d\alpha = 0$ or, since $d\sigma/d\alpha = -(\partial f/\partial \alpha)/(\partial f/\partial \sigma)$, by

$$\frac{\partial}{\partial \alpha} f(\sigma_x, \sigma_y, \tau_{xy}) = 0,$$

where $\quad \sigma_x = \sigma\cos^2\alpha, \quad \sigma_y = \sigma\sin^2\alpha, \quad \tau_{xy} = \sigma\sin\alpha\cos\alpha.$

† A theorem stated by the author in a letter (10 Dec. 1948) to Prof. W. M. Baldwin, Case Institute of Technology.

This is
$$\frac{\partial f}{\partial \sigma_x}\frac{\partial \sigma_x}{\partial \alpha}+\frac{\partial f}{\partial \sigma_y}\frac{\partial \sigma_y}{\partial \alpha}+2\frac{\partial f}{\partial \tau_{xy}}\frac{\partial \tau_{xy}}{\partial \alpha}=0,$$
or
$$2\sigma\left[-\frac{\partial f}{\partial \sigma_x}\sin\alpha\cos\alpha+\frac{\partial f}{\partial \sigma_y}\sin\alpha\cos\alpha+\frac{\partial f}{\partial \tau_{xy}}(\cos^2\alpha-\sin^2\alpha)\right]=0.$$

After rearrangement this becomes
$$\frac{\cos^2\alpha-\sin^2\alpha}{2\sin\alpha\cos\alpha}=\frac{\dfrac{\partial f}{\partial \sigma_x}-\dfrac{\partial f}{\partial \sigma_y}}{2\dfrac{\partial f}{\partial \tau_{xy}}},$$

which is identical with the condition for the coincidence of the principal axes of stress and strain-increment.

6. Variation of the anisotropic parameters during cold-work

The problem of relating the values of the six parameters to the strain-history is obviously extremely complicated. Here we shall restrict our attention to a metal in which a pronounced preferred orientation is already present, and to a range of strain such that changes in the orientation can be neglected. Since the state of anisotropy then remains effectively the same, the yield stresses must increase in *strict proportion* as the material work-hardens. We may write $X = hX_0$, $Y = hY_0$, etc., where the subscript zero denotes the initial value, and h is a parameter increasing monotonically from unity and expressing the amount of hardening. The anisotropic parameters decrease in strict proportion, since $F = F_0/h^2$ etc., from (2). Thus the relations between the strain-increment *ratios* and the stress *ratios* remain invariant; for example, in uniaxial tension the strain-increment ratios remain constant.† To complete the system of equations we need only to specify the way in which h varies with the strain. We may regard

$$\bar{\sigma} = \sqrt{\frac{3}{2}}\frac{h}{(F_0+G_0+H_0)^{\frac{1}{2}}}$$
$$= \sqrt{\frac{3}{2}}\left[\frac{F_0(\sigma_y-\sigma_z)^2+G_0(\sigma_z-\sigma_x)^2+H_0(\sigma_x-\sigma_y)^2+}{F_0+G_0+H_0}\right.$$
$$\left.\phantom{= \sqrt{\frac{3}{2}}\Big[}\frac{+2L_0\tau_{yz}^2+2M_0\tau_{zx}^2+2N_0\tau_{xy}^2}{F_0+G_0+H_0}\right]^{\frac{1}{2}} \quad (17)$$

as a non-dimensional measure of the equivalent stress. By analogy with the isotropic theory it is natural to assume that $\bar{\sigma}$ is a function of the

---

† This has been observed in tensile specimens cut from rolled aluminium plate; L. J. Klingler and G. Sachs, *Journ. Aero. Sci.* **15** (1948), 599.

plastic work.† Now the increment of plastic work per unit volume is

$$dW = \sigma_{ij}\,d\epsilon_{ij} = \sigma_{ij}\frac{\partial f}{\partial \sigma_{ij}}\,d\lambda = 2f\,d\lambda = d\lambda, \qquad (18)$$

with the use of equation (1) and Euler's theorem on homogeneous functions. From (5) we have also

$$G\,d\epsilon_y - H\,d\epsilon_z = (FG+GH+HF)(\sigma_y-\sigma_z)\,d\lambda,$$

together with two similar equations obtained by cyclic permutation. Hence

$$\sum\left[F\left(\frac{G\,d\epsilon_y-H\,d\epsilon_z}{FG+GH+HF}\right)^2 + \frac{2d\gamma_{yz}^2}{L}\right]$$
$$= (d\lambda)^2 \sum\left[F(\sigma_y-\sigma_z)^2 + 2L\tau_{yz}^2\right] = (d\lambda)^2.$$

This suggests the following definition of an equivalent strain-increment for anisotropic material:

$$\overline{d\epsilon} = \sqrt{\tfrac{2}{3}}(F_0+G_0+H_0)^{\frac{1}{2}}\left[F_0\left(\frac{G_0\,d\epsilon_y-H_0\,d\epsilon_z}{F_0G_0+G_0H_0+H_0F_0}\right)^2 + \dots + \frac{2d\gamma_{yz}^2}{L_0} + \dots\right]^{\frac{1}{2}}. \qquad (19)$$

From (18) and (19) we find

$$\overline{d\epsilon} = \sqrt{\tfrac{2}{3}}(F_0+G_0+H_0)^{\frac{1}{2}}\frac{dW}{h}, \qquad \text{and} \qquad dW = \bar{\sigma}\,\overline{d\epsilon}. \qquad (20)$$

Thus, if there is a functional relation between $\bar{\sigma}$ and W (this has yet to be demonstrated), there must also be one between $\bar{\sigma}$ and $\int \overline{d\epsilon}$. This is the analogue of the equivalent stress-strain curve for isotropic material, the area under the curve being again equal to the work per unit volume.

In uniaxial tension X parallel to the x-axis,

$$\bar{\sigma} = \sqrt{\frac{3}{2}\left(\frac{G_0+H_0}{F_0+G_0+H_0}\right)^{\frac{1}{2}}}X, \qquad \overline{d\epsilon} = \sqrt{\frac{2}{3}\left(\frac{F_0+G_0+H_0}{G_0+H_0}\right)^{\frac{1}{2}}}d\epsilon_x.$$

There are two similar expressions, obtained by cyclic permutation, for tensions Y and Z parallel respectively to the y- and z-axes. A comparison of the stress-strain curves in the x, y, and z directions provides an immediate test of the proposed $(\bar{\sigma}, \int \overline{d\epsilon})$ relation. Thus, if $\bar{\sigma} = F(\int \overline{d\epsilon})$, then $\alpha X, \beta Y, \gamma Z$ are the same functions F of ϵ_x/α, ϵ_y/β, ϵ_z/γ, respectively, where

$$\alpha = \sqrt{\frac{3}{2}\left(\frac{G_0+H_0}{F_0+G_0+H_0}\right)^{\frac{1}{2}}}, \qquad \beta = \sqrt{\frac{3}{2}\left(\frac{H_0+F_0}{F_0+G_0+H_0}\right)^{\frac{1}{2}}},$$

$$\gamma = \sqrt{\frac{3}{2}\left(\frac{F_0+G_0}{F_0+G_0+H_0}\right)^{\frac{1}{2}}} \qquad (\alpha^2+\beta^2+\gamma^2 = 3).$$

† L. R. Jackson, K. F. Smith, and W. T. Lankford, *Metals Technology*, Tech. Pub. 2440 (1948). J. E. Dorn, *Journ. App. Phys.* **20** (1949), 15.

α, β, γ are calculable from the ratios F_0/H_0 and G_0/H_0, which are given by the strain ratios in tensile tests in the x and y directions. If the material is in the form of a thin sheet the stress-strain curve for the z direction (through the thickness) can conveniently be obtained by the application of balanced biaxial tension in the plane of the sheet; this is equivalent to a uniaxial compression normal to the sheet, together with an equal hydrostatic tension. Such a biaxial state of stress may be approximately realized in the bulge test where a circular sheet, clamped at the circumference, is plastically deformed by pressure applied to one side.[†]

7. Theory of plane strain for anisotropic metals[‡]

(i) *Fundamental equations.* Let the state of plane strain be such that the z principal axis of anisotropy is normal to the planes of flow. Setting $d\epsilon_z = 0$ in (5) we have

$$\sigma_z = \frac{G\sigma_x + F\sigma_y}{G + F}. \tag{21}$$

In the compression of a chrome-vanadium steel block under conditions of plane strain, Bridgman[§] has found that σ_z becomes increasingly greater than $\tfrac{1}{2}\sigma_x$ (20 per cent. more for a strain of 0·4), where x is the direction of compression; this implies $G > F$. A cylindrical specimen cut from the compressed block, with its axis in the original z direction becomes elliptical when compressed, the major axis being in the former x direction; this also implies $G > F$.

If (21) is substituted in (1), with $\tau_{yz} = \tau_{zx} = 0$, there results

$$\left(\frac{FG + GH + HF}{F + G}\right)(\sigma_x - \sigma_y)^2 + 2N\tau_{xy}^2 = 1.$$

For reasons that will appear later it is convenient to write

$$c = 1 - \frac{N(F+G)}{2(FG + GH + HF)} \quad (-\infty < c < 1). \tag{22}$$

If N is greater than both $F + 2H$ and $G + 2H$, c is negative, while if N is less than both $F + 2H$ and $G + 2H$, c is positive. c is zero if the material is isotropic, and also when the anisotropy is rotationally symmetrical about the z-axis so that

$$N = F + 2H = G + 2H \quad (F \neq H).$$

[†] C. C. Chow, A. W. Dana, and G. Sachs, *Journal of Metals*, **1** (1949), 49. L. R. Jackson, K. F. Smith, and W. T. Lankford, op. cit., p. 333.
[‡] R. Hill, *Proc. Roy. Soc.* A, **198** (1949), 428.
[§] P. W. Bridgman, *Journ. App. Phys.* **17** (1946), 225. See also R. Hill, *Proc. Inst. Mech. Eng.* **159** (1948), 157.

With the use of (22) the yield criterion can be written as

$$\frac{(\sigma_x-\sigma_y)^2}{4(1-c)}+\tau_{xy}^2 = T^2, \tag{23}$$

where T is the yield stress in shear with respect to the x-, y-axes, defined in (3).

The plane-strain tensile yield stress σ in the direction making an angle θ with the x-axis is found by substituting

$$\sigma_x = \sigma\cos^2\theta, \qquad \sigma_y = \sigma\sin^2\theta, \qquad \tau_{xy} = \sigma\sin\theta\cos\theta,$$

in (23), leading to

$$\sigma = 2T\left(\frac{1-c}{1-c\sin^2 2\theta}\right)^{\frac{1}{2}}. \tag{24}$$

It is evident that σ has equal values in any set of four directions

$$\pm\theta, \qquad \pm(\tfrac{1}{2}\pi-\theta),$$

and hence that the angular variation of σ is symmetrical about the axes of anisotropy and about the 45° directions. The corresponding values of σ_z are, however, different unless $F = G$. If c is positive, σ has a minimum value $2T\sqrt{(1-c)}$ in the directions of the axes of anisotropy, and a maximum value $2T$ in the 45° directions; if c is negative, σ has a maximum $2T\sqrt{(1-c)}$ in the directions of the axes of anisotropy, and a minimum $2T$ in the 45° directions.

If u_x and v_y are the components of velocity referred to the anisotropic axes, the stress-strain relations (5) reduce to

$$\frac{\partial u_x}{\partial x}+\frac{\partial v_y}{\partial y} = 0; \qquad (1-c)\frac{\dfrac{\partial u_x}{\partial x}-\dfrac{\partial v_y}{\partial y}}{\dfrac{\partial u_x}{\partial y}+\dfrac{\partial v_y}{\partial x}} = \frac{\sigma_x-\sigma_y}{2\tau_{xy}}. \tag{25}$$

These should be compared with the equations of Saint-Venant for an isotropic material ((4) and (7) of Chap. VI). If ψ is the orientation of a principal stress direction with respect to the x-axis, and ψ' is the orientation of a principal strain-rate direction,

$$\tan 2\psi' = (1-c)\tan 2\psi. \tag{26}$$

Hence $\psi' = \psi$ if and only if $\psi = 0°, 45°,$ or $90°$; this is a consequence of the fourfold symmetry of the angular variation of the tensile yield stress.

The equations (23) and (25) together with the equilibrium equations, between the five unknowns σ_x, σ_y, τ_{xy}, u_x, and v_y, involve only two parameters, namely T, which is a measure of the average resistance to deformation, and c, which specifies the state of anisotropy in the planes of flow. T and c can be experimentally determined by two measurements, for example in compression tests (under conditions of plane strain) at $0°$ and $45°$ to the axes of anisotropy. It is necessary to know the separate magnitudes of the four parameters F, G, H, N (or X, Y, Z, T) only when it is required to calculate σ_z.

(ii) *The existence of characteristics.* As in the theory of isotropic plane strain it may be shown that there exist curves (characteristics) across which certain derivatives of the stress and velocity components may be discontinuous under suitable boundary conditions. The proof is analogous to that in Section 3 of Chapter VI. It is found that the characteristics for the stresses and velocities are the same, and that they are the slip-lines, or directions of maximum shear strain-rate; these are not, in general, maximum shear-stress directions. If the slip-lines at some point P are taken as (ξ, η) axes of reference, we find (by the method leading to equation (9) of Chap. VI) that

$$\frac{\partial \sigma_\xi}{\partial \xi} = 0 = \frac{\partial \sigma_\eta}{\partial \eta} \tag{27}$$

at the point P, where σ_ξ, σ_η, $\tau_{\xi\eta}$ are the stress components referred to the (ξ, η) axes. Similarly, if u_ξ and v_η are the velocity components referred to these axes,

$$\frac{\partial u_\xi}{\partial \xi} = 0 = \frac{\partial v_\eta}{\partial \eta} \tag{28}$$

at the point P. The inclinations dy/dx of the slip-lines relative to the x-axis of anisotropy may easily be shown to be the roots of the equation

$$(\sigma_x - \sigma_y)(dx^2 - dy^2) + 4(1-c)\tau_{xy}\, dx\, dy = 0. \tag{29}$$

(iii) *Relations along the characteristics.* It is now assumed that the anisotropy is uniformly distributed in magnitude and direction. We introduce the stress components σ_α, σ_β, $\tau_{\alpha\beta}$, referred to the slip-lines as curvilinear axes, where the two families are distinguished by the convention that $\tau_{\alpha\beta}$ shall be a positive quantity, to preserve the analogy with the isotropic theory. If ϕ is the anti-clockwise orientation of an α-line to the x-axis, the yield criterion (23) becomes

$$f(\sigma_\alpha, \sigma_\beta, \tau_{\alpha\beta}) \equiv \frac{1}{(1-c)}[(\sigma_\alpha - \sigma_\beta)\cos 2\phi - 2\tau_{\alpha\beta}\sin 2\phi]^2 +$$
$$+ [(\sigma_\alpha - \sigma_\beta)\sin 2\phi + 2\tau_{\alpha\beta}\cos 2\phi]^2 = 4T^2,$$

on transforming the components of stress. Since the slip-lines are

XII. 7]   PLANE STRAIN FOR ANISOTROPIC METALS   337

directions of zero extension $\partial f/\partial \sigma_\alpha = -\partial f/\partial \sigma_\beta = 0$, or

$$\frac{\cos 2\phi}{(1-c)}[(\sigma_\alpha-\sigma_\beta)\cos 2\phi - 2\tau_{\alpha\beta}\sin 2\phi] + \\ + \sin 2\phi[(\sigma_\alpha-\sigma_\beta)\sin 2\phi + 2\tau_{\alpha\beta}\cos 2\phi] = 0.$$

Solving these two equations for $\sigma_\alpha - \sigma_\beta$ and $\tau_{\alpha\beta}$:

$$\left.\begin{aligned}\frac{\tau_{\alpha\beta}}{T} &= (1-c\sin^2 2\phi)^{\frac{1}{2}}, \\ \frac{\sigma_\alpha-\sigma_\beta}{T} &= \frac{2c\sin 2\phi \cos 2\phi}{(1-c\sin^2 2\phi)^{\frac{1}{2}}},\end{aligned}\right\} \quad (30)$$

where the positive square root is to be taken. It may be verified that

$$\frac{d\tau_{\alpha\beta}}{d\phi} + (\sigma_\alpha-\sigma_\beta) = 0.$$

Hence $\quad \sigma_\alpha = -p - \tfrac{1}{2}T\dfrac{dh}{d\phi}, \quad \sigma_\beta = -p + \tfrac{1}{2}T\dfrac{dh}{d\phi}, \quad \tau_{\alpha\beta} = Th,$ $\bigg\}$ (31)

where $\quad h(\phi) = (1-c\sin^2 2\phi)^{\frac{1}{2}},$

and $p = -\tfrac{1}{2}(\sigma_\alpha+\sigma_\beta)$ is the mean compressive stress in the plane.

To express the relations (27) along the slip-lines in terms of the variation of p and ϕ, we transform the stress components according to the equations

$$\sigma_\xi = -p + \tfrac{1}{2}(\sigma_\alpha-\sigma_\beta)\cos 2(\phi-\phi_0) - \tau_{\alpha\beta}\sin 2(\phi-\phi_0),$$
$$\sigma_\eta = -p - \tfrac{1}{2}(\sigma_\alpha-\sigma_\beta)\cos 2(\phi-\phi_0) + \tau_{\alpha\beta}\sin 2(\phi-\phi_0),$$

where ϕ_0 is the value of ϕ at the point P under consideration. Substituting in (27), and using (31), we obtain

$$\left[\frac{\partial}{\partial \xi}\left\{p + \tfrac{1}{2}T\frac{dh}{d\phi}\cos 2(\phi-\phi_0) + Th\sin 2(\phi-\phi_0)\right\}\right]_{\phi=\phi_0} = 0,$$

$$\left[\frac{\partial}{\partial \eta}\left\{p - \tfrac{1}{2}T\frac{dh}{d\phi}\cos 2(\phi-\phi_0) - Th\sin 2(\phi-\phi_0)\right\}\right]_{\phi=\phi_0} = 0.$$

Hence

$$\left(\frac{\partial p}{\partial \xi}\right)_P + \left(\tfrac{1}{2}T\frac{d^2h}{d\phi^2}+2Th\right)\left(\frac{\partial \phi}{\partial \xi}\right)_P = 0,$$

$$\left(\frac{\partial p}{\partial \eta}\right)_P - \left(\tfrac{1}{2}T\frac{d^2h}{d\phi^2}+2Th\right)\left(\frac{\partial \phi}{\partial \eta}\right)_P = 0.$$

Since P is a general point,

$$\left.\begin{aligned}\frac{p}{2T}+g &= \text{constant on an } \alpha\text{-line}, \\ \frac{p}{2T}-g &= \text{constant on a } \beta\text{-line},\end{aligned}\right\} \quad (32)$$

where
$$g(\phi) = \int_0^\phi \left(\frac{1}{4}\frac{d^2h}{d\phi^2}+h\right) d\phi = \frac{1}{4}\left[\frac{dh}{d\phi}\right]_0^\phi + \int_0^\phi h\, d\phi.$$

Inserting the value (31) of $h(\phi)$, we immediately obtain

$$g(\phi) = -\frac{\tfrac{1}{2}c\sin 2\phi \cos 2\phi}{(1-c\sin^2 2\phi)^{\frac{1}{2}}} + \tfrac{1}{2}E(u,k), \qquad (33)$$

where
$$\sin 2\phi = \operatorname{sn}(u,k); \qquad k^2 = c.$$

$\operatorname{sn}(u,k)$ is the Jacobian elliptic function with modulus k (which takes complex values when c is negative), and $E(u,k)$ is the standard elliptic integral of the second kind:†

$$E(u,k) = \int_0^u \operatorname{dn}^2(u,k)\, du = \int_0^{2\phi} (1-k^2\sin^2\theta)^{\frac{1}{2}}\, d\theta.$$

For values of ϕ in the range $(-\tfrac{1}{4}\pi, \tfrac{1}{4}\pi)$ u lies in the range $(-K, K)$, where K is the quarter-period of the elliptic function. When ϕ lies in the range $(\tfrac{1}{4}\pi, \tfrac{1}{2}\pi)$ u must be taken in the range $(K, 2K)$, and so on for other values of ϕ. When $c = 0$, equations (32) reduce to the Hencky relations, (12) of Chapter VI. If u and v are the components of velocity along the α and β slip-lines, equations (28) transform to

$$\left. \begin{array}{l} du - v\, d\phi = 0 \text{ along an } \alpha\text{-line,} \\ dv + u\, d\phi = 0 \text{ along a } \beta\text{-line,} \end{array} \right\} \qquad (34)$$

exactly as in the isotropic theory (equations (14) of Chap. VI).

(iv) *Properties of the slip-line field.* We now examine whether geometrical theorems, analogous to those of Hencky, exist in the anisotropic theory. It is immediately evident, by the method used to prove Hencky's first theorem, that the difference in the values of g (or p), where two given slip-lines of one family are cut by any member of the other family, is a constant. Conversely, any two orthogonal families of curves possessing this property constitute a slip-line field for a plastic state under certain boundary conditions. In Hencky's theorem it is the difference in the values of ϕ that is constant along two given slip-lines. This is not true for anisotropic metals except when the difference is $\tfrac{1}{2}\pi$.

If, now, a section of an α-line, say, is straight, ϕ is constant along it; hence g and p are also constant along the section. It follows from the previous result that the corresponding sections of all α-lines are also straight, and so, as a simple consequence, that they are of equal length.

† Tabulated in E. Jahnke and F. Emde, *Funktionentafeln*, 2nd Edition, p. 141 (Teubner, Leipzig, 1933).

An example of such a field is that consisting of radii and concentric circular arcs, whose common centre is a point singularity for the stress distribution.

The analogy of Hencky's second theorem is too complicated to be useful.

(v) *Indentation by a flat rigid die.* It is evident that the present theory is only applicable throughout a process of plastic deformation so long as the state of anisotropy does not change appreciably, or changes in such a way as to remain uniformly distributed. The indentation of the

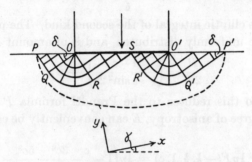

FIG. 90. Slip-line field in a semi-infinite anisotropic medium indented by a flat die.

plane surface of a block of metal by a flat rigid die satisfies the first condition. At the moment when indentation becomes possible it is assumed, in accordance with the discussion in Sections 3 and 5 of Chapter IX, that the plastic zone covers the area between the broken curve and the section PP' of the surface (Fig. 90). The state of stress is then uniquely determined within the triangles OPQ and $O'P'Q'$ formed by the intersecting pairs of slip-lines through O, P, and O', P', respectively. The state of stress in each of these regions is a uniform compression parallel to the surface. Let γ ($0 \leqslant \gamma \leqslant \tfrac{1}{2}\pi$) be the angle between the surface and the x-axis of anisotropy (in the sense indicated in Fig. 90), and let angles POQ and $O'P'Q'$ be denoted by δ. Then according to (24) and (29),

$$p_0 = T\left(\frac{1-c}{1-c\sin^2 2\gamma}\right)^{\tfrac{1}{2}}, \qquad \delta = \gamma + \tfrac{1}{2}\cot^{-1}\{(1-c)\tan 2\gamma\}, \qquad (35)$$

where the inverse cotangent is an angle in the interval $(-\tfrac{1}{2}\pi, \tfrac{1}{2}\pi)$. It is easy to show that, regardless of the value of γ, δ lies between $\cot^{-1}\sqrt{(1-c)}$ and $\tfrac{1}{2}\pi - \cot^{-1}\sqrt{(1-c)}$.

The slip-lines in the regions OQR and $O'Q'R'$, defined respectively by the singularities O and O' and the positions of the slip-lines OQ and

$O'Q'$, are radii and concentric circular arcs. The positions of OR and $O'R'$ are determined by the condition that the surface of the die is frictionless. The regions ORS and $O'R'S$ are therefore uniformly stressed, the principal axes of stress being parallel and perpendicular to the surface. Hence the orientation of the slip-lines is the same as in regions OPQ and $O'P'Q'$, the angles QOR and $Q'O'R'$ being $\frac{1}{4}\pi$. According to (32) and (33), the value of p in ORS and $O'R'S$ is $p_0 + 2TE$, where
$$E = E(K, k) = \int_0^{\frac{1}{2}\pi} (1 - k^2 \sin^2\theta)^{\frac{1}{2}} d\theta$$
is the complete elliptic integral of the second kind. The pressure on the die is therefore uniformly distributed, and is of amount P, where
$$\frac{P}{2T} = \left(\frac{1-c}{1-c\sin^2 2\gamma}\right)^{\frac{1}{2}} + E. \tag{36}$$
When c is zero this reduces to the Prandtl formula $P = 2T(1 + \frac{1}{2}\pi)$. For a small degree of anisotropy, E can conveniently be calculated from the series
$$E = \tfrac{1}{2}\pi F(-\tfrac{1}{2}, \tfrac{1}{2}, 1, c) = \tfrac{1}{2}\pi \left(1 - \frac{c}{4} - \frac{3c^2}{64} - \frac{5c^3}{256} - \ldots\right), \tag{37}$$
where F is the hypergeometric function. The series expansion for P is
$$\frac{P}{2T} = 1 + \tfrac{1}{2}\pi - \frac{c}{2}(1 + \tfrac{1}{4}\pi + \sin^2 2\gamma)\ldots \,. \tag{38}$$
Tables for E as a function of c or k^2 are available† for the calculation of P for finite degrees of anisotropy. If c is positive, P is less than
$$2T(1 + \tfrac{1}{2}\pi);$$
if c is negative, P is greater than $2T(1 + \tfrac{1}{2}\pi)$. Furthermore, P and the slip-line field are the same for orientations γ and $\tfrac{1}{2}\pi - \gamma$ of the axes of anisotropy; this is due to the symmetry of the anisotropy about directions making 45° with the axes of anisotropy.

The position of the point S, hitherto not specified, depends upon the state of stress in the non-plastic material, for it is this that controls the development of the plastic zone. As in the isotropic theory the flow streamlines coincide with the slip-lines parallel to $PQRS$ and $P'Q'R'S$, respectively, and the plastic material below these boundaries is constrained to remain rigid. The velocity is of magnitude $V \sec\delta$ in $O'P'Q'R'S$, and of magnitude $V \cosec\delta$ in $OPQRS$, where V is the downward speed of the die.

† E. Jahnke and F. Emde, *Funktionentafeln*, 2nd edition, p. 150 (Teubner, Leipzig, 1933).

APPENDIX I
SUFFIX NOTATION AND THE SUMMATION CONVENTION

(i) *Suffix notation.* In particular applications, the notation used in this book for the nine components of stress at a point is σ_x, σ_y, σ_z, τ_{yz}, τ_{zy}, τ_{zx}, τ_{xz}, τ_{xy}, τ_{yx}, referred to rectangular axes x, y, z; only six components are independent since $\tau_{yz} = \tau_{zy}$, etc. The components of an increment of strain are denoted by $d\epsilon_x$, $d\gamma_{yz}$, etc. In discussions of relations between stress and strain-increment the normal components of stress are denoted by σ_{11}, σ_{22}, σ_{33}, where $\sigma_{11} \equiv \sigma_x$, etc., and the shear components by σ_{23}, σ_{31}, σ_{12}, where $\sigma_{23} \equiv \tau_{yz}$, etc. A general component is denoted by σ_{ij}, where letters are used as suffixes instead of numbers; a particular component is obtained by giving i and j the appropriate values 1, 2, or 3; any other letters can, of course, be used instead of i and j. For brevity a stress is referred to simply by writing its general component. Similarly, the components of an increment of strain are denoted by $d\epsilon_{ij}$ $(i,j = 1, 2, 3)$.

(ii) *Summation convention.* In a sum such as

$$\sum_i \sigma_{ii} = \sigma_{11} + \sigma_{22} + \sigma_{33},$$

taken over the values $i = 1, 2, 3$, the summation sign is omitted and the whole expression is written simply σ_{ii}. This is the summation convention, according to which a recurring *letter* suffix indicates that the sum must be formed of all terms obtainable by assigning to the suffix the values 1, 2, and 3. The convention does not apply to numerical suffixes. Similarly, in a quantity containing two repeated suffixes, such as $\sigma_{ij} d\epsilon_{ij}$, the summation must be carried out for all values 1, 2, 3 of both i and j. Thus,

$$\sigma_{ij} d\epsilon_{ij} = \sigma_{11} d\epsilon_{11} + \sigma_{22} d\epsilon_{22} + \sigma_{33} d\epsilon_{33} + 2(\sigma_{23} d\epsilon_{23} + \sigma_{31} d\epsilon_{31} + \sigma_{12} d\epsilon_{12}),$$

where the symmetric property $\sigma_{ij} = \sigma_{ji}$, $d\epsilon_{ij} = d\epsilon_{ji}$, has been used. Repeated letter suffixes are called 'dummy' suffixes, and non-repeated ones 'free' suffixes. For example, in

$$\sigma_{ij} \sigma_{jk} = \sigma_{i1} \sigma_{1k} + \sigma_{i2} \sigma_{2k} + \sigma_{i3} \sigma_{3k}$$

j is a dummy suffix, while i and k are free. Evidently a dummy suffix can be replaced by any other letter which does not occur elsewhere in the same term.

The use of brackets must be carefully noted. Thus,

$$(\sigma_{ii})^2 = (\sigma_{11} + \sigma_{22} + \sigma_{33})^2$$

but

$$\sigma_{ii}^2 = \sigma_{11}^2 + \sigma_{22}^2 + \sigma_{33}^2.$$

Summation must be carried out inside a bracket before any other operation is performed.

In an equation such as

$$p_{ij} q_{jk} = r_{ik}$$

the free suffixes (i and k) must be the same on both sides, and the equation is to be understood to hold for all values of the free suffixes. The above expression stands therefore for nine equations of the type

$$p_{11} q_{11} + p_{12} q_{21} + p_{13} q_{31} = r_{11} \quad (i = 1, k = 1),$$
$$p_{21} q_{13} + p_{22} q_{23} + p_{23} q_{33} = r_{23} \quad (i = 2, k = 3).$$

Summation is also to be understood in an expression involving a derivative, such as $\partial u_i/\partial x_i$, which stands for

$$\frac{\partial u_1}{\partial x_1}+\frac{\partial u_2}{\partial x_2}+\frac{\partial u_3}{\partial x_3}.$$

Similarly, the equations of equilibrium

$$\frac{\partial \sigma_{11}}{\partial x_1}+\frac{\partial \sigma_{12}}{\partial x_2}+\frac{\partial \sigma_{13}}{\partial x_3}=0,$$

$$\frac{\partial \sigma_{21}}{\partial x_1}+\frac{\partial \sigma_{22}}{\partial x_2}+\frac{\partial \sigma_{23}}{\partial x_3}=0,$$

$$\frac{\partial \sigma_{31}}{\partial x_1}+\frac{\partial \sigma_{32}}{\partial x_2}+\frac{\partial \sigma_{33}}{\partial x_3}=0,$$

can be written as
$$\frac{\partial \sigma_{ij}}{\partial x_j}=0.$$

Here summation is over the dummy suffix j, and the equation holds for all values 1, 2, 3 of the free suffix i.

(iii) *The delta symbol.* It is occasionally advantageous to use a symbol δ_{ij} which, by definition, is equal to unity when $i=j$, and to zero when $i \neq j$. For example, the deviatoric or reduced stress, which is the part of the stress σ_{ij} remaining after the hydrostatic tension $\frac{1}{3}(\sigma_{11}+\sigma_{22}+\sigma_{33})$ has been removed, has components

$$\sigma_{11}-\tfrac{1}{3}(\sigma_{11}+\sigma_{22}+\sigma_{33}), \qquad \sigma_{23},$$

and two other similar pairs obtained by cyclic permutation of the suffixes. These components can be written as $\sigma_{ij}-\frac{1}{3}\sigma_{kk}$ when $i=j$ and as σ_{ij} when $i\neq j$, or, still more shortly and uniformly, as $\sigma_{ij}-\frac{1}{3}\delta_{ij}\sigma_{kk}$.

(iv) *Tensors.* A Cartesian tensor of the kind occurring in this book is a quantity comprising nine components p_{ij} ($i,j=1,2,3$). It is implied that there exists some rule (depending on the particular field of application) whereby definite components may be associated with any given set of three-dimensional Cartesian axes x_i through the point under consideration. The three components associated with the ith axis are p_{ij} ($j=1,2,3$). The nine components can be visualized as a 3×3 array in which p_{ij} is the element in the ith row and jth column. A tensor is said to be symmetrical when $p_{ij}=p_{ji}$. The defining property of a tensor is that its components p_{ij} and p'_{ij}, corresponding to *any* two different sets of axes x_i and x'_i, are related by the equations

$$p'_{ij}=a_{ik}a_{jl}p_{kl}. \tag{1}$$

Here a_{ik} ($k=1,2,3$) are the direction-cosines of the x'_i-axis with respect to the x set of axes; the direction-cosines of the x_i-axis with respect to the x' set are therefore a_{ki} ($k=1,2,3$). The direction-cosines satisfy the well-known geometrical relations
$$a_{ik}a_{jk}=\delta_{ij}, \qquad a_{ki}a_{kj}=\delta_{ij}. \tag{2}$$

Equations (1) are precisely the equations of transformation between stress components referred to different sets of axes. In the usual notation these are

$$\sigma=\lambda^2\sigma_x+\mu^2\sigma_y+\nu^2\sigma_z+2(\mu\nu\tau_{yz}+\nu\lambda\tau_{zx}+\lambda\mu\tau_{xy}),$$

$$\tau=\lambda'\lambda''\sigma_x+\mu'\mu''\sigma_y+\nu'\nu''\sigma_z+(\mu'\nu''+\mu''\nu')\tau_{yz}+(\nu'\lambda''+\nu''\lambda')\tau_{zx}+(\lambda'\mu''+\lambda''\mu')\tau_{xy},$$

where σ is the normal stress component in the direction (λ,μ,ν) and τ is the shear

SUFFIX NOTATION AND THE SUMMATION CONVENTION

component with respect to the directions (λ', μ', ν') and $(\lambda'', \mu'', \nu'')$. Thus the nine components of stress (and similarly of strain-increment) constitute a tensor which, moreover, is symmetrical since $\sigma_{ij} = \sigma_{ji}$ (and $d\epsilon_{ij} = d\epsilon_{ji}$). In this notation equations (2) are of the type

$$\lambda^2 + \mu^2 + \nu^2 = 1, \qquad \lambda'\lambda'' + \mu'\mu'' + \nu'\nu'' = 0.$$

The following theorem on tensors is drawn on in Chapters II and III. If the scalar quantity $f(p_{ij})$ is a function of the components of a tensor p_{ij}, then the assembly of derivatives $q_{ij} = \partial f / \partial p_{ij}$ constitute a tensor, its components referred to another set of axes x'_i being formed according to the rule $q'_{ij} = \partial f(p_{ij})/\partial p'_{ij}$. Now

$$\frac{\partial f(p_{ij})}{\partial p'_{ij}} = \frac{\partial f}{\partial p_{kl}} \frac{\partial p_{kl}}{\partial p'_{ij}} = \frac{\partial f}{\partial p_{kl}} a_{ik} a_{jl},$$

since

$$p_{kl} = a_{ik} a_{jl} p'_{ij}.$$

Hence

$$q'_{ij} = a_{ik} a_{jl} \frac{\partial f}{\partial p_{kl}} = a_{ik} a_{jl} q_{kl},$$

and the theorem is proved.

(v) *Invariants of a tensor.* A function of tensor components which retains the same form when any other axes of reference are taken is known as an invariant. Thus a function $f(p_{ij})$ is invariant if

$$f(p_{ij}) = f(a_{ki} a_{lj} p'_{kl}) \equiv f(p'_{ij}).$$

It may be proved that tensors of the kind considered here have only three *independent* invariants, any other invariant function being expressible in terms of them. The simplest independent invariants are

$$p_{ii}, \quad p_{ij} p_{ji}, \quad p_{ij} p_{jk} p_{ki},$$

or any scalar multiples of them. Taking these in turn, the proofs of invariance are as follows. From (2) we have

$$p'_{ii} = a_{ik} a_{il} p_{kl} = \delta_{kl} p_{kl} = p_{kk} = p_{ii},$$

by the definition of the delta symbol and the interchangeability of dummy suffixes. Similarly,

$$p'_{ij} p'_{ji} = (a_{ik} a_{jl} p_{kl})(a_{jm} a_{in} p_{mn}) = (a_{ik} a_{in} p_{kl})(a_{jl} a_{jm} p_{mn})$$
$$= (\delta_{kn} p_{kl})(\delta_{lm} p_{mn}) = p_{nl} p_{ln} = p_{ij} p_{ji};$$
$$p'_{ij} p'_{jk} p'_{ki} = (a_{il} a_{jm} p_{lm})(a_{jn} a_{kr} p_{nr})(a_{ks} a_{it} p_{st})$$
$$= (a_{il} a_{it} p_{lm})(a_{jm} a_{jn} p_{nr})(a_{kr} a_{ks} p_{st})$$
$$= (\delta_{lt} p_{lm})(\delta_{mn} p_{nr})(\delta_{rs} p_{st})$$
$$= p_{tm} p_{mr} p_{rt} = p_{ij} p_{jk} p_{ki}.$$

The summation convention is virtually indispensable in proofs such as these, not only in saving time and space but also in displaying the structure of the equations and suggesting the correct grouping of the terms.

APPENDIX II
COORDINATE SYSTEMS

(i) *Cylindrical coordinates* (r, θ, z). The equations of equilibrium (body-forces being disregarded) are

$$\frac{\partial \sigma_r}{\partial r} + \frac{1}{r}\frac{\partial \tau_{r\theta}}{\partial \theta} + \frac{\partial \tau_{rz}}{\partial z} + \frac{\sigma_r - \sigma_\theta}{r} = 0,$$

$$\frac{\partial \tau_{r\theta}}{\partial r} + \frac{1}{r}\frac{\partial \sigma_\theta}{\partial \theta} + \frac{\partial \tau_{\theta z}}{\partial z} + \frac{2\tau_{r\theta}}{r} = 0,$$

$$\frac{\partial \tau_{rz}}{\partial r} + \frac{1}{r}\frac{\partial \tau_{\theta z}}{\partial \theta} + \frac{\partial \sigma_z}{\partial z} + \frac{\tau_{rz}}{r} = 0,$$

where $\sigma_r, \sigma_\theta, \sigma_z, \tau_{\theta z}, \tau_{rz}, \tau_{r\theta}$ are the stress components referred to these coordinates.

The components of the strain-rate tensor are

$$\dot{\epsilon}_r = \frac{\partial u}{\partial r}, \qquad \dot{\gamma}_{\theta z} = \frac{1}{2}\left(\frac{\partial v}{\partial z} + \frac{1}{r}\frac{\partial w}{\partial \theta}\right),$$

$$\dot{\epsilon}_\theta = \frac{1}{r}\left(\frac{\partial v}{\partial \theta} + u\right), \qquad \dot{\gamma}_{rz} = \frac{1}{2}\left(\frac{\partial u}{\partial z} + \frac{\partial w}{\partial r}\right),$$

$$\dot{\epsilon}_z = \frac{\partial w}{\partial z}, \qquad \dot{\gamma}_{r\theta} = \frac{1}{2}\left(\frac{\partial v}{\partial r} - \frac{v}{r} + \frac{1}{r}\frac{\partial u}{\partial \theta}\right),$$

where u, v, w are the components of velocity in the radial, circumferential, and axial directions, respectively.

(ii) *Spherical polar coordinates* (r, θ, ϕ). θ is the angle between a radius and the positive z-axis; ϕ is measured round the z-axis in a right-handed sense. The equations of equilibrium (body-forces being disregarded) are

$$\frac{\partial \sigma_r}{\partial r} + \frac{1}{r}\frac{\partial \tau_{r\theta}}{\partial \theta} + \frac{1}{r\sin\theta}\frac{\partial \tau_{r\phi}}{\partial \phi} + \frac{1}{r}(2\sigma_r - \sigma_\theta - \sigma_\phi + \tau_{r\theta}\cot\theta) = 0,$$

$$\frac{\partial \tau_{r\theta}}{\partial r} + \frac{1}{r}\frac{\partial \sigma_\theta}{\partial \theta} + \frac{1}{r\sin\theta}\frac{\partial \tau_{\theta\phi}}{\partial \phi} + \frac{1}{r}\{(\sigma_\theta - \sigma_\phi)\cot\theta + 3\tau_{r\theta}\} = 0,$$

$$\frac{\partial \tau_{r\phi}}{\partial r} + \frac{1}{r}\frac{\partial \tau_{\theta\phi}}{\partial \theta} + \frac{1}{r\sin\theta}\frac{\partial \sigma_\phi}{\partial \phi} + \frac{1}{r}(3\tau_{r\phi} + 2\tau_{\theta\phi}\cot\theta) = 0,$$

where $\sigma_r, \sigma_\theta, \sigma_\phi, \tau_{\theta\phi}, \tau_{r\phi}, \tau_{r\theta}$ are the stress components referred to these coordinates.

The components of the strain-rate tensor are

$$\dot{\epsilon}_r = \frac{\partial u}{\partial r}, \qquad \dot{\gamma}_{\theta\phi} = \frac{1}{2r\sin\theta}\left(\sin\theta\frac{\partial w}{\partial \theta} - w\cos\theta + \frac{\partial v}{\partial \phi}\right),$$

$$\dot{\epsilon}_\theta = \frac{1}{r}\left(\frac{\partial v}{\partial \theta} + u\right), \qquad \dot{\gamma}_{r\phi} = \frac{1}{2}\left(\frac{1}{r\sin\theta}\frac{\partial u}{\partial \phi} + \frac{\partial w}{\partial r} - \frac{w}{r}\right),$$

$$\dot{\epsilon}_\phi = \frac{1}{r\sin\theta}\left(\frac{\partial w}{\partial \phi} + u\sin\theta + v\cos\theta\right), \qquad \dot{\gamma}_{r\theta} = \frac{1}{2}\left(\frac{\partial v}{\partial r} - \frac{v}{r} + \frac{1}{r}\frac{\partial u}{\partial \theta}\right),$$

where u, v, w are the components of velocity in the r, θ, ϕ directions, respectively.

APPENDIX III
CHARACTERISTICS

We begin with the first-order linear equation

$$P\frac{\partial z}{\partial x} + Q\frac{\partial z}{\partial y} = R, \qquad (1)$$

where z is the unknown dependent variable and x, y are the independent variables, and where P, Q, R are continuous single-valued functions of x, y, z. The variation of z in the direction dy/dx in the (x, y) plane is

$$dz = \frac{\partial z}{\partial x} dx + \frac{\partial z}{\partial y} dy. \qquad (2)$$

Consider the family of curves with slope

$$\frac{dx}{P} = \frac{dy}{Q}, \qquad (3)$$

one, and only one, of which passes through every point (except possibly a singularity of z). The variation of z along a curve of this family is

$$dz = \left(P\frac{\partial z}{\partial x} + Q\frac{\partial z}{\partial y}\right)\frac{dx}{P} = \frac{R}{P} dx.$$

Hence
$$\frac{dx}{P} = \frac{dy}{Q} = \frac{dz}{R}. \qquad (4)$$

These equations, a statement of the variation of z along the family of curves (3), are completely equivalent to the original differential equation (1). The curves (3) are known as *characteristics*.

If the value of z is given at one point O in the (x, y) plane, its values are uniquely determined at all points along the characteristic through O. In a numerical solution we simply replace (4) by the finite difference relations

$$\frac{\Delta x}{\bar{P}} = \frac{\Delta y}{\bar{Q}} = \frac{\Delta z}{\bar{R}},$$

where $\Delta x, \Delta y, \Delta z$ are small increments between two neighbouring points on the characteristic and $\bar{P}, \bar{Q}, \bar{R}$ are mean values over the interval. If P and Q do not involve z, the characteristic through O can be calculated first and the values of z afterwards; if z enters into the functions P and Q the calculations of z and the characteristic have to be carried out together. In the first case the shape of the characteristic does not depend on the given value of z at O; in the second case it does. Similarly, if values of z are assigned along a section JK of some curve Σ, which is not a characteristic, the values of z are uniquely determined at all points within the strip of the (x, y) plane covered by characteristics cutting JK. (It is evident that JK must be such that no characteristic cuts it twice since the value of z can only be *arbitrarily* assigned at one point of a characteristic; for example, there is no solution of (1) taking arbitrary values on a closed curve.) In particular, if the boundary values in the neighbourhood of a point O on JK are slightly altered, the solution is only affected in the corresponding neighbourhood of the characteristic through O; we may say therefore that small disturbances are propagated along characteristics.

The boundary values of z on JK need not necessarily be continuous, unless this is required by extraneous considerations in the particular problem. Also, the boundary values may be such that the space derivative of z along JK is discontinuous at some point O. If this happens, the discontinuity in derivative is propagated along the characteristic C through O; in other words, the derivative in any direction except along C has different values on opposite sides of C. It will be seen that this possibility stems directly from the circumstance that the original differential equation is equivalent to a specification of the variation of z only along the characteristics. It follows, also, that if the region of the (x, y) plane covered by the solution consists of several domains in each of which z is analytic (i.e., possesses continuous derivatives) then the domains can adjoin only along characteristics. For these reasons the field of characteristics must be regarded as the fundamental element to be calculated.

When dealing with more complicated equations it is convenient to take the basic defining property of a characteristic to be that it is a curve across which certain derivatives may be discontinuous under suitable boundary conditions. To investigate the characteristics of (1) from this standpoint, suppose that values of z are given along some curve C, so that the differential dz is known at all points on C. There are then two equations, (1) and (2), available for the calculation of $\partial z/\partial x$ and $\partial z/\partial y$. In general the equations have a unique solution and the derivatives are then identical on both sides of C. However, when the determinant

$$\begin{vmatrix} dx & dy \\ P & Q \end{vmatrix}$$

vanishes, the equations do not have a unique solution and the derivatives cannot be evaluated without additional information; this may be such that the derivatives are discontinuous across C. The determinant vanishes at all points of the curve when C is one of the family (3), which are therefore, by definition, the characteristics. The condition for the compatibility of (1) and (2) is then

$$\begin{vmatrix} dx & dz \\ P & R \end{vmatrix} = 0,$$

which is just the equation (4), governing the variation of z along a characteristic.

Consider, now, two simultaneous first-order equations

$$\left.\begin{aligned} P\frac{\partial u}{\partial x}+Q\frac{\partial u}{\partial y}+R\frac{\partial v}{\partial x}+S\frac{\partial v}{\partial y} &= T, \\ P'\frac{\partial u}{\partial x}+Q'\frac{\partial u}{\partial y}+R'\frac{\partial v}{\partial x}+S'\frac{\partial v}{\partial y} &= T', \end{aligned}\right\} \tag{5}$$

where P, P' etc., are functions of x, y, u, v. Suppose that u and v are given along some curve C, so that the differentials

$$du = \frac{\partial u}{\partial x}dx + \frac{\partial u}{\partial y}dy, \qquad dv = \frac{\partial v}{\partial x}dx + \frac{\partial v}{\partial y}dy,$$

are known on C. The first derivatives are uniquely determined by the above four equations at any point on C unless

$$\begin{vmatrix} dx & dy & 0 & 0 \\ 0 & 0 & dx & dy \\ P & Q & R & S \\ P' & Q' & R' & S' \end{vmatrix} = 0. \tag{6}$$

This is a quadratic form in dx and dy, having two roots dy/dx.

If the roots are real and distinct, there are two families of characteristics, a member of each family passing through any point. The equations are then said to be *hyperbolic*, and they may be reduced to a pair of differential relations expressing the variation of u and v along the characteristics. These are solved numerically by replacing them by difference relations across a network of characteristics. If values of u and v are given along some section JK of a curve Σ, which is not a characteristic, it may be proved that u and v are uniquely determined in the curvilinear quadrilateral bounded by the two pairs of intersecting characteristics through J and K. An example of hyperbolic equations is

$$\frac{\partial u}{\partial x} = \frac{\partial v}{\partial y}, \quad \frac{\partial v}{\partial x} = \frac{\partial u}{\partial y},$$

of which the characteristics are $dy^2 - dx^2 = 0$, that is, the lines $x \pm y =$ constant. Along the family $x+y =$ constant,

$$du = \frac{\partial u}{\partial x}dx + \frac{\partial u}{\partial y}dy = dx\left(\frac{\partial u}{\partial x} - \frac{\partial u}{\partial y}\right) = dx\left(\frac{\partial u}{\partial x} - \frac{\partial v}{\partial x}\right),$$

$$dv = \frac{\partial v}{\partial x}dx + \frac{\partial v}{\partial y}dy = dx\left(\frac{\partial v}{\partial x} - \frac{\partial v}{\partial y}\right) = dx\left(\frac{\partial v}{\partial x} - \frac{\partial u}{\partial x}\right),$$

whence $du + dv = 0.$

Thus, $u+v$ is constant along each line $x+y =$ constant, and so $u+v = 2f(x+y)$, say, where f is an arbitrary function to be determined by the boundary conditions. Similarly, $u-v = 2g(x-y)$, and the final solution is

$$\left.\begin{matrix}u\\v\end{matrix}\right\} = f(x+y) \pm g(x-y).$$

In this example the (x, y) equations of the characteristics could be expressed explicitly, but this, of course, is not usually possible.

If the roots of (6) are complex the equations (5) are said to be *elliptic*. An example is

$$\frac{\partial u}{\partial x} = \frac{\partial v}{\partial y}, \quad \frac{\partial v}{\partial x} = -\frac{\partial u}{\partial y}.$$

Since $\quad \dfrac{\partial^2 u}{\partial x^2} = \dfrac{\partial^2 v}{\partial x \partial y} = -\dfrac{\partial^2 u}{\partial y^2} \quad$ and $\quad \dfrac{\partial^2 v}{\partial x^2} = -\dfrac{\partial^2 u}{\partial x \partial y} = -\dfrac{\partial^2 v}{\partial y^2},$

u and v are related solutions of Laplace's equation. It will be realized from this that the boundary-value problems and methods of solution of hyperbolic and elliptic equations are completely different (in the real plane).

If the roots of (6) are equal at all points the equations (5) are said to be *parabolic*. An example is

$$\frac{\partial u}{\partial x} = v, \quad \frac{\partial v}{\partial x} = u,$$

the characteristics being lines parallel to the x-axis.

The same method is used for finding whether characteristics exist for a set of more than two simultaneous linear equations. The labour of expanding the determinant can be reduced by the following artifice. Let $\phi \equiv \phi(x, y)$ and $\psi \equiv \psi(x, y)$ be the parametric equations of two families of curves in the (x, y) plane. Taking equations (5) as an example, u and v may be considered as functions of ϕ and ψ instead of x and y. Suppose, now, that u and v are given along one of the curves

ϕ = constant, so that $\partial u/\partial \psi$ and $\partial v/\partial \psi$ are also known on this curve. Equations (5) can therefore be written in the form:

$$\left(P\frac{\partial \phi}{\partial x}+Q\frac{\partial \phi}{\partial y}\right)\frac{\partial u}{\partial \phi}+\left(R\frac{\partial \phi}{\partial x}+S\frac{\partial \phi}{\partial y}\right)\frac{\partial v}{\partial \phi}+\text{known terms} = T,$$

$$\left(P'\frac{\partial \phi}{\partial x}+Q'\frac{\partial \phi}{\partial y}\right)\frac{\partial u}{\partial \phi}+\left(R'\frac{\partial \phi}{\partial x}+S'\frac{\partial \phi}{\partial y}\right)\frac{\partial v}{\partial \phi}+\text{known terms} = T'.$$

The derivatives $\partial u/\partial \phi$ and $\partial v/\partial \phi$, and hence the derivatives in any direction, are uniquely determined unless

$$\begin{vmatrix} P\dfrac{\partial \phi}{\partial x}+Q\dfrac{\partial \phi}{\partial y} & R\dfrac{\partial \phi}{\partial x}+S\dfrac{\partial \phi}{\partial y} \\ P'\dfrac{\partial \phi}{\partial x}+Q'\dfrac{\partial \phi}{\partial y} & R'\dfrac{\partial \phi}{\partial x}+S'\dfrac{\partial \phi}{\partial y} \end{vmatrix} = 0. \quad (7)$$

The equivalence of (6) and (7) can easily be verified, remembering that

$$d\phi = \frac{\partial \phi}{\partial x}dx+\frac{\partial \phi}{\partial y}dy = 0 \quad \text{or} \quad dy/dx = -\frac{\partial \phi}{\partial x}\bigg/\frac{\partial \phi}{\partial y}$$

along a characteristic ϕ = constant. The order of the determinant is halved by this method.

We come finally to the linear second-order equation

$$A\frac{\partial^2 z}{\partial x^2}+2B\frac{\partial^2 z}{\partial x \partial y}+C\frac{\partial^2 z}{\partial y^2} = D\left(x,y,z,\frac{\partial z}{\partial x},\frac{\partial z}{\partial y}\right).$$

Appropriate boundary conditions are the specification of z and its first derivatives $\partial z/\partial x$, $\partial z/\partial y$, along some curve Σ. The variations

$$d\left(\frac{\partial z}{\partial x}\right) = \frac{\partial^2 z}{\partial x^2}dx+\frac{\partial^2 z}{\partial x \partial y}dy, \quad d\left(\frac{\partial z}{\partial y}\right) = \frac{\partial^2 z}{\partial x \partial y}dx+\frac{\partial^2 z}{\partial y^2}dy,$$

are therefore known along Σ. The three equations are sufficient to determine the second derivatives $\partial^2 z/\partial x^2$, $\partial^2 z/\partial x \partial y$, $\partial^2 z/\partial y^2$ uniquely unless

$$\begin{vmatrix} dx & dy & 0 \\ 0 & dx & dy \\ A & 2B & C \end{vmatrix} = 0$$

or

$$A\,dy^2-2B\,dxdy+C\,dx^2 = 0. \quad (8)$$

The equation is therefore hyperbolic (with two families of characteristics) if the roots of the quadratic are real and distinct ($B^2 > AC$), elliptic if they are complex ($B^2 < AC$), and parabolic if they are equal ($B^2 = AC$).

APPENDIX IV

TRANSFORMATION EQUATIONS FOR STRESS COMPONENTS IN A PLANE

LET $\sigma_x, \sigma_y, \tau_{xy}$ be the components of stress in the (x, y) plane. Let (ξ, η) be another pair of axes in the same plane, and let θ be the inclination of the ξ-axis to the x-axis, measured anti-clockwise. Then the (ξ, η) components of stress are related to the (x, y) components according to the equations

$$\sigma_\xi = \sigma_x \cos^2\theta + \sigma_y \sin^2\theta + 2\tau_{xy} \sin\theta \cos\theta,$$
$$\sigma_\eta = \sigma_x \sin^2\theta + \sigma_y \cos^2\theta - 2\tau_{xy} \sin\theta \cos\theta,$$
$$\tau_{\xi\eta} = -(\sigma_x - \sigma_y)\sin\theta \cos\theta + \tau_{xy}(\cos^2\theta - \sin^2\theta).$$

Alternatively, these may be written

$$\sigma_\xi = \tfrac{1}{2}(\sigma_x + \sigma_y) + \tfrac{1}{2}(\sigma_x - \sigma_y)\cos 2\theta + \tau_{xy} \sin 2\theta,$$
$$\sigma_\eta = \tfrac{1}{2}(\sigma_x + \sigma_y) - \tfrac{1}{2}(\sigma_x - \sigma_y)\cos 2\theta - \tau_{xy} \sin 2\theta,$$
$$\tau_{\xi\eta} = -\tfrac{1}{2}(\sigma_x - \sigma_y)\sin 2\theta + \tau_{xy} \cos 2\theta.$$

The following Table gives the Cartesian coordinates of nodal points in a slip-line field, with a 15° mesh, defined by two intersecting equal circular arcs of unit radius (Fig. 30). The coordinates are referred to rectangular axes (x', y') passing through 0 and directed along, and perpendicular to, the axis of symmetry, and such that

$$\sqrt{2}x' = x+y, \qquad \sqrt{2}y' = y-x,$$

where x and y are as in Fig. 30. Since the values of α and β at nodal points are integer multiples of 15°, we can write

$$\alpha° = -m \times 15°, \qquad \beta° = n \times 15°,$$

and refer to a particular point by a pair of positive integers (m, n). This is the convention adopted in the Table. The computation was carried out for a 5° mesh and it is thought that the error nowhere exceeds one unit in the third figure; the accuracy is likely to be least good in x' for large values of m and n.

TABLE 1

(m, n)	x'	y'	(m, n)	x'	y'
(1, 1)	0·428	0	(3, 3)	1·867	0
(1, 2)	0·647	0·286	(3, 4)	2·338	0·615
(1, 3)	0·792	0·639	(3, 5)	2·673	1·429
(1, 4)	0·845	1·040	(3, 6)	2·804	2·414
(1, 5)	0·789	1·464			
(1, 6)	0·617	1·880	(4, 4)	3·13(4)	0
			(4, 5)	3·85	0·937
			(4, 6)	4·37	2·205
(2, 2)	1·018	0			
(2, 3)	1·333	0·413			
(2, 4)	1·552	0·944	(5, 5)	5·06	0
(2, 5)	1·634	1·571	(5, 6)	6·17	1·453
(2, 6)	1·545	2·256	(6, 6)	8·04	0

TABLE 2 (see Chap. XI, equation (43))

ψ degrees	ψ radians	λ radians	θ degrees	θ radians	λ radians
0	0	0	0	0	0
10	0·175	0·252	10	0·175	0·151
20	0·349	0·455	20	0·349	0·300
30	0·524	0·595	30	0·524	0·447
40	0·698	0·685	40	0·698	0·583
50	0·873	0·737	50	0·873	0·706
60	1·047	0·766	60	1·047	0·785
70	1·222	0·780			
80	1·396	0·784			
90	1·571	0·785			

AUTHOR INDEX

Authors' names are followed by the numbers of pages on which papers or books are explicitly mentioned. Where essentially distinct parts of the same source are quoted on different pages, all page numbers are given. If the same material is quoted more than once, only the page where it first occurs is given.

Aldous, C. W., 257.
Alkins, W. E., 27.
Allen, D. N. de G., 125, 245, 307.
Ansoff, H. I., 162, 172.
Aul, E. L., 287.

Baldwin, W. M., Jr., 17, 269, 271, 282, 317, 323, 330.
Baranski, G., 78.
Barrett, C. S., 4, 329.
Bauer, F. B., 58, 64, 66.
Becker, A. J., 23.
Belayev, N. M., 124.
Beltrami, E., 20.
Bishop, R. F., 104, 127, 218.
Bitter, F., 325.
Bland, D. R., 189, 197, 201.
Boas, W., 4.
Bohlen, E. C., 330.
Bourne, L., 323, 330.
Boussinesq, J., 19, 135.
Bragg, W. L., 14.
Brick, R. M., 329.
Bridgman, P. W., 12, 16, 77, 273, 334.
Briggs, G. C., 168.
Burghoff, H. L., 330.

Cairns, W. J., 14, 251.
Carathéodory, C., 153.
Carrier, G. F., 160, 162, 172.
Chow, C. C., 334.
Christianovitch, S., 143.
Christopherson, D. G., 92.
Coffin, L. F., Jr., 125.
Cook, G., 23, 119, 123, 124, 125.
Cook, M., 199, 277, 317, 322, 329.
Cottrell, A. H., 4, 11.
Coulomb, C. A., 19.
Cox, H. L., 23.
Cunningham, D. M., 32.

Dana, A. W., 334.
Davidenkov, N. N., 273.
Davies, R. M., 258.
Davis, E. A., 22, 32, 45, 172.
Davis, H. E., 32.
Dehlinger, U., 23.
Dokos, S. J., 172.
Dorn, J. E., 14, 32, 320, 333.
Drucker, D. C., 47.

Eddy, R. P., 96.
Edwards, S. H., 22.
Eichinger, A., 23, 29, 32, 171.

Eisbein, W., 185.
Elam, C. F., 4, 14.
Ernst, H., 206, 207.

Fangmeier, E., 187.
Fisher, J. C., 125.
Ford, H., 78, 189, 197, 201, 258.
Foster, P. Field, 213.
Fraenkel, S. J., 32, 35, 45.
Francis, E. L., 178.
Fried, M. L., 248.

Galin, L. A., 252.
Geiringer, H., 88, 136, 157, 233.
Gensamer, M., 32.
Gleyzal, A., 48.
Goldberg, A., 14.
Greenberg, H. J., 58, 64, 66, 67.
Guest, J. J., 23.

Haar, A., 67, 280.
Haigh, B. P., 17, 20.
Handelman, G. H., 33, 45.
Hankins, G. A., 257.
Hanson, D., 27.
Hencky, H., 20, 45, 135, 281.
Hermite, R. L., 218.
Heyer, R. H., 261.
Hill, R., 41, 59, 64, 67, 77, 100, 115, 119, 125, 142, 163, 178, 181, 182, 203, 215, 218, 219, 221, 223, 228, 242, 248, 256, 263, 265, 278, 284, 287, 312, 318, 323, 329, 330, 334.
Hitchcock, J. H., 189.
Hodge, P. G., 64, 89, 118, 220.
Hoff, H., 325.
Hohenemser, K., 72.
Hollomon, J. H., 14, 29, 30.
Howald, T. S., 271, 282, 323, 330.
Howard, J. V., 15.
Huber, M. T., 20.
Hume-Rothery, W., 4.

Ilyushin, A. A., 46, 47, 48.
Ishlinsky, A., 281.
Iterson, F. K. Th. van., 162, 181.

Jackson, K. L., 287.
Jackson, L. R., 30, 320, 333.
Jacobs, J. A., 252.

Karman, Th. von., 67, 195, 280.
Klingler, L. J., 172, 318, 322, 323.
Köchendörfer, A., 25.

AUTHOR INDEX

Körber, F., 29, 171, 256, 325.
Köster, W., 317.
Kötter, F., 136, 299.

Lankford, W. T., 32, 320, 333.
Larke, E. C., 199, 277.
Lee, E. H., 115, 119, 142, 215, 228.
Lessells, J. M., 22.
Lévy, M., 38.
Lin, C. C., 33, 45.
Linicus, W., 177.
Lode, W., 18, 22, 44, 45.
Low, J. R., 32.
Lubahn, J. D., 14, 181, 287.
Ludwik, P., 12, 14, 29.
Lueg, W., 201.
Lunt, R. W., 176, 177.

McAdam, D. J., Jr., 15.
MacGregor, C. W., 12, 22, 125.
MacLellan, G. D. S., 176, 177.
Malaval, M. P., 29.
Mandel, J., 297, 299.
Marin, J., 22.
Markov, A. A., 68.
Mason, W., 23.
Maxwell, J. Clerk, 20.
Mebs, R. W., 15.
Melan, E., 34, 57.
Merchant, M. E., 207, 208.
Meyer, O. E., 261.
Miller, C. P., 22.
Mises, R. von, 20, 21, 38, 51, 141.
Mohr, O., 296.
Morkovin, D., 23.
Morrison, J. L. M., 23, 72, 94, 123, 124.
Mott, N. F., 104, 127, 218.

Nadai, A., 20, 32, 41, 45, 46, 76, 84, 87, 93, 94, 117, 122, 136, 192, 197, 211, 230, 234, 253, 255, 256, 269, 308.
Nye, J. F., 14, 228, 251.

Odquist, F. K. G., 30.
Orowan, E., 14, 189, 192, 201, 202, 251.
Osgood, W. R., 32.

Palm, J. H., 13.
Palmer, E. W., 322.
Parasyuk, O. S., 253.
Parker, E. R., 32.
Pearson, C. E., 185.
Pearson, K., 38.
Philippidis, A. H., 67.
Piispanen, V., 208.
Polanyi, M., 16.
Pope, J. A., 23.
Prager, W., 33, 41, 45, 48, 49, 64, 65, 72, 88, 93, 136, 139, 142, 157, 158.
Prandtl, L., 39, 41, 86, 233, 235, 255, 256, 297.
Pumphrey, S. L., 275.
Putnam, W. J., 23.

Quinney, H., 22, 27, 45, 171.

Reiner, M., 13.
Reuss, A., 39, 41, 100.
Robertson, A., 23.
Roderick, J. W., 84.
Ros, M., 23, 32.
Rosenhain, W., 206.
Ross, A. W., 323, 330.

Sachs, G., 14, 23, 171, 172, 177, 181, 185, 248, 255, 269, 287, 318, 322, 323, 334.
Sadowsky, M. A., 67, 135.
Saint-Venant, B. de, 38, 117, 131.
Sangdahl, G. S., Jr., 287.
Scheu, A., 29.
Schleicher, F., 21.
Schmid, E., 4, 16.
Schmidt, E., 153.
Schmidt, R., 27, 45.
Schroeder, W., 277.
Scoble, W. A., 23.
Seely, F. B., 23.
Seitz, F., 4.
Shaw, F. S., 92, 96.
Shepherd, W. M., 29, 72, 123.
Shevchenko, K. H., 162, 172.
Shoji, H., 14.
Sidebotham, O., 23.
Siebel, E., 162, 181, 187, 201, 256, 277.
Sinitsky, A. K., 124.
Smith, C. S., 322.
Smith, K. F., 320, 333.
Smith, S. L., 15.
Sneddon, I. N., 253.
Snitko, N. K., 23.
Sokolnikoff, I. S., 83, 84, 86.
Sokolovsky, W. W., 17, 48, 84, 90, 93, 96, 124, 141, 162, 169, 172, 235, 250, 256, 258, 299, 303, 304, 307.
Sopwith, D. G., 23, 125.
Southwell, R. V., 83, 84, 92, 97, 131, 245, 307.
Spiridonova, N. I., 273.
Stanley, R. L., 22.
Stevenson, A. C., 253.
Sturney, A. C., 206.
Swainger, K. H., 48.
Swift, H. W., 94, 168, 269, 282, 286, 328.
Symonds, P. S., 141, 250, 263.

Tabor, D., 258, 260.
Taylor, G. I., 22, 24, 25, 27, 45, 52, 171, 307.
Thompson, F. C., 178.
Thomsen, E. G., 32.
Tietz, T. E., 14.
Timoshenko, S., 83, 84, 94, 97, 131, 190, 258.
Todhunter, I., 38.
Tracy, D. P., 181.
Trefftz, E., 92.
Tresca, H., 19.
Trinks, W., 197.
Tselikov, A. T., 195.
Tupper, S. J., 115, 119, 142, 163, 215, 228.
Turner, L. B., 110, 121.

AUTHOR INDEX

Underwood, L. R., 195, 197.

Voce, E., 13.

Warren, A. G., 121, 124.
Webster, D. A., 277.
Weiss, L., 178.
Westergaard, H. M., 17.

Wheeler, M. A., 27.
White, G. N., Jr., 118.
Williams, H. A., 84.
Wilson, F. H., 329.
Winzer, A., 48, 160.
Wistreich, J. G., 178.

Zener, C., 14, 29, 30.

SUBJECT INDEX

Anisotropy, 29, 44, 45, Chap. XII.
Annealing, 5.
Autofrettaged tube, axial expansion of, 121–2; axial stress distribution in, 116, 120; under plane strain, 115–18; with closed ends, 118–24; with open ends, 124.

Bauschinger effect, 8, 16, 24, 30, 72, 121.
Bending, of a sheet, 79–81, 287–92; of a beam, 81–4.
— under tension, 292–4.

Cavity, expansion in plate of circular, 307–13; expansion in surface of semi-cylindrical, 223–6; expansion of cylindrical, 125–7, 252; expansion of spherical, 103–6.
Channel, flow of plastic material through, 209–12.
Characteristics, theory of, 345–8; in deep-drawing, 285; in expansion of a tube, 113; in plane strain, 132, 296–7, 336; in plane stress, 300–1; in torsion, 89, 95–6.
Compressibility during plastic distortion, 26–7.
Compression, of block between rough plates, 226–36; of block between smooth plates, 77–9; of cylinder between rough plates, 277; of cylinder under distributed load, 265–7; of wedge by a flat die, 221–2, 351.
Consistent strain-increment for plastic-rigid body, 59.
Constraint factor in notched-bar test, 250.
Creep, physical mechanism of, 5.
Criterion of yielding, definition of, 15; dependence on cold-work of, 23–32; experimental investigations of, 19, 22–3; for anisotropic metals, 318; geometrical representation of, 17, 295, 301; influence of hydrostatic stress on, 16.
Crystal, lattice structure of, 4; plastic glide in, 6.
Cycloidal slip-line field, 232.
Cylindrical cavity in infinite medium, 125–7, 252.
— tube, under internal pressure, 115–25; under tension and internal pressure, 267–9; under tension and torsion, 71–5.

Deep-drawing, 282–7, 328–32.
Deviatoric, strain-increment, 27; stress, 16.
Discontinuity, in stress, 93, 157; in stress gradient, 133; in stress rate, 55, 239; in velocity, 150, 160; in velocity gradient, 134.
Drawing, deformation in, 171; die pressure and load in, 167, 172–3; efficiency of, 170; influence of back-pull in, 175;

of sheet, 163–76; of wire, 176–8; standing wave in, 168.

Earing of deep-drawn cups, 328–32.
Elastic limit, *see* Criterion of yielding.
— moduli, influence of cold-work on, 6, 15.
Elastic-plastic boundary, conditions along, 55, 132, Chaps. IV, V, IX.
Envelope of Mohr circles, 295–6.
Equivalent, strain, 30; stress, 26.
Extremum principles, for elastic body, 60–3; for plastic-elastic body, 63–6; for plastic-rigid body, 66–8.
Extrusion, deformation in, 185; direct, 186; efficiency of, 185; from contracting container, 263–5; inverted, 182; pressure of, 185.

Forward slip in rolling, 193.

Glide system, 6, 34, 50.
Grain boundaries, 7.

Hardness test, significance of, 258–61.
Hole, *see* Cavity.
Hydrostatic stress, influence on necking, 12; influence on yielding, 16, 294–5.
Hysteresis loop, 10.

Indentation, by cone, 218; by flat cylinder 281; by rectangular die, 254–8, 339–40; by sphere, 281; by wedge, 215–20.
Internal stress, 24, 35, 37, 50, 328.
Invariants, 15, 30, 343.
Ironing, 178–81.
Isotropy, conditions for, 7.

Lattice imperfections, 5.
Lode's variables, 18, 36, 44.
Lüders bands, 10, 52, 255–6.

Machining, 206–9.
Maximum plastic work, principle of, 66.
Mild steel, yield-point in, 10, 23, 52, 123–4.

Necking in tension, of cylindrical bar, 272–7; of thin strip, 323–5.
Neutral loading, 33.
— surface in bending, 79, 82, 288–91.
Notched bar under tension, 245–52.

Orientation, preferred, 8, 317.

Piercing, 106, 186–8.
Plane plastic strain, 77–8, Chap. VI.
— — stress, 300–7.
Plastic potential, 50.
Plastic-elastic boundary, conditions along, 55, 132, Chaps. IV, V, IX.
Plastic-rigid body, 38–9, 58, 128, Chap. IX.

SUBJECT INDEX

Prismatic beam, bending of, 81–4; tension and torsion of, 75–6; torsion of, 84–94; torsion, tension, and bending of, 313–6.

Reduced stress, 16.
Residual stresses in overstrained shell, 102–3, 121.
Riemann method, 153–4.
Rolling, efficiency of, 199; empirical formulae in theory of, 199; influence of strip-tension on, 202; influence of work-hardening on, 200–1; pressure distribution in, 199–201; roll-distortion in, 189–90.

Shear, deformation in, 326–7.
Size effects, 23, 124.
Slip bands, 6.
Slip-lines, definition of, 134; Hencky's theorems on, 136–8; in anisotropic metals, 336; in axially-symmetric state, 278; numerical calculation of, 140–9; variation of curvatures of, 138–9; variation of pressure along, 135; variation of velocity along, 136.
Soil mechanics, 294–300.
Spherical cavity in infinite medium, 103–4, 106.
Spherical shell, under internal pressure, 97–103, 104–5.
Statically determined problems, 87, 100, 111, 131, 242–5.
Strain, engineering, 9; equivalent, 30; increment of, 26; logarithmic or natural, 9, 28.

Strain-hardening, see Work-hardening.
Stress, deviatoric or reduced, 16; equivalent, 26.
Stress-strain curves, comparison of, 27–31; significance of, 8–13.
— relations, experimental, 44–5, 71–2; geometrical representation of, 41, 301; inversion of, 68; theoretical, 38–40.
Suffix notation and summation convention, 342.

Thermal phenomena, 14.
Torsion, membrane-roof analogy for, 86–7, 93; of anisotropic tube, 325–8; of annealed bar, 93; of non-uniform bar, 94–6; of uniform bar, 85–92; warping of section in, 88–9.
Tube drawing and sinking, 269–72.

Uniqueness theorems, for plastic-elastic body, 53–8, 242–5; for plastic-rigid body, 58–60.
Unit diagram, 213.

Variational principles, see Extremum principles.
Viscous fluid, contrast to plastic solid, 38.

Work-hardening, dependence on plastic strain, 26–32; physical mechanism of, 5.

Yielding, see Criterion of.
Yield-point, in annealed mild steel, 10, 23, 52, 123–4.